Preface

This text was prepared for a first course in fluid dynamics for engineering students. It is the result of an effort to develop an introductory treatment having increased relevance to the educational process in a modern engineering curriculum and to the needs of engineers of the future. This objective has led to some important viewpoints regarding content and method of treatment.

In content, the text emphasizes the *dynamics* of *real* fluids, both compressible and incompressible. Moreover, it attempts to distill the significant fundamental ideas from the conventional introductory treatment and to combine these with some important and stimulating topics normally untreated or reserved for later courses. This changes the accustomed pattern in certain respects and offers some challenges in coordination, but it is believed that only by such steps can today's education meet tomorrow's needs.

In treatment, the text reflects the awareness that the sophistication of modern technical problems requires an appreciation of the general, as well as the specific, basis for solutions. It is increasingly more important that the limitations of "special" solutions must be understood. The authors believe that today's students are better served by an approach that stresses fundamental relations in their general forms and treats particular examples as special cases derivable from the general. Thus a feature of this text is the introduction of the Navier-Stokes general equations of motion at an early stage, followed by their application to a variety of special cases. For the development of the general equations of motion the authors have chosen a method which draws upon the student's background in solid mechanics. This provides a useful correlation between the two basic branches of mechanics. A more coherent presentation of the specialized concepts of fluid dynamics is possible from the common analytical background of the general equations of motion. Indeed, it is difficult to imagine successful alternatives for presenting the ideas of creeping motions and boundary layer flows. In addition, this approach avoids the undue repetition that is often necessary to establish rigor and generality in the usual introductory treatment, which proceeds from the specialized or simplified equations toward the general. The resulting efficiency in the development of new material thus makes it possible to include additional topics in an introductory course. The authors recognize that the basic needs in the subject-area of fluids are the same for all fields of technology. Consequently, every effort has been made to present the subject on an interdisciplinary basis.

In a real sense, fluid dynamics involves the distribution and diffusion of fluid matter and its various properties, and the motion of the fluid throughout a system. Thus it is concerned with the flux and transport of matter, energy, momentum, and other properties from point to point in a system. Depending on the kind of problem, two approaches are useful. One is the finite control-volume method by which bulk or gross effects are related to end conditions. The other employs infinitesimal elements and control volumes to examine the motion of fluid particles and to obtain differential and integral relations. The text uses both methods to derive fundamental equations. A special effort is made to distinguish clearly the basis and the limitations of these equations and the relations derived from them. This distinction is especially important because in a few cases relations having the same terms can be obtained by either route, even though the underlying premises are different. The confusion between the various derivations of the so-called Bernoulli equation is an example.

The text conveniently divides into two parts. The first eight chapters are concerned with fundamental concepts and fundamental equations of fluid dynamics. The remaining eight chapters are devoted to specific topics and applications. The manuscript has profited from extensive classroom use, first in the Civil Engineering Department at the Massachusetts Institute of Technology, and later in the Engineering Mechanics Department at the University of Michigan. This has led to important revisions in both text and problems, and has provided the opportunity to evaluate a reasonable pace for effective class learning. The authors have found that the equivalent of eleven chapters can be covered efficiently by a class meeting three days a week throughout a normal semester. The entire volume can be covered conveniently in something less than two semesters, leaving time for additional special topics or discussion in greater depth of some of the topics presented. For a one-semester terminal exposure, a course offering satisfactory coverage without being superficial could include Chapters 1 through 8 and selected articles from Chapters 9 through 13.

The authors wish to acknowledge the support of the Massachusetts Institute of Technology and the Ford Foundation during the development of the manuscript. The former was given through the continued encouragement and cooperation of both Dean G. S. Brown of the School of Engineering and Professor C. L. Miller of the Civil Engineering Department, including a period of released time at a key stage. The latter was provided through funds made available from the grant of the Ford Foundation to the School of Engineering at MIT for the development of undergraduate engineering curricula. The authors are also indebted to all their colleagues, both at MIT and Michigan, who have been the source of constructive ideas and criticisms. At MIT, particular acknowledgment is due Professor A. T. Ippen for his advice and counsel and to Professors P. A. Drinker, C. Elata, J. F. Kennedy, F. Raichlen, R. R. Rumer, Jr., and F. E. Perkins, who participated in the instruction and contributed to the improvement of the text in its formative stages. At later stages, Professors P. S. Eagleson, R. T. McLaughlin, L. W. Gelhar, and E. Partheniades cooperated in the teaching program with preliminary editions. At Michigan, Professors W. R. Debler, J. D. Murray, H. J. Smith, R. A. Yagle, and Drs. C-C Hsu, S. P. Lin, and H. F. Keedy taught from a preliminary edition and provided many helpful suggestions. The assembly and the solution of problems are major efforts in programming a textbook. In the present case sincere thanks for major help are due to Messrs. J. R. Wallace, D. A. Haith, J. H. Nath, V. J. Siciunas, and C. H. Popelar. The aid of Miss H. Waechter and Miss E. von Oppolzer in preparing the manuscript is gratefully acknowledged. A special debt is owed to our wives and families for their patience during the writing of this book.

Ann Arbor, Michigan   J. W. D.
Cambridge, Massachusetts   D. R. F. H.
November, 1965

JAMES W. DAILY
Engineering Mechanics Department
University of Michigan

DONALD R. F. HARLEMAN
Civil Engineering Department
Massachusetts Institute of Technology

Fluid Dynamics

ADDISON-WESLEY PUBLISHING COMPANY, INC., READING, MASSACHUSETTS, U.S.A.
ADDISON-WESLEY (CANADA) LIMITED, DON MILLS, ONTARIO

This book is in the

ADDISON-WESLEY SERIES IN MECHANICS AND THERMODYNAMICS

Howard W. Emmons and Bernard Budiansky

Consulting Editors

Library of Congress Catalog Card No. 65-23029

ADDISON-WESLEY PUBLISHING COMPANY, INC.
READING, MASSACHUSETTS · Palo Alto · London
NEW YORK · DALLAS · ATLANTA · BARRINGTON, ILLINOIS

ADDISON-WESLEY (CANADA) LIMITED
DON MILLS, ONTARIO

Contents

Acknowledgments

In addition to the references included in the text, credit is due for the figures listed below.

Fig. 9–4 (p. 175); from H. SCHLICTING, *Boundary Layer Theory*, McGraw-Hill Book Co., New York, 4th ed., 1960, p. 96.

Fig 9–5 (p. 177); from H. SCHLICTING, *op. cit.*, p. 16.

Fig. 10–7 (p. 201); from H. SCHLICTING, *Nat. Advisory Comm. Aeron.* Tech. Memo. 1217 (1949).

Fig. 10–9 (p. 203); from H. SCHLICTING, *Boundary Layer Theory*, McGraw-Hill Book Co., New York, 4th ed., 1960, Figs. 12.10, 12.11, 12.12.

Fig. 12–13 (p. 249); from H. SCHLICTING, *op. cit.*, Fig. 21.12.

Fig. 15–9 (p. 383); from H. SCHLICTING, *op. cit.*

Fig. 16–4 (p. 420); from H. SCHLICTING, *op. cit.*

CHAPTER 1

Fluid Characteristics

1–1 INTRODUCTION

In general, matter can be classified by the physical form of its occurrence. These forms, known as *phases*, are solid, liquid, and gas or vapor. *Fluids* comprise matter in the *liquid* and *gas* or *vapor* phases. We are all familiar with the distinguishing features of these phases versus the solid phase. But we also know that liquids and gases have quite distinct features, so we must look for a common characteristic that allows us to put them in the single classification of fluid. In discussing fluid dynamics, we are interested in how a fluid behaves when in motion and how this behavior is related to applied forces and moments. Liquids, gases, and vapors have in common a distinct manner of reacting when subjected to shear stresses, which explains their "fluidity" and provides the key basis for developing the principles of fluid dynamics. This common and distinguishing feature is stated as follows:

A fluid will deform continuously under shearing (tangential) stresses, no matter how small the stress. As will be discussed later on, the magnitude of the stress depends on the **rate of angular deformation.**

A solid, on the other hand, will deform by an amount proportional to the stress applied, after which static equilibrium will result. The magnitude of a shear stress depends on the **magnitude of the angular deformation.**

Not all fluids show exactly the same relation between stress and rate of deformation. Fluids are called *Newtonian** if shear stress is *directly* proportional to rate of angular deformation starting with zero stress and zero deformation. In these cases, the constant of proportionality is defined as μ, the absolute or dynamic viscosity. Thus Newtonian fluids have the property of a dynamic viscosity which is independent of the motion the fluid happens to undergo. The more common fluids like air and water are Newtonian. There is an analogy between Newtonian fluids having a constant viscosity relating stress and rate of deformation and solids obeying Hooke's law of a constant modulus of elasticity relating stress and magnitude of deformation.

---

* Named after Sir Isaac Newton, who studied fluid motion, with assumptions corresponding to constant absolute viscosity.

Fluids having a variable proportionality between stress and deformation rate are called *non-Newtonian*. In such cases, the proportionality may depend on the length of time of exposure to stress as well as on the magnitude of the stress. A vast number of fluids which are not commonly encountered but which are extremely important, nevertheless, are non-Newtonian. Some substances, notably some of the plastics, exhibit a yield stress below which they behave like a solid and above which they have fluidlike behavior. *Rheology* is the subject which treats plastics and non-Newtonian fluids. Recently the increasing importance of non-Newtonian fluids in engineering applications has resulted in more attention being accorded them in engineering literature. In Fig. 1–1 types of fluid and plastic behaviors are shown on diagrams of rate of deformation versus stress and time versus stress.

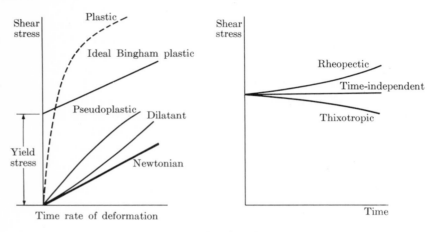

FIG. 1–1. Types of rheological behavior.

A subdivision of fluids into two main classes, either compressible or incompressible, can be made on the basis of reaction to pressure (normal) stresses. All gases and vapors are highly compressible. Liquids by comparison are only slightly compressible. As we shall see, compressibility introduces thermodynamic considerations into fluid flow problems. If incompressibility can be assumed, it is much easier to describe the state of the fluid and its behavior in motion. With some important exceptions, liquids usually are treated as incompressible for practical purposes. Gases, on the other hand, can be treated as incompressible only if the change in pressure is small throughout the flow system.

All fluids are composed of molecules discretely spaced and in continuous motion. In the previous definitions and distinctions used to describe fluids, this discrete molecular structure was ignored and the fluid was considered as a *continuum*. This means that all dimensions in a fluid space are taken as large compared to the molecular spacing—an assumption we shall make throughout this text even when considering the limit of zero distance from a bounding wall.

It also means that all properties of the fluid such as density and viscosity will be continuous from point to point throughout a given volume of fluid.

Finally, an important behavior of continuum-type viscous fluids is the no-slip condition at rigid boundaries. Experimentally, we observe that real fluids tend to adhere to boundaries, which results in zero relative velocity at the boundary surface. Thus, in analyzing motions of fluids with viscosity, we see that this physical condition must always be satisfied.

The treatment in this text will be concerned primarily with single-phase fluid systems, although certain multiphase systems will be discussed. Multiphase systems include combinations of liquid-gas, liquid-vapor, liquid-solid, and gas- or vapor-solid. Examples are air slugs in water lines, condensed water in steam lines, cavitation, and suspended sediment in streams of water or air.

1–2 UNITS OF MEASUREMENT

All physical quantities are described in terms of standard units of measurement, and in this text we will use the English system.

The magnitudes of the units for mechanical properties and effects are established by use of Newton's second law of motion. This law states that a mass moving by virtue of an applied force will be accelerated and that the component of force in the direction of the acceleration is proportional to the product of the mass and the acceleration. If we select the units such that *one* unit of force acting on *one* unit of mass will produce *one* unit of acceleration, the proportionality constant will be *unity*, and we may write force *equals* mass multiplied by acceleration. Adopting this and using the pull of gravity on a mass, i.e. its weight, as a measure of force, we have the vector equation

$$\mathbf{F} = m\mathbf{a} = \frac{W}{g}\,\mathbf{a}, \tag{1–1}$$

where

$\mathbf{F}$ = force, lb,

m = mass, slug,

$\mathbf{a}$ = acceleration, ft/sec^2,

W = weight, lb,

g = local gravitational acceleration, ft/sec^2.

Generally, we will use the English Gravitational system which relates force in pounds (lb), mass in slugs, and acceleration in feet per second per second (ft/sec^2), such that the force to accelerate a mass in slugs is given by

$$1 \text{ lb} = 1 \text{ slug} \times 1 \text{ ft/sec}^2. \tag{1–2}$$

In some cases it is convenient to measure mass in pounds-mass (lbm), the unit for the English Absolute system. The slug and the pound-mass magnitudes are related by the conversion factor

$$g_c = 32.1740 \ \frac{\text{lbm}}{\text{slug}},$$

or

$$1 \text{ slug} = 32.1740 \text{ lbm}. \tag{1–3}$$

In a mixed system of units, the force to accelerate a mass in pounds-mass is given by

$$\mathbf{F}(\text{lb}) = m(\text{lbm})/g_c \times \mathbf{a}(\text{ft/sec}^2). \tag{1–4}$$

The conversion factor g_c is *numerically* equal to the standard acceleration of gravity, but it *is not* an acceleration. It is solely a proportionality constant relating the magnitudes of the pound-mass and the slug due to our departure from a single system of units. The resulting magnitude relation between pound and pound-mass is

$$1 \text{ lb} = 1 \text{ lbm} \times \frac{g}{g_c}. \tag{1–4a}$$

Certain quantities depend on thermostatic and thermodynamic effects and involve temperature and the heat equivalent of mechanical energy. We will measure temperature in degrees fahrenheit (ice point = 32°F) and degrees rankine (°R = °F + 459.69°). Heat energy is usually given in British thermal units (Btu). The conversion to mechanical energy is

$$J = 778.26 \ \frac{\text{ft-lb}}{\text{Btu}},$$

or

$$1 \text{ Btu} = 778.26 \text{ ft-lb}. \tag{1–5}$$

Table 1–1 gives the dimensions and units of the several quantities which will be used extensively in this text.

1–3 PROPERTIES AND STATES OF FLUIDS

A *property* is a characteristic of a substance which is invariant when the substance is in a particular state. In each *state* the condition of the substance is unique and is described by its properties. A distinction is made in thermodynamics [1] between *extensive* properties, whose values depend on the amount of the substance present, and *intensive* properties, which have values independent of the amount present. For example, total volume, total energy, and total weight of a substance are extensive properties. The corresponding *specific* values, namely, volume per unit mass, energy per unit mass, and weight per unit volume, are intensive properties. Temperature, pressure, viscosity, and surface tension are also independent of the amount of the substance and are intensive properties. The intensive properties are the values which apply to a "particle" of the fluid.

Over any finite volume or mass in a system, the intensive values may change from particle to particle. Thus we make the distinction between the *state of a substance* and the *state of a system*. We will use the concept of the state of a system in later chapters. In this chapter, however, we will confine our attention to the state of substances and to some of their corresponding properties.

We use two kinds of relations in determining the states of a substance. First, for every substance, *equations of state* relate the minimum number of properties or conditions necessary to determine both the state and the other properties. Second, the variation of the conditions and properties of a substance as it changes from one state to another depends on how the change is controlled, that is, on the *process*. For example, the process may be at constant temperature, constant volume, constant pressure, zero heat transfer, or some combination of these or other special cases.

1–3.1 Properties of importance in fluid dynamics

Definitions and descriptions of properties and quantities of importance in fluid dynamics are given in the following paragraphs. Numerical values are listed in Tables 1–2 through 1–8 at the end of this chapter.

Pressure, p. Pressure is force/area. If a volume of matter is isolated as a free body, the force system acting on the volume includes *surface forces* over every element of area bounding the volume. In general, a surface force will have components perpendicular and parallel to the surface. At any point, the perpendicular component per unit area is called the *normal stress*. If this is a compression stress, it is called *pressure intensity*, or simply *pressure*. Pressure is a scalar quantity, and the force associated with a given pressure acting on a unit of area is $p\,d\mathbf{A}$ and has the direction of the normal to the unit area $d\mathbf{A}$. Thus, at a point in the interior of a fluid mass, the direction of the *pressure force* depends on the orientation of the plane or "cut" through the point.

Pressure is measured relative to an absolute zero value (called *absolute* pressure) or relative to the atmospheric pressure at the location of the measurement (called *gauge* pressure). Thus,

$$p\text{ (gauge)} = p\text{ (absolute)} - p_{\text{atm}}\text{ (absolute)}.$$

The *standard atmosphere* is a pressure defined as

$$1\text{ standard atmosphere} = 14.6959\text{ (absolute) lb/in}^2. \qquad (1\text{–}6)$$

In unusual cases liquids will support tensile stresses which would be denoted as negative absolute pressures. As indicated in Table 1–1, pressure intensity is commonly given in units of lb/in². This is often written as psig and psia for gauge and absolute pressure, respectively.

Temperature, T. Two bodies in thermal equilibrium exhibit the same value of the property that we call temperature. Changes in temperature cause changes in other properties of matter and afford us methods of measurement. One example is the expansion of mercury with temperature increase, while another is the increase in pressure of a constant volume of gas as its temperature rises.

Density, ρ. Density is mass/volume. A given amount of matter is said to have a certain mass which is treated as an invariant. Thus it follows that density is a constant so long as the volume of the given amount of matter is unaffected (i.e., for a gas, so long as pressure and temperature conditions are the same).

Specific weight, γ. Specific weight is weight/volume. Weight depends on the gravitational field. (In the field of the earth, it is the force of gravity extended on a given mass at a given locality.) Consequently, specific weight, in contrast to density, depends on the gravitational field.

Viscosity (dynamic molecular), μ. Due to molecular mobility, a property which is called *viscosity* is evident whenever a fluid moves such that a relative motion exists between adjacent volumes. This leads to the common method of defining the magnitude of viscosity for measurement in terms of a simple flow situation. Consider the two-dimensional parallel shear field described by velocity u in the x-direction whose magnitude is a function of the normal y-direction only. For this case, the relation between shear stress and rate of angular deformation of the fluid is simply

$$\tau_{yx} = \mu \frac{du}{dy}, \tag{1–7}$$

where

$$\tau_{yx} = \text{shear stress acting in the } x\text{-direction on a}$$
$$\text{plane whose normal is the } y\text{-direction, lb/ft}^2,$$

$$u = u(y) \text{ only} = \text{velocity, ft/sec},$$

$$du/dy = \text{rate of angular deformation, 1/sec},$$

$$\mu = \text{dynamic molecular viscosity, lb-sec/ft}^2.$$

The factor μ is called *dynamic molecular* (or simply dynamic) viscosity because Eq. (1–7) is a dynamic relation between a force and the fluid motion. Its units contain the dynamic quantity of force (or mass in M, L, T units). For Newtonian fluids, μ has a unique value depending only on the fluid state, and hence it is one of the fluid properties. As already pointed out, many real fluids approximate this Newtonian assumption, but there are important exceptions.*

---

* Note also that the definition of μ by Eq. (1–7) does not show the possibility of a stress as the result of a pure dilation with zero shear. For compressible fluids, certain experiments have indicated such an effect, which for an isotropic fluid requires a *second coefficient* of viscosity. The effect is secondary, however, and in the present text discussions will be in terms of the above definition unless otherwise noted.

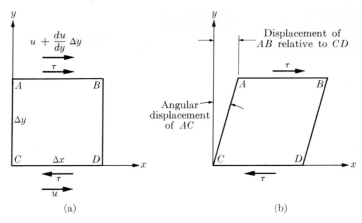

FIG. 1–2. Angular displacement due to shear deformation.

That du/dy equals rate of angular deformation, or displacement, can be seen by considering the simple case of an elementary volume of fluid under shear. In Fig. 1–2, let shear be caused by the upper surface AB of the element $\Delta x \, \Delta y$ moving faster than the lower surface CD. The element $\Delta x \, \Delta y$ in (a) will deform as shown in (b). The angular velocity of the left-hand vertical edge of the element relative to the lower edge is the rate of angular deformation. This is measured as the rate at which the top edge is displaced with respect to the lower edge divided by the length of the left-hand edge. Thus

$$\frac{du}{dy} \Delta y = \text{rate of displacement of } AB \text{ relative to } CD,$$

$$\left(\frac{du}{dy} \Delta y \right) \frac{1}{\Delta y} = \frac{du}{dy} = \text{rate of angular displacement of } AC$$

$$= \text{rate of angular deformation of the}$$
$$\text{elementary volume.}$$

For solids subjected to the same two-dimensional shear, let $d\xi$ be the relative static displacement of AB due to a given shear stress. Then the relation analogous to Eq. (1–7) is

$$\tau_{yx} = G \frac{d\xi}{dy}, \qquad (1\text{–}8)$$

where

$$\tau_{yx} = \text{shear stress, lb/ft}^2,$$
$$d\xi/dy = \text{angular deformation (shear strain)},$$
$$G = \text{modulus of elasticity in torsion (shear), lb/ft}^2.$$

Equation (1–7) is the basis for evaluating μ in most viscosity-measuring instruments. For example, two-dimensional motion can be approximated with a

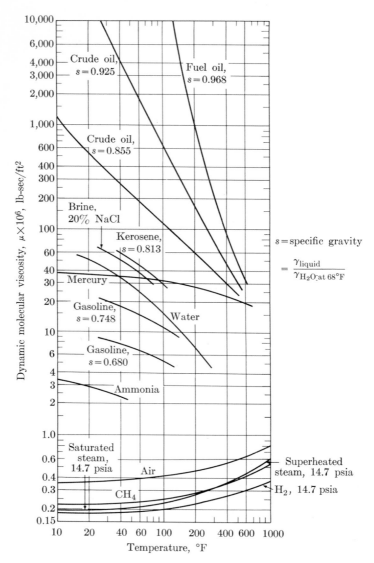

FIG. 1–3. Dynamic molecular viscosity [2].

fluid in a narrow annulus between two large cylinders. When one cylinder rotates, the fluid, which tends to stick to solid boundaries, is subjected to a shear measured by the applied torque. For very narrow annuli, the fluid velocity varies linearly from the rotating to the stationary boundary, giving

$$\frac{du}{dy} \propto \frac{\text{radius} \times \text{angular velocity}}{\text{annulus thickness}}.$$

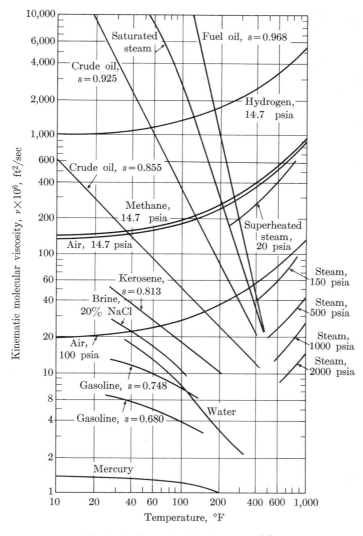

FIG. 1–4. Kinematic molecular viscosity [2].

The dynamic molecular viscosity is a function of temperature and pressure. The pressure dependence is practically negligible for liquids and is small or negligible for most gases and vapors unless the pressure becomes very high. Curves of dynamic viscosity in Fig. 1–3 [2] show that the viscosities of all liquids decrease with temperature and those of all gases increase. These different temperature effects are the consequence of the structural difference between liquids and gases. The liquid state can be viewed as having a relatively stable lattice structure within which molecules vibrate about an equilibrium position.

Under a shear force, fluid layers will slip relative to one another, and the vibrating particles in the layer will jump at intervals into new equilibrium positions. An increase in molecular vibration which is manifested as a temperature rise will ease the rigidity so that particles are more easily displaced. The "fluidity" is increased and the viscosity decreased. For gases, viscosity is explained by the kinetic theory as being a consequence of the momentum transfer accompanying the exchange of molecules between fluid layers. Temperature increases molecular activity and hence increases both the momentum exchange and the viscosity.

Viscosity (kinematic molecular), ν. The ratio μ/ρ appears frequently when dealing with fluid dynamics. It has only kinematic dimensions and units which explains the reason for its name. Thus

$$\nu = \frac{\mu}{\rho}. \tag{1–9}$$

Figure 1–4 [2] gives the values of kinematic viscosity for various liquids and gases.

Specific heat, c. *Specific heat* or *specific heat capacity* is the ratio of the quantity of heat flowing into a substance per unit mass to the change in temperature. Specific heats must be determined experimentally or must be computed from a molecular theory. Table 1–2 shows values of the specific heat capacity for common liquids.

The specific heats for gases and vapors depend on how the change in state is effected. Table 1–6 gives specific heats at constant volume, or constant density, c_v, and at constant pressure, c_p for some common gases. Note that the values shown are mean values for the range of 32°F to 400°F. In general, both c_v and c_p are functions of temperature.

Internal energy, u. The *specific internal energy* is measured as energy per unit mass, usually in units of Btu/lbm. Internal energy is due to the kinetic and potential energies bound into the substance by its molecular activity and depends primarily on temperature. Table 1–8 gives values of u for air at different temperatures.

For a perfect gas [a gas which obeys the equation of state, Eq. (1–17), described in Section 1–3.2], internal energy is a function solely of temperature according to the relation

$$du = c_v \, dT, \tag{1–10a}$$

or if c_v is constant, by the relation

$$u_2 - u_1 = c_v(T_2 - T_1). \tag{1–10b}$$

Equations (1–10a) and (1–10b) hold for any process of change of state between the temperatures T_1 and T_2. Air follows perfect gas laws closely.

Enthalpy, $(u + p/\rho)$. The sum $(u + p/\rho)$ is called *specific enthalpy*.* Since u, p, and ρ are all properties, enthalpy is a property. Specific enthalpy is measured as energy per unit mass, usually in units of Btu/lbm.

For a perfect gas, enthalpy is a function of temperature only. It can be calculated by the relation

$$d\left(u + \frac{p}{\rho}\right) = c_p \, dT, \tag{1-11a}$$

or, if c_p is constant, by

$$\left(u + \frac{p}{\rho}\right)_2 - \left(u + \frac{p}{\rho}\right)_1 = c_p(T_2 - T_1). \tag{1-11b}$$

This is an equation that can be applied to any process of change of state between temperatures T_1 and T_2. Table 1–8 shows enthalpy versus temperature for air.

All the quantities in Eqs. (1–10b) and (1–11b) must be in consistent units. The proper conversions are illustrated by the multipliers in Table 1–1(c).

Bulk modulus of elasticity, E_v, and compressibility. Compressibility is the measure of change of volume and density when a substance is subjected to normal pressures or tensions. It is defined as

compressibility = % change in volume (or density)
for a given pressure change

$$= -\frac{d\upsilon}{\upsilon}\frac{1}{dp} = +\frac{d\rho}{\rho}\frac{1}{dp}, \tag{1-12a}$$

where the negative sign indicates a decrease in volume, υ, with an increase in pressure. The change in volume equals

(compressibility) × (original volume) × (pressure increment),

and the change in density equals

(compressibility) × (original density) × (pressure increment).

The reciprocal of compressibility is known as the *bulk modulus of elasticity*,

$$E_v = -\frac{dp}{d\upsilon/\upsilon} = +\frac{dp}{d\rho/\rho}. \tag{1-12b}$$

Compressibility data for liquids are usually given in terms of E_v as determined experimentally. Theoretically, E_v should depend on the manner or process in which the volume or density change is effected, e.g., *adiabatically, isothermally,*

---

* Enthalpy per unit mass is denoted by the letter h in thermodynamics literature. However, in this text h will be used to denote elevation measured above a horizontal datum, and specific enthalpy will be expressed always as the sum $(u + p/\rho)$.

etc. For the common gases such as oxygen, these two processes* give

$$\text{adiabatic } E_v = kp, \tag{1–13}$$

$$\text{isothermal } E_v = p, \tag{1–14}$$

where

$k = c_p/c_v = $ ratio of specific heat at constant pressure to that at constant volume,

$p = $ absolute pressure.

The effect of the process is smaller for liquids than for gases.

The value of E_v for water at normal atmospheric pressure and temperature is 312,000 psi. From Eq. (1–14) we see that the isothermal value of E_v for atmospheric air is of the order of 15 psi. Thus atmospheric air is about 20,000 times more compressible than water. Water is about 100 times more compressible than steel. Nevertheless, as can be seen from Table 1–5, increasing the pressure on water by 1000 psi will increase the density by only about 0.3%. For practical purposes, the assumption of incompressibility is satisfactory unless very large pressure ranges are involved.

Velocity of sound, c. Associated with each state of a substance is a definite *velocity of sound* given by

$$c = \sqrt{dp/d\rho}, \tag{1–15a}$$

or, using the bulk modulus, Eq. (1–12b), we have

$$c = \sqrt{E_v/\rho}. \tag{1–15b}$$

In these two equations, the velocity is given in ft/sec if p and E_v are in units of lb/ft^2 and ρ is in slug/ft^3. The magnitude of c is the velocity with which small-amplitude pressure signals will be transmitted through a fluid of infinite extent or through a fluid confined by completely rigid walls. The density change caused by an infinitely small pressure wave occurs almost frictionlessly and adiabatically.

For liquids the speed of sound is usually determined from experimental values of E_v and ρ, such as those given in Table 1–3.

Using the adiabatic bulk modulus of Eq. (1–13), we obtain for the common gases a relation which can be applied quite accurately. Then

$$c = \sqrt{kp/\rho}, \tag{1–16}$$

where we again express p in lb/ft^2 and ρ in slug/ft^3 to give c in ft/sec. The quantities p and ρ are related by an equation of state and for gases may be replaced by a function of the single variable, temperature. This step is shown by Example 1–3 in Section 1–3.3.

---

* See Examples 1–1 and 1–2 for the derivation of Eqs. (1–13) and (1–14).

Vapor pressure, p_v. The pressure at which liquids boil is called the *vapor pressure*. This is the equilibrium partial pressure which escaping liquid molecules will exert above any free surface; its magnitude increases with temperature. For boiling to occur, the equilibrium must be upset either by raising the temperature to cause the vapor pressure to equal or exceed the total pressure applied at the free surface, or by lowering the total pressure at the free surface until it is equal to or less than the vapor pressure. The more volatile the liquid, the higher its vapor pressure, as can be seen from Table 1–2.

Surface energy and surface tension, σ. At boundaries between gas and liquid phases or between different immiscible liquids, molecular attraction introduces forces which cause the interface to behave like a membrane under tension. In any liquid body, every surface molecule suffers a strong attraction perpendicular to the surface which results in a tendency to deform any but a plane surface. For example, water drops in air will spontaneously contract into spheres, the shape for minimum surface for a given volume of liquid. Conversely, work must be done to extend such a contracted surface by drawing more molecules into it. This work equals the *free energy of the surface*, the additional potential energy per unit surface due to the surface molecules lacking like neighbors opposite the interface. This free energy is measured by the *surface tension* and has the units

$$\sigma = \frac{\text{force} \times \text{distance}}{\text{area}} = \frac{\text{work}}{\text{area}} = \frac{\text{force}}{\text{length}},$$

which leads to the concept of a tension force and to an analogy to membrane behavior.

The magnitude of the surface tension depends on the two fluids in contact and is very nearly independent of temperature. The surface tensions of a liquid in contact with only its own vapor or with a gas are very nearly the same. Interfacial tensions between two liquids fall in between the values of each liquid in the presence of its vapor. Tables 1–2 through 1–4 give typical surface tension values.

1–3.2 Equations of state

Liquids. The equations of state for most physical substances are complex and are expressible in simple forms only for limited ranges of conditions. This is true for liquids if all the several effects of temperature, pressure, volume, etc., are to be expressed. Thus it is both customary and more practical to use tabulations and graphical curves of liquid properties rather than equations of state as such. Fortunately, for a wide range of pressures most liquids are nearly incompressible, and their states can be defined closely by temperature alone. Only when subjected to very large pressures or extreme temperatures are additional conditions significant in fixing the state. This is reflected in several of the tables and curves of liquid properties in this chapter.

Gases. A gas in equilibrium with its liquid is usually called a *saturated vapor.* If a gas at a given pressure has a temperature above the value for equilibrium with the liquid phase it is called a *superheated vapor.* The common gases, such as those which form our air, are highly superheated. As for liquids, the equations of state for vapors at or near saturation conditions are complex. On the other hand, a useful approximation for gases in a highly superheated condition is the theoretical equation of state for the perfect gas. This *perfect gas law* is given by the relation

$$\frac{p}{\rho} = RT, \tag{1–17}$$

where

$T =$ absolute temperature, °R,

$R =$ gas constant, ft-lb/mass-°R,

$p =$ absolute pressure, lb/ft^2,

$\rho =$ density, mass/ft^3.

The gas constant R is proportional to the universal constant 1545.4 divided by the molecular weight. Its numerical value depends on the units used for density. For density in slug/ft^3 and pressure in lb/ft^2,

$$R = R_{\text{slug}} = \frac{1545.4\, g_c}{\text{molecular weight}}, \tag{1–18a}$$

expressed in units of ft-lb/slug-°R. For density in lbm/ft^3 and pressure in lb/ft^2,

$$R = R_{\text{lbm}} = \frac{1545.4}{\text{molecular weight}}, \tag{1–18b}$$

expressed in units of ft-lb/lbm-°R. The deviations from Eq. (1–17) shown by actual gases are small at ordinary temperature but may become marked if the gases are highly compressed or if their temperatures are reduced to near-liquifica-tion.

The pressure defined through an equation of state such as Eq. (1–17) is the value obtained under thermodynamic equilibrium and is sometimes called the *thermodynamic pressure.*

For a perfect gas, the constant R is related to the specific heats as follows:

$$c_p = c_v + R = \frac{k}{k-1}\,R, \qquad c_v = \frac{R}{k-1}, \qquad k = \frac{c_p}{c_v}. \tag{1–19}$$

For perfect gases, R and $(c_p - c_v)$ are each constant, even though c_p and c_v need not be. Specific heats are usually given in units of Btu/lbm-°R. In Eq. (1–19) they should be converted to ft-lb/lbm-°R (by multiplying by 778) if R has the units of Eq. (1–18b).

1–3.3   Change of state processes for gases

If we use the perfect gas law, several *processes of changing state* are given by simple relations as follows:

Isothermal process

$$\frac{p}{\rho} = RT = \text{const.} \tag{1-20}$$

During compression, work done on the gas would increase its temperature unless the heat equivalent of the work were allowed to flow from the gas to its surroundings.

Constant-pressure process

$$p = \rho RT = \text{const.} \tag{1-21}$$

Thus the volume of a given mass is proportional to temperature for a constant-pressure process.

Isentropic adiabatic process. For the case of zero heat transfer (the adiabatic process), we can obtain the following relation from thermodynamic theory if there is no friction (isentropic):

$$\frac{p}{\rho^k} = C_s = \text{const,} \tag{1-22a}$$

where

$$k = \frac{c_p}{c_v}, \text{ the ratio } \frac{\text{specific heat at constant pressure}}{\text{specific heat at constant volume}}$$

$$= \text{adiabatic constant for the gas.}$$

Using Eq. (1–17) with Eq. (1–22a), we can obtain two additional equations:

$$\frac{T}{\rho^{k-1}} = \frac{C_s}{R} = \text{const,} \qquad \frac{T}{p^{(k-1)/k}} = \frac{C_s^{1/k}}{R} = \text{const.} \tag{1-22b}$$

Ideally, the exponent k is a true constant and for actual gases it is nearly constant over a wide range of states. In adiabatic compression (with or without internal friction), the external work done on the gas increases its internal energy and raises its temperature. If the process is adiabatic but not frictionless, it is described by an equation like Eqs. (1–22) but with an exponent slightly different from k. The exponent is smaller than k for expansion and is larger than k for compression.

The heating or cooling effects and all the intermediate states may also be determined by Eqs. (1–22). These equations were the basis for establishing the adiabatic bulk modulus of a gas, Eq. (1–13).

Example 1–1: Adiabatic bulk modulus for gases

Take the logarithm of Eq. (1–22a) for the isentropic change of state

$$\ln p - k \ln \rho = \ln C_s.$$

Differentiating, we have

$$\frac{dp}{p} - k \frac{d\rho}{\rho} = 0.$$

Then

$$\frac{dp}{d\rho/\rho} = kp = E_v \quad \text{(adiabatic)}.$$

This is Eq. (1–13).

Example 1–2: Isothermal bulk modulus for gases

If we take the derivative of Eq. (1–20) for the isothermal change of state, we find

$$dp/d\rho = \text{const} = p/\rho,$$

so that

$$\frac{dp}{d\rho/\rho} = p = E_v \quad \text{(isothermal)}.$$

This is Eq. (1–14).

Example 1–3: Velocity of sound in a gas

Assume that the density change accompanying a sound wave occurs isentropically. Using Eq. (1–22a) as shown in Example 1–1 gives $dp/d\rho = k(p/\rho)$. Substituting into Eq. (1–15a), we have

$$c = \sqrt{k(p/\rho)},$$

which is Eq. (1–16). For p in lb/ft^2 and ρ in slug/ft^3, the velocity c is in ft/sec. Using Eqs. (1–17) and the definition of Eq. (1–18a), we obtain

$$c = \sqrt{k R_{\text{slug}} T},$$

which shows that the velocity of sound in a perfect gas depends only on its absolute temperature.

1–4 EQUILIBRIUM OF FLUIDS

A fluid volume may be in *equilibrium* either when in a condition of rest or when moving as a solid body. The equilibrium is the result of the force field in which the fluid is placed and the restraints exerted on the boundaries of the fluid volume. A distribution of pressure (and density if the fluid body is compressible or of variable density) results throughout the volume, depending on the force field. Important examples of force fields are gravity and centrifugal fields. Restraints leading to equilibrium include the normal pressures of confining rigid boundaries and the surface tension forces.

1–4.1   Stress at a point for fluids in equilibrium

When a volume of a real fluid is subjected to external forces, an internal stress condition develops. This can be illustrated by assuming that such a stressed volume suffers an internal cut. In order to avoid slip or parting of the fluid at the cut, it would be necessary to apply a surface force over the cut equal and opposite to the original stresses on the uncut surface, as illustrated in Fig. 1–5. In general, this force will be at an angle to the cut, thus having both normal and tangential components. These forces per unit area will be *pressure* (or *tension*) and *shear stresses*. It is clear that the magnitudes of the stress will depend on the orientation of the cut surface, and thus we conclude that a definite relationship must exist between the stresses on arbitrarily oriented surfaces passing through a given point in a fluid.

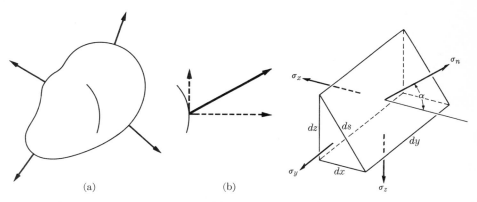

FIG. 1–5. Internal stresses in a fluid volume: (a) a cut in a stressed volume; (b) stresses at the cut face.

FIG. 1–6. Diagram for stress at a point in equilibrium.

However, as we can deduce from Eq. (1–7), fluids in equilibrium will not develop shear stresses since internal velocity gradients will not exist. The relation for the stresses at a point has a unique solution in this case. Consider an elementary volume of fluid in the prismatic form shown in Fig. 1–6. The fluid volume is taken to be in equilibrium with surface forces composed of stresses balancing the field forces which are proportional to the volume or mass. The equilibrium equations in the x- and z-coordinate directions are

$$-\sigma_x \, dy \, dz + \sigma_n \, ds \, dy \cos \alpha = 0, \quad -\sigma_z \, dx \, dy + \sigma_n \, ds \, dy \sin \alpha - \gamma \frac{dx \, dy \, dz}{2} = 0.$$

Noting that $\cos \alpha = dz/ds$ and $\sin \alpha = dx/ds$, we find that these equations reduce to

$$-\sigma_x + \sigma_n = 0, \quad -\sigma_z + \sigma_n - \gamma \frac{dz}{2} = 0.$$

By allowing the volume to shrink to zero, we get $\sigma_x = \sigma_z = \sigma_n$. By reorienting the volume, we can show that $\sigma_y = \sigma_n$, and therefore

$$\sigma_x = \sigma_y = \sigma_z = \sigma_n = \sigma. \tag{1–23}$$

Thus the normal stress at a point in a fluid in equilibrium without shear is independent of the direction. It is a scalar quantity. The tension σ is the negative of the pressure p, that is,

$$\sigma = -p. \tag{1–24}$$

If the fluid is not in equilibrium, the normal stresses are not identical. However, as we shall see in Chapter 5, use is made of an average normal stress which becomes equal to the above value for equilibrium.

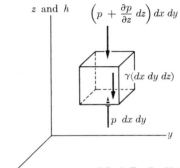

FIG. 1–7.   Equilibrium in a gravity field.

1–4.2   Equilibrium in a gravity field

Consider the equilibrium of forces on an elementary volume $dx\,dy\,dz$ in a gravity field. The forces are gravity in balance with pressures acting over the faces of the element. Selecting z to coincide with the vertical coordinate direction, h (Fig. 1–7), we have in the z-direction

$$p\,dx\,dy - \left[p\,dx\,dy + \left(\frac{\partial p}{\partial z}\,dz\right)dx\,dy\right] = \gamma\,dx\,dy\,dz, \quad \text{or} \quad -\frac{\partial p}{\partial z} = \gamma = -\frac{\partial p}{\partial h}. \tag{1–25}$$

In the other directions, we have

$$\frac{\partial p}{\partial x} = 0, \qquad \frac{\partial p}{\partial y} = 0. \tag{1–26}$$

Therefore, we have merely

$$\frac{dp}{dz} = \frac{dp}{dh} = -\gamma. \tag{1–27}$$

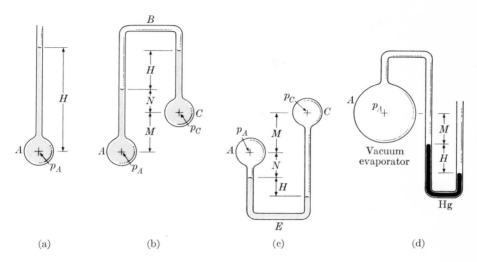

FIG. 1–8. Manometer examples.

Homogeneous liquids. By "homogeneous" we mean one species, and since liquids are essentially incompressible, a *homogeneous liquid* implies a constant density fluid. Then we can integrate Eq. (1–27) between elevations h_1 and h_2 and obtain

$$p_2 - p_1 = \gamma(h_1 - h_2), \quad \text{or} \quad \frac{p}{\gamma} + h = \text{const.} \quad (1\text{–}28)$$

This is the equation for *hydrostatic pressure distribution.* As elevation increases within a body of liquid, p decreases. If $p = p_{\text{atm}}$ at and above a free surface and $(h_1 - h_2) = d$, the depth below the free surface, we have

$$p - p_{\text{atm}} = \gamma d. \quad (1\text{–}29)$$

We note from this that everywhere over a plane of $d = \text{const}$, i.e., any horizontal plane, p is constant.

All the above equations give *pressure differences.* When we use Eqs. (1–27) or (1–28), either *gauge pressure*, measured relative to atmospheric pressure, or *absolute pressure*, measured relative to absolute vacuum, may be employed.

Devices known as *manometers* use the principle expressed by Eqs. (1–27) and (1–28) for measuring pressure differences in terms of heights of liquid columns. For example, applying Eq. (1–28) to Fig. 1–8(a), we find that the absolute pressure p_A at the center line of vessel A differs from p_{atm} at the open end of the manometer tube by the amount γH, giving

$$p_A = \gamma H + p_{\text{atm}}.$$

In Fig. 1–8(b), Eq. (1–28) is applied successively between the interface levels separating the three immiscible fluids. In this example, A and C are liquids and B is a gas or immiscible liquid of γ_B less than γ_A or γ_C. Then

$$p_A - M\gamma_A - N\gamma_A - H\gamma_B + H\gamma_C + N\gamma_C = p_C,$$

or

$$p_A - (M + N)\gamma_A + N\gamma_C + H(\gamma_C - \gamma_B) = p_C.$$

If $\gamma_A = \gamma_C$, we have

$$p_A + H(\gamma_C - \gamma_B) - M\gamma_C = p_C.$$

For $M = 0$ and $\gamma_B \lll \gamma_C$, we have simply

$$p_A + H\gamma_C = p_C.$$

Note that p_A and p_C may be gauge or absolute so long as both are in the same terms. In Fig. 1–8(c), γ_E exceeds both γ_A and γ_C so that we compute

$$p_A + N\gamma_A + H\gamma_E - H\gamma_C - (M + N)\gamma_C = p_C,$$

or

$$p_A - (M + N)\gamma_C + N\gamma_A + H(\gamma_E - \gamma_C) = p_C.$$

Example 1–4: Measuring vacuums with manometers

Consider the vacuum evaporator in Fig. 1–8(d) to which a mercury U-tube manometer is attached. When the manometer deflection H is 12 in. and the distance M is 18 in., find the absolute pressure and the vacuum in the evaporator.

Apply Eq. (1–28),

$$p_A + M\gamma_A + H\gamma_{Hg} = p_{atm}.$$

However, fluid A in the connection from the evaporator is a vapor so that $\gamma_A \lll \gamma_{Hg}$. Therefore $M\gamma_A$ is negligible and

$$p_A = p_{atm} - H\gamma_{Hg}.$$

From Table 1–2, we have $\gamma_{Hg} = 847$ lb/ft^3, so that

$$\text{Absolute pressure} = p_A = 14.7 - \frac{1 \times 847}{144} = 14.7 - 5.9 = 8.8 \text{ psia,}$$

$$\text{Vacuum} = p_{atm} - p_A = 5.9 \text{ psi}$$

$$= 12 \text{ in. Hg.}$$

The surface of a submerged object, such as shown in Fig. 1–9(a), is subject to forces due to the hydrostatic pressure. Letting p denote this hydrostatic pressure, we find that the resultant force will be

$$\mathbf{F} = \int d\mathbf{F} = \int p \, d\mathbf{A}. \tag{1–30}$$

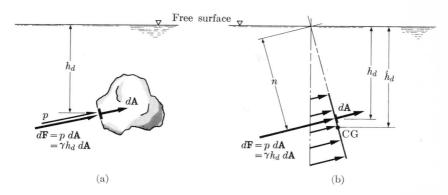

FIG. 1-9. Force due to hydrostatic pressure on submerged surfaces: (a) submerged object; (b) submerged plane.

Using Eq. (1–29), we have

$$\mathbf{F} = \gamma \int h_d \, d\mathbf{A}, \tag{1-30a}$$

where

$d\mathbf{F}$ = unit force acting in the direction of vector area element $d\mathbf{A}$,

h_d = submergence depth to the element $d\mathbf{A}$.

Taking the scalar product with the unit vectors $\mathbf{i}$, $\mathbf{j}$, and $\mathbf{k}$ gives the components

$$F_x = \gamma \int h_d \, dA_x, \qquad F_y = \gamma \int h_d \, dA_y, \qquad F_z = \gamma \int h_d \, dA_z, \tag{1-30b}$$

where A_x, A_y, A_z are the projections A on to planes normal to x, y, z.

For a plane surface (e.g., Fig. 1–9b), h_d is proportional to the distance n measured parallel to the surface, and each of the force components can be put in the form

$$F \propto \gamma \int n \, dA \propto \gamma \dot{n} A,$$

where $\dot{n}$ is the distance to the centroid (CG) of the plane area A. Thus, for the side of the plane which is subject to the hydrostatic pressure, we find the components

$$F_x = \gamma \dot{h}_d A_x, \qquad F_y = \gamma \dot{h}_d A_y, \qquad F_z = \gamma \dot{h}_d A_z, \tag{1-31}$$

where $\gamma \dot{h}_d$ is the hydrostatic pressure at the centroid of A. The resultant force is merely

$$\mathbf{F} = \gamma \dot{h}_d \mathbf{A}, \tag{1-31a}$$

so that we may compute the magnitude F directly without first determining its components.

The point of application of a single force of magnitude F which will produce the same equilibrium as the distributed pressure loading is called the center of

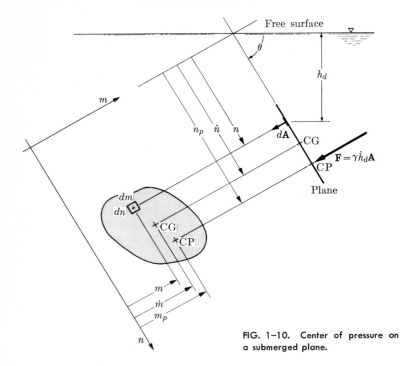

FIG. 1–10. Center of pressure on a submerged plane.

pressure (CP). To find this point for the general case of a plane at angle θ to the free surface, we use the notation of Fig. 1–10, in which the m-axis is taken as the intersection of the plane projected to the free surface and the n-axis lies in the plane.   Then $h_d = n \sin \theta$ and $dF = p \, dA = \gamma n \sin \theta \, dA$.   The coordinates m_p and n_p to the center of pressure are found by taking moments

$$m_p F = m_p \int dF = \int m \, dF, \qquad n_p F = n_p \int dF = \int n \, dF.$$

Substituting for dF and integrating gives

$$m_p = \frac{\gamma \sin \theta \int mn \, dA}{\gamma \sin \theta \int n \, dA} = \frac{I_{mn}}{\dot{n} A} = \frac{\dot{I}_{mn} + \dot{m}\dot{n}A}{\dot{n} A}, \tag{1–32}$$

$$n_p = \frac{\gamma \sin \theta \int n^2 \, dA}{\gamma \sin \theta \int n \, dA} = \frac{I_m}{\dot{n} A} = \frac{\dot{I}_m + \dot{n}^2 A}{\dot{n} A}, \tag{1–33}$$

where

I_{mn} = product of inertia about the axes $m = 0, n = 0$,

$\dot{I}_{mn}$ = product of inertia about the centroidal axes $m = \dot{m}, n = \dot{n}$,

I_m = moment of inertia (second moment) about the axis $n = 0$,

$\dot{I}_m$ = moment of inertia about the centroidal axis $n = \dot{n}$.

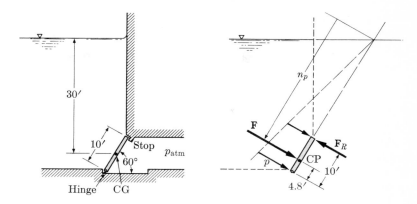

FIG. 1–11.   Forces on a submerged gate.

Note that the second moment is always positive so that the center of pressure always falls below the centroid.

Example 1–5: Forces on a submerged gate

A 10-ft by 20-ft rectangular gate hinged at its lower edge closes the entrance to a tunnel as shown in Fig. 1–11.   Determine the total hydrostatic pressure load on the gate and the force on the stop along the upper edge of the gate.   The depth h_d to the center of gravity of the gate is 30 ft.   Using Eq. (1–31a), we find that the magnitude of the total hydrostatic force is

$$F = 62.4 \times 30 \times 10 \times 20 = 37{,}500 \text{ lb,}$$

with horizontal and vertical components

$$F_H = \sin 60° \times 37{,}500 = 32{,}400 \text{ lb,} \qquad F_V = \cos 60° \times 37{,}500 = 18{,}750 \text{ lb.}$$

This load F has the same effect as a concentrated force acting at the center of pressure. Extending the line of the gate and the line of free surface level until they intersect gives an origin for the measurement of the n-coordinate used in deriving Eq. (1–33). The inclined distance to the centroid of the gate is

$$\dot{n} = \frac{30}{\sin 60°} = 34.7 \text{ ft.}$$

When we use Eq. (1–33), the inclined distance to the center of pressure is

$$n_p = \frac{\dot{I}_m + \dot{n}^2 A}{\dot{n} A} = \frac{(20 \times 10^3)/12 + (34.7)^2 \times 200}{34.7 \times 200} = 34.9 \text{ ft.}$$

The distance from the lower edge of the gate to the center of pressure is

$$(34.7 + 5) - 34.9 = 4.8 \text{ ft.}$$

Now taking moments about the lower edge we find the reacting force F_R at the stop on the upper edge as

$$F_R = \frac{37,500 \times 4.8}{10} = 18,000 \text{ lb.}$$

The resultant hydrostatic force on a curved or irregular surface is found by combining its components, because for all nonplanes the expression for the integral of $h_d \, dA$ depends on the coordinate direction. Consider a surface with curvature in all three coordinate directions. Let z coincide with the vertical direction and x and y be horizontal directions. From Eqs. (1–30b) we see that since h_d is a function only of the elevation of dA, $h_d \, dA_x$ and $h_d \, dA_y$ are both independent of the horizontal position of dA. Hence the integral of $h_d \, dA_x$, as well as that of $h_d \, dA_y$, is the same as the integral over a vertical plane of projected area A_x or A_y. Thus the horizontal force components are found by the same relations used for planes, namely

$$F_{Hx} = \gamma \dot{h}_d A_x, \qquad F_{Hy} = \gamma \dot{h}_d A_y.$$
$$(1-34)$$

This is illustrated for a cylindrical surface in Fig. 1–12, where the coordinates are chosen so that $F_{Hy} = 0$. On the other hand, the integral in the vertical direction is

$$\int h_d \, dA_z = \int z \, dA_z,$$

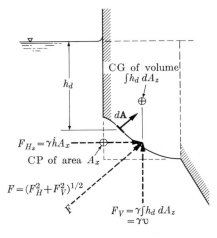

FIG. 1–12. Horizontal and vertical components on a curved surface.

which is merely the volume between the submerged surface and the free surface. Hence the vertical component equals the weight of a liquid mass of the same volume. Calling this volume $\mathcal{V}$, we have

$$F_V = \gamma \mathcal{V}. \qquad (1-35)$$

The sense of F_V depends only on whether the upper or lower side of the irregular surface is "wetted." As illustrated in Fig. 1–12, the horizontal force components act at the center of pressure of their respective projected areas A_x or A_y. The vertical component F_V acts through the centroid of the volume $\mathcal{V}$. For surfaces curved in all three dimensions, the three force components may not intersect so that the resultant of the distributed pressure loading may be a force plus a couple.

An extension of the computation of the vertical component is Archimedes' principle, which states that a submerged body is subject to a buoyant upthrust equal to the weight of the fluid displaced. The integral of $h_d \, dA_z$ taken over the surface of a completely submerged object is the volume of the object. Calling the volume $\mathcal{V}_B$, then, for the *buoyant force* F_B we have

$$F_B = \gamma \mathcal{V}_B. \tag{1–36}$$

The net weight of a submerged object is its true weight less the buoyant force, or

$$\text{net weight} = \text{true weight} - \gamma \mathcal{V}_B.$$

If the object is free and in equilibrium, the net weight is zero.

Example 1–6: Hydrometer

The buoyancy principle is the basis of the *hydrometer* shown in Fig. 1–13.
(a) The weighted bulb displaces a volume of 1.0 in^3 and the attached stem which has a cross-sectional area of 0.01 in^2 submerges to a depth of 2.0 in. in distilled water. Determine the total weight of the hydrometer.

For equilibrium the weight of the body equals the buoyancy upthrust. Hence

$$\text{hydrometer weight} = \gamma_{\text{H}_2\text{O}} \mathcal{V}_B = 62.4 \, \frac{1 + 2 \times 0.01}{1728} = 0.0368 \text{ lb.}$$

(b) If the hydrometer is placed in an oil having a specific gravity $(\gamma_{\text{oil}}/\gamma_{\text{water}})$ of 0.95, determine how far it will sink compared to its position when in water.

Again equating the hydrometer weight to the weight of displaced liquid, we have

$$\text{hydrometer weight} = \gamma_{0.95} \mathcal{V}_B,$$

$$0.0368 = 59.3 \, \frac{1 + s \times 0.01}{1728},$$

from which

$$s = \left[\frac{1728 \times 0.0368}{59.3} - 1 \right] \frac{1}{0.01} = 7.3 \text{ in.}$$

Therefore, the extra submergence is 5.3 in.

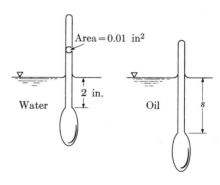

FIG. 1–13.   Hydrometer in water and oil.

Heterogeneous or variable liquids. Liquid bodies such as salt solutions of variable concentration or homogeneous liquids of variable temperature are found in estuaries, in lakes, and reservoirs, and are used in industrial processes. These are often stratified, and the equations for homogeneous liquids apply over each stratum. Considering a field of liquid or liquids of variable density, we can conclude that equilibrium is not possible unless the specific weight is constant in each horizontal layer. In addition, equilibrium will be stable only if the more dense liquids lie below the less dense. These conclusions also apply to cases of continuous variation in density.

Gases. Equations (1–28) apply whenever γ can be considered constant. More generally, however, it is necessary to integrate Eq. (1–27) with $\gamma = \gamma(h)$. For gases obeying Eq. (1–17), we can compute the pressure variation for any arbitrary temperature-elevation relation.

A useful temperature-altitude relation is defined for the *standard atmosphere* used when comparing the performance of flight vehicles. Beginning with the standard atmosphere at sea level, the air temperature is assumed to vary linearly with altitude to 35,000 ft and remain constant at $-67°$F above that elevation. The standard sea-level atmosphere is

$$p_0 = 14.7 \text{ psia,} \qquad T_0 = 59°\text{F} = 519°\text{R.} \tag{1–37}$$

The linear decrease with altitude h is taken to be

$$T = T_0 + Ch$$
$$= (519 - 0.00357\ h)°\text{R.}$$

For the linear change, we have a *constant gradient of temperature* so that

$$dT = C\ dh.$$

Combining this with Eqs. (1–17) and (1–27), we obtain

$$\frac{dp}{p} = -\frac{g\ dT}{CRT}.$$

Integrating from T_0 to T, we have

$$\ln \frac{p}{p_0} = -\frac{g}{CR} \ln \frac{T}{T_0}, \qquad \text{or} \qquad p = p_0 \left(\frac{T_0}{T_0 + Ch}\right)^{g/CR}. \tag{1–38}$$

For *isothermal* conditions above 35,000 ft, we have by Eq. (1–20),

$$\frac{p}{\gamma} = \frac{C'}{g} = K,$$

where $\gamma = \rho g$, $K = $ const. Then, from Eq. (1–27), we have

$$\frac{dp}{dh} = -\frac{p}{K},$$

and integrating from a known p_1, we obtain

$$\left.\begin{array}{l} \ln \dfrac{p_2}{p_1} = \dfrac{-1}{K}\ (h_2 - h_1) = \dfrac{-\gamma_1}{p_1}\ (h_2 - h_1), \\[3mm] p_2 = p_1 \exp\left[\dfrac{-\gamma_1}{p_1}\ (h_2 - h_1)\right]. \end{array}\right\} \tag{1–39}$$

or

Table 1–7 gives a tabulation of standard-atmosphere properties.

1–4.3 Capillarity effects

The property of surface tension causes special effects at liquid-liquid or liquid-gas interfaces or whenever we find mutual contact between three immiscible liquids; two immiscible liquids and a solid; or a liquid, a gas, and a solid. At simple interfaces there is the tendency of surfaces to wrinkle and form capillary waves when disturbed. Flat plane interfaces are in unstable equilibrium. The extreme is demonstrated by a drop of liquid which assumes a spherical shape when projected into the air, the common method of making lead "shot." The sphere is the stable configuration. When mutual contact occurs among three substances, stability and equilibrium depend on the magnitudes of the interfacial tensions and, if a solid surface is involved, on the degree of wetting.

Stable lenses at interfaces. Consider the case of two immiscible liquids, A and B, and a third fluid, either gas or immiscible liquid, C, in contact as shown in Fig. 1–14. To be in equilibrium, the interfacial surface tensions and angles of contact must be

$$\sigma_{AC} = \sigma_{AB} \cos \phi_{AB} + \sigma_{BC} \cos \phi_{BC},$$

$$\sigma_{AB} \sin \phi_{AB} = \sigma_{BC} \sin \phi_{BC}.$$

These equations can be satisfied if $\sigma_{AC} < (\sigma_{AB} + \sigma_{BC})$, and a lens of B will form. If $\sigma_{AC} > (\sigma_{AB} + \sigma_{BC})$, equilibrium is not possible, and liquid B will spread out between A and C.

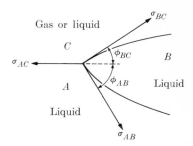

FIG. 1–14. Immiscible liquids in common contact.

Wetting of solid surfaces. At the boundaries separating gas-liquid-solid phases, the resulting free-surface configuration depends on the relative strengths of "cohesive" forces between liquid molecules and the "adhesive" forces between liquid and solid molecules. If the resulting affinity of the solid molecule for the liquid causes wetting, the solid surface is called *hydrophilic;* otherwise, *hydrophobic.* The contact angle θ at the liquid-solid-gas contact point is a measure of wetting (θ measured *in the liquid* from the solid surface). With reference to Fig. 1–15, this can be shown as follows: The work required to effect separation of liquid from a solid face depends upon their separate interfacial surface tensions, according to the relation

W_{SL} = work per unit area to separate solid and liquid,

$$= \sigma_{SG} + \sigma_{LG} - \sigma_{SL}.$$

For equilibrium, the surface tension forces must satisfy

$$\sigma_{SG} = \sigma_{SL} + \sigma_{LG} \cos \theta.$$

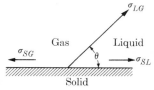

FIG. 1–15. Liquid-gas-solid common contact.

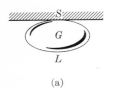
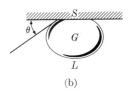
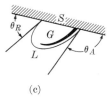

(a)   (b)   (c)

FIG. 1–16. Examples of wetting and contact angles: (a) stable bubble on hydrophilic surface, perfect wetting, $\theta = 0°$; (b) stable bubble on hydrophobic surface, imperfect wetting, $\theta > 0°$; (c) bubble at incipient motion on hydrophobic surface, $\theta_A > \theta_R$ (always).

By combining the two, we obtain

$$W_{SL} = \sigma_{LG}(1 + \cos \theta). \tag{1–40}$$

For zero contact, angle $W_{SL} = 2\sigma_{LG}$ and the liquid's attractions for the solid and for itself are the same (adhesion equals cohesion). When the liquid's attraction for the solid exceeds its own cohesive force, the angle will also be zero. For $\theta = 180°$, there would be no adhesion. Actually, there is always some adhesion, so θ is always less than 180°. Thus for perfect wetting, the contact angle measured *in the liquid* as shown in Fig. 1–16(a) is zero degrees. For complete nonwetting, $\theta = 180°$. All intermediate angles correspond to imperfect wetting as Fig. 1–16(b) shows.

The contact angle should be measured when the solid-gas interface is covered with a very thin film of liquid. If the solid surface is dry and wetting imperfect, θ depends on whether the liquid approaches an equilibrium position by advancing or by receding over the surface (giving θ_A and θ_R, respectively). As shown for the example in Fig. 1–16(c), $\theta_A > \theta_R$, a condition which is always true.

Capillarity in tubes. The behavior of liquids in narrow tubes (capillary tubes, from the Latin *capillus*, a hair) depends on surface tension and the wetting of solids. For example, as illustrated in Fig. 1–17 when a glass tube is dipped in water, the water in the tube rises above the hydrostatic level, while if the tube is dipped in mercury, a depression results. Wetting (or nonwetting) causes the

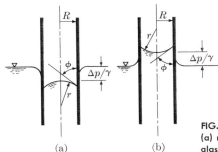

(a)   (b)

FIG. 1–17. Capillarity in tubes: (a) mercury in glass; (b) water in glass.

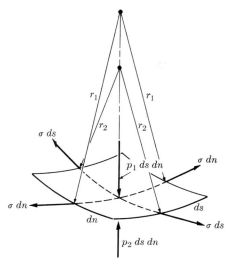

FIG. 1–18. Equilibrium under surface tension.

liquid surface to become curved. As a consequence, surface tension causes a pressure difference to develop across the surface. This pressure difference between convex to concave sides for a surface warped in two directions as shown in Fig. 1–18 is

$$\Delta p = p_1 - p_2 = \sigma \left(\frac{1}{r_1} + \frac{1}{r_2} \right), \tag{1–41}$$

where

$$\sigma = \text{interfacial surface tension,}$$
$$r_1, r_2 = \text{radii of curvature of the surface.}$$

For circular tubes, $r_1 = r_2 = r$, r being the radius of curvature of the *meniscus*, the name given to the interfacial surface. For tubes less than about 0.1 in. in diameter, the meniscus is nearly spherical and $r = R/\cos \phi$, where

$$R = \text{radius of tube,}$$
$$\phi = \text{acute angle measured at the liquid-solid contact point.}$$

(*Note:* For wetting, ϕ equals the contact angle θ as previously defined. For non-wetting ϕ is the supplement of θ.) Then the capillary rise (or depression) is given by

$$\frac{\Delta p}{\gamma} = \frac{2\sigma}{\gamma R} \cos \phi. \tag{1–42}$$

For clean water and clean glass, ϕ (and θ) is close to zero, while for mercury and glass, ϕ is about 50° (θ about 130°).

TABLE 1–1

DIMENSIONS AND UNITS OF PHYSICAL QUANTITIES

A. Mechanical Quantities

English Gravitational System:
Foot-pound(force)-second (ft-lb-sec)

Item	Symbol	Dimensions	Units
Length	L	$L = \dfrac{Ft^2}{m}$	$ft = \dfrac{lb\text{-}sec^2}{slug}$
Force	F	$\dfrac{mL}{t^2} = F$	$\dfrac{slug\text{-}ft}{sec^2} = lb$
Time	t	$t = \left(\dfrac{mL}{F}\right)^{1/2}$	$sec = \left(\dfrac{ft\text{-}slug}{lb}\right)^{1/2}$
Mass	m	$m = \dfrac{Ft^2}{L}$	$slug = \dfrac{lb\text{-}sec^2}{ft}$
Acceleration	a	$\dfrac{F}{m} = \dfrac{L}{t^2}$	$\dfrac{lb}{slug} = \dfrac{ft}{sec^2}$
Pressure*	p	$\dfrac{m}{Lt^2} = \dfrac{F}{L^2}$	$\dfrac{slug}{ft\text{-}sec^2} = \dfrac{lb}{ft^2}$
Density	ρ	$\dfrac{m}{L^3} = \dfrac{Ft^2}{L^4}$	$\dfrac{slug}{ft^3} = \dfrac{lb\text{-}sec^2}{ft^4}$
Specific weight	γ	$\dfrac{m}{L^2 t^2} = \dfrac{F}{L^3}$	$\dfrac{slug}{ft^2\text{-}sec^2} = \dfrac{lb}{ft^3}$
Viscosity (dynamic)	μ	$\dfrac{m}{Lt} = \dfrac{Ft}{L^2}$	$\dfrac{slug}{ft\text{-}sec} = \dfrac{lb\text{-}sec}{ft^2}$
Viscosity (kinematic)	ν	$\dfrac{L^2}{t}$	$\dfrac{ft^2}{sec}$
Bulk modulus of elasticity*	E_v	$\dfrac{m}{Lt^2} = \dfrac{F}{L^2}$	$\dfrac{slug}{ft\text{-}sec^2} = \dfrac{lb}{ft^2}$
Surface tension	σ	$\dfrac{m}{t^2} = \dfrac{FL}{L^2}$	$\dfrac{slug}{sec^2} = \dfrac{lb\text{-}ft}{ft^2}$

* These items are usually reported in pounds per square inch (psi).

TABLE 1–1 (Cont'd.)

DIMENSIONS AND UNITS OF PHYSICAL QUANTITIES

B.   Thermodynamic Quantities

Item	Symbol	Dimensions	Units
Temperature	T	T	°F or °R
Gas constant	R	$\dfrac{LF}{mT} = \dfrac{L^2}{t^2 T}$	$\dfrac{\text{ft-lb}}{\text{mass-°R}}$
Specific heat: Constant pressure Constant volume	$\left.\begin{array}{c} c_p \\ c_v \end{array}\right\}$		$\dfrac{\text{Btu}}{\text{mass-°R}}$
Specific internal energy	u		$\dfrac{\text{Btu}}{\text{mass}}$

C.   Conversion of Thermodynamic Quantities

Quantity	Given units	Desired units	Conversion multiplier
u	$\dfrac{\text{Btu}}{\text{lbm}}$	$\dfrac{\text{Btu}}{\text{slug}}$	$\dfrac{g_c}{(\text{lbm/slug})}$
$\dfrac{p}{\rho}$	$\dfrac{\text{lb/in}^2}{\text{slug/ft}^3}$	$\dfrac{\text{Btu}}{\text{lbm}}$	$\dfrac{144}{778.26\,g_c}$ $\dfrac{(\text{in}^2/\text{ft}^2)}{(\text{ft-lb/Btu}) \times (\text{lbm/slug})}$
$\dfrac{p}{\rho}$	$\dfrac{\text{lb/in}^2}{\text{slug/ft}^3}$	$\dfrac{\text{Btu}}{\text{slug}}$	$\dfrac{144}{778.26}$ $\dfrac{(\text{in}^2/\text{ft}^2)}{\text{ft-lb/Btu}}$
$c_v,\ c_p$	$\dfrac{\text{Btu}}{\text{lbm-°R}}$	$\dfrac{\text{Btu}}{\text{slug-°R}}$	$\dfrac{g_c}{(\text{lbm/slug})}$

TABLE 1–2

PROPERTIES OF COMMON LIQUIDS AT 68°F ATMOSPHERIC PRESSURE AND STANDARD GRAVITY

Substance	Density (ρ), slug/ft³	Specific weight (γ), lb/ft³	Dynamic viscosity ($\mu \times 10^5$), lb-sec/ft²	Kinematic viscosity ($\nu \times 10^5$), ft²/sec	Surface* tension (σ), lb/ft	Vapor pressure at 68°F (p_v), psia	Bulk modulus of elasticity (E_v), psi	Specific heat (c), Btu/lbm-°F
Alcohol (ethyl)	1.53	49.3	2.51	1.64	0.00153	0.85		0.581(77°F)
Benzene	1.71	54.9	1.37	0.80	0.00198	1.45	150,000	0.406
Carbon tetrachloride	3.09	99.5	2.035	0.66	0.00183	1.74		0.201
Freon-12 (at 66.3°F)	2.59	83.3				80.0		
Glycerin	2.45	78.8	3140.0	1280.0		2×10^{-6}	630,000	0.55
Kerosene	1.51–1.59	49–51			0.0016–0.0022			
Mercury	26.3	847	3.24	0.123	0.0352–0.0333†	2.3×10^{-5}	3,800,000	0.0332
Oil: crude	1.65–1.80	53–58						
fuel	1.80–1.90	58–61			0.0016–0.0026			
linseed	1.83	58.7	92.0	50.3	0.0023			
lubricating	1.65–1.70	53–55			0.0024–0.0026			
Water: fresh	1.937	62.31	2.11	1.09	0.00499	0.339	320,000	0.999
sea	1.99	64.0	2.28	1.14				
20% NaCl	2.23	71.6						

* In contact with air. † In contact with own vapor.

TABLE 1–3

PROPERTIES OF WATER AT ATMOSPHERIC PRESSURE AND STANDARD GRAVITY

Temperature (T), °F	Density (ρ), slug/ft^3	Specific weight (γ), lb/ft^3	Dynamic viscosity $(\mu \times 10^5)$, lb-sec/ft^2	Kinematic viscosity $(\nu \times 10^5)$, ft^2/sec	Surface* tension (σ), lb/ft	Vapor pressure (p_v), psia	Bulk modulus of elasticity (E_v), psi
32	1.940	62.42	3.75	1.93	0.00518	0.0885	292,000
40	1.940	62.43	3.23	1.66	0.00514	0.122	
50	1.940	62.41	2.74	1.41	0.00509	0.178	
60	1.938	62.37	2.36	1.22	0.00504	0.256	
68	1.937	62.31	2.11	1.09	0.00499	0.339	320,000
80	1.934	62.22	1.80	0.930	0.00492	0.507	
90	1.931	62.11	1.60	0.826	0.00486	0.698	
100	1.927	62.00	1.42	0.739	0.00480	0.949	
120	1.918	61.71	1.17	0.609	0.00467	1.69	332,000
140	1.908	61.38	0.981	0.514	0.00454	2.89	
160	1.896	61.00	0.838	0.442	0.00440	4.74	
180	1.883	60.58	0.726	0.385	0.00427	7.51	
200	1.868	60.12	0.637	0.341	0.00413	11.53	308,000
212	1.860	59.83	0.593	0.319	0.00404	14.70	

* In contact with air.

TABLE 1–4

SURFACE TENSIONS

Locus	Temperature (T), °F	Surface tension (σ), lb/ft
In presence of air:		
Silver-air	1778	0.0548
Sodium chloride-air	1824.8	0.00683
Water-air	68	0.00499
Zinc-air	1094	0.00485
Interfacial:		
Benzene-mercury	68	0.0257
Water-benzene	68	0.00240
Water-carbon tetrachloride	68	0.00308
Water-mercury	68	0.0257

TABLE 1–5

BULK MODULUS OF ELASTICITY FOR WATER E_v(psi) [2]

Pressure, psi	Temperature, °F				
	32	68	120	200	300
15	292,000	320,000	332,000	308,000	
1500	300,000	330,000	342,000	319,000	248,000
4500	317,000	348,000	362,000	338,000	271,000
15000	380,000	410,000	426,000	405,000	350,000

TABLE 1-6

PROPERTIES OF COMMON GASES AT ATMOSPHERIC PRESSURE AND 60°F

Gases	Density (ρ), slug/ft³	Specific weight (γ), lb/ft³	Dynamic viscosity (μ), lb-sec/ft²	Kinematic viscosity (ν), ft²/sec	Gas constant (R_{lbm}), $\dfrac{\text{ft-lb}}{\text{lbm-°R}}$	Specific heat mean values 32°F–400°F		Specific heat ratio (k)
						(c_v), Btu/(lbm-°R)	(c_p), Btu/(lbm-°R)	
Air	0.00237	0.0763	3.77×10^{-7}	1.6×10^{-4}	53.3	0.171	0.241	1.40
Ammonia	0.00141	0.0455	$2 \ \times 10^{-7}$	1.4×10^{-4}	90.8	0.406	0.523	1.29
Argon					38.6	0.075	0.124	1.66
Carbon dioxide	0.00363	0.117	3.06×10^{-7}	8.5×10^{-5}	35.1	0.16	0.205	1.28
Helium	0.000329	0.0106			386.2	0.746	1.24	1.66
Hydrogen	0.000165	0.00531	$2 \ \times 10^{-7}$	1.2×10^{-3}	766.8	2.435	3.42	1.40
Methane	0.00132	0.0424	$2.4 \ \times 10^{-7}$	1.8×10^{-4}	96.4	0.469	0.593	1.26
Oxygen	0.00262	0.0844			48.3	0.155	0.217	1.40
Nitrogen	0.00229	0.0739			55.2	0.177	0.248	1.40
Sulphur dioxide	0.00537	0.173			23.6			1.26
Water vapor (saturated steam)	0.00148	0.0475	$2.1 \ \times 10^{-7}$	7.7×10^{-3}	85.8	0.36	0.46	1.28

TABLE 1–7

Properties of Standard Atmosphere

Altitude, ft	Temperature (T), °F	Pressure (p), psfa	Density (ρ), slugs/ft³	Kinematic viscosity $(\nu \times 10^4)$, ft²/sec	Velocity of sound (c), ft/sec
0	59	2116.2	0.00238	1.56	1,117
1,000	55.44	2040.9	0.00231	1.60	1,113
2,000	51.87	1967.7	0.00224	1.64	1,109
3,000	48.31	1896.7	0.00218	1.68	1,105
4,000	44.74	1827.7	0.00211	1.72	1,104
5,000	41.18	1760.8	0.00205	1.77	1,098
10,000	23.36	1455.4	0.00176	2.00	1,078
15,000	5.54	1194.3	0.00150	2.28	1,058
20,000	−12.28	972.6	0.00127	2.61	1,037
25,000	−30.10	785.3	0.00107	3.00	1,016
30,000	−48.0	628.4	0.000880		995
35,000	−65.8	498.0	0.000737		973
36,000	−67.6	474.8	0.000707		971
37,000	−67.6	452.5	0.000676		971

TABLE 1–8

Thermodynamic Properties of Air [3]

Temperature (Abs) (T), °R	Enthalpy $\left(u + \dfrac{p}{\rho}\right)$, Btu/lbm	Internal energy (u), Btu/lbm	Temperature (Abs) (T), °R	Enthalpy $\left(u + \dfrac{p}{\rho}\right)$, Btu/lbm	Internal energy (u), Btu/lbm
100	23.74	16.38	600	143.47	102.34
140	33.31	23.71	620	148.24	105.78
180	42.89	30.55	640	153.09	109.21
220	52.46	37.38	660	157.92	112.67
260	62.03	44.21	680	162.73	116.12
300	71.61	51.04	700	167.56	119.58
320	76.40	54.46	720	172.39	123.04
340	81.18	57.87	740	177.23	126.51
360	85.97	61.29	760	182.08	129.99
380	90.75	64.70	780	186.94	133.47
400	95.53	68.11	800	191.81	136.97
420	100.32	71.52	840	201.56	143.98
440	105.11	74.93	880	211.35	151.02
460	109.90	78.36	920	221.18	158.12
480	114.69	81.77	960	231.06	165.26
500	119.48	85.20	1000	240.98	172.43
520	124.27	88.62	1040	250.95	179.66
540	129.06	92.04	1080	260.97	186.93
560	133.86	95.47	1120	271.03	194.25
580	138.66	98.90	1160	281.14	201.63
			1200	291.30	209.05

PROBLEMS

1-1. (a) Write the relation for shear stress for an ideal Bingham plastic as a function of the time rate of deformation. (b) An ideal Bingham plastic is placed between two parallel closely spaced flat plates. One plate is moved in a direction parallel to its face at 10 ft/sec when the spacing is 0.01 ft. If the shear stress which develops is 0.075 lb/ft^2 and the yield stress is 0.05 lb/ft^2, find the rate at which the stress increases with rate of deformation. (c) What would be the dynamic viscosity if the fluid were Newtonian instead of non-Newtonian?

1-2. (a) Given the density of water at atmospheric pressure and 68°F, compute the density at 1000 atm and 200°F, assuming the velocity of sound remains constant. Carry calculations to three significant figures. (b) Compute the specific gravity at 1000 atm and 200°F.

1-3. A Newtonian fluid is contained between parallel plates 0.04 in. apart. (a) Find the rate of angular deformation in radians/sec if one plate is moving relative to the other at a linear velocity of 4 ft/sec. (b) Find the dynamic viscosity if the shear stress on one plate is 0.50 lb/in^2.

1-4. Compute the change in specific internal energy of standard air as given in Table 1–7 for an altitude increase from zero to 37,000 ft. Assume air behaves as a perfect gas.

1-5. Assume that the perfect gas relations of Eqs. (1–10b), (1–11b), and (1–19) hold for the air data in Table 1–8. Determine how the specific heat ratio k and the gas constant R change between temperatures of 100°R and 1000°R.

1-6. The specific heats for chlorine gas in the temperature range of $T = 60 - 360°F$ are $c_p = 0.124$ Btu/lbm-°R and $c_v = 0.094$ Btu/lbm-°R. Consider a mass of the gas at $p = 14.7$ psia, $T = 300°F$. Assume that chlorine behaves as a perfect gas. (a) Compute the gas constant R_{lbm}. (b) Compute the change in specific internal energy when the temperature drops from 300°F to 100°F. (c) Compute the specific enthalpy change for the same change of conditions.

1-7. Compare the internal energy release of the same masses of bromine gas and methane gas each undergoing a temperature drop of 250°F, given the specific heat data below.

	c_p Btu/lbm-°R	c_p/c_v
Bromine	0.056	1.29
Methane	0.593	1.32

1-8. (a) Compute the gas constants R_{lbm} for bromine and methane using the data of Problem 1–7. (b) Compute the adiabatic elastic modulus E_v and the velocity of sound for these two gases at 250°F and $p = 14.7$ psia.

1-9. Compute and compare the velocity of sound in air at 59°F and 14.7 psia with that at 500°F and 100 psia.

1-10. Compute and compare the velocity of sound at 68°F and 14.7 psia in air, water, and mercury.

1-11. Compute the dynamic viscosity for standard air as given in Table 1–7. What conclusions can you draw regarding the effect of pressure on the dynamic viscosity of air?

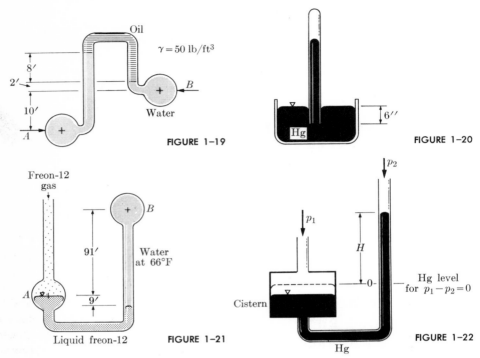

FIGURE 1–19

FIGURE 1–20

FIGURE 1–21

FIGURE 1–22

1–12. Develop the proof for the statement that a liquid of variable density can stand in equilibrium in a gravity field only if the specific weight is constant in each horizontal layer. Show that the equilibrium will be stable only if the more dense portions of the liquid lie below the less dense.

1–13. Determine the pressure intensity at A, if the pressure at B is 20 psig (Fig. 1–19).

1–14. A glass tube with a closed end is 48 in. long. It is completely filled with mercury after which its open end is immersed in a mercury-filled dish and the tube is placed in a vertical position as shown in Fig. 1–20. Determine the pressure intensity within the tube at its highest point and at a point 30 in. above the open end of the tube.

1–15. Freon-12 gas is entering the top of tank A at 66°F and condensing. Find the absolute pressure in the tank B. (See Fig. 1–21.)

1–16. A single-leg manometer is useful since only one reading is necessary to measure a pressure differential. For the manometer shown in Fig. 1–22, determine the necessary ratio of tube-bore diameter to the cistern diameter if the indicated height H of the mercury column is to be within 0.1% of the height corresponding to the actual pressure difference ($p_1 - p_2$).

1–17. The length of the liquid column for a given pressure differential is magnified by inclining the manometer leg. For the manometer shown in Fig. 1–23, the ratio of the cistern to manometer tube diameters is 10:1. Determine the angle α if the true pressure difference is 2.5 psf when $L = 1$ ft, where L is measured from the zero pressure position of the manometer fluid in the inclined tube.

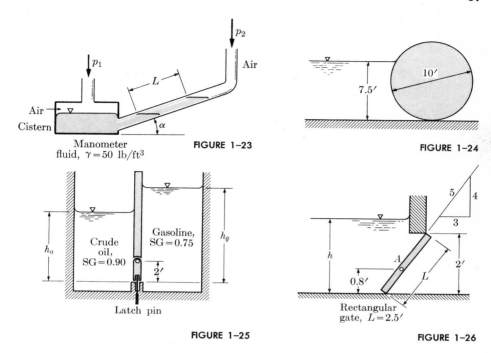

FIGURE 1-23

FIGURE 1-24

FIGURE 1-25

FIGURE 1-26

1-18. Determine the horizontal, vertical, and resultant forces acting on a drum gate 10 ft in diameter and 30 ft long when water stands to a depth of 7.5 ft as shown in Fig. 1-24. What are the direction and line of action of the resultant?

1-19. The storage tank illustrated in Fig. 1-25 is divided into two compartments, which are separated by a 2 × 2-ft square gate hinged at the top with a sill at the bottom of the tank. The left-hand side contains crude oil of SG = 0.90 and the right-hand side, gasoline of SG = 0.75. The oil side is filled to a depth h_0 = 5 ft. Determine the depth of gasoline, h_g, such that there will be no force on the latch pin.

1-20. For what value of the depth, h, will the rectangular pivot gate, A (Fig. 1-26), open? Neglect the weight of the gate.

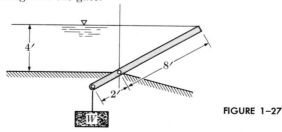

FIGURE 1-27

1-21. The flow from a canal is controlled by a counterweighted pivot gate as shown in Fig. 1-27. The rectangular gate is 8 ft high by 10 ft wide. Determine the value of the weight, W, such that the water will just spill when the depth of water in the canal is 4 ft.

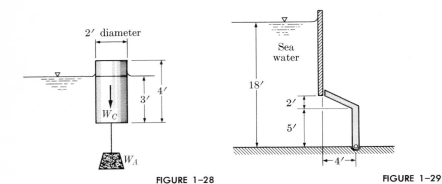

FIGURE 1–28   FIGURE 1–29

1–22. A cylinder 2 ft in diameter, 4 ft long, and weighing 75 lb (Fig. 1–28) floats in water with its axis vertical. An anchor weighing 150 lb/ft³ is hanging from the lower end. Determine the total weight of the anchor, given that the bottom of the cylinder is submerged 3 ft below the water surface.

1–23. A 1 foot by 1 foot square open tank holds water to a depth of 1 ft. It is placed on a platform weighing scale. A 6-in. × 6-in. wooden timber is lowered into the tank to a depth of 6 in. The upper end of this 6 in. × 6 in. timber is rigidly attached to the ceiling above the tank and scale. Determine the reading of the platform scale. Neglect the weight of the tank.

1–24. Determine the moment per foot of length of the gate shown in Fig. 1–29 which must be applied to hold the gate closed.

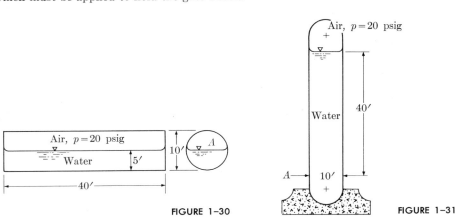

FIGURE 1–30   FIGURE 1–31

1–25. Determine the total force on the end wall A of the cylindrical tank (Fig. 1–30). Find the location of the center of pressure.

1–26. Refer to Fig. 1–31. (a) Determine the circumferential stress (hoop stress) at the level A for the circular steel water tank given that the wall thickness is $\frac{1}{2}$ in. (b) Determine the longitudinal stress at A.

REFERENCES

1. LEE, J.F., and F. W. SEARS, *Thermodynamics*, Addison-Wesley Publishing Co., Inc., Reading, Mass., 2nd ed., 1963, Chapter 1.
2. DAUGHERTY, R. L., "Some Properties of Water and Other Fluids," *Trans. ASME*, **57,** 5 (July, 1935).
3. KEENAN, J. H., and J. KAYE, *Gas Tables*, John Wiley and Sons, Inc., New York, 1948, Table 1.

CHAPTER 2

Kinematics

2–1 THE VELOCITY FIELD

Different portions of a moving fluid usually have different velocities and accelerations. The *field of motion* must then be described in terms of the velocities and accelerations of fluid particles at various points throughout the fluid-filled space. Both velocities and accelerations are vector quantities which we denote by the terms **q** and **a**, respectively. In Cartesian coordinates, the x-, y-, z-components are u, v, w and a_x, a_y, a_z. In general, **q** and **a** are functions of time and also depend on position in space at any instant.

There are two methods of describing the motion of a group of particles in a continuum. In the first, or Lagrangian method, the *coordinates* of moving particles are represented as functions of time. This implies that at some arbitrary time, t_0, the coordinates of a particle, (a, b, c), are identified and that thereafter we follow *that particle* through the flow field. The position of the particle at any other instant is given by a set of equations of the form:

$$x = f_1(a, b, c, t), \qquad y = f_2(a, b, c, t), \qquad z = f_3(a, b, c, t). \qquad (2\text{–}1)$$

The corresponding velocities and accelerations are

$$u = \left(\frac{\partial x}{\partial t}\right)_0, \qquad v = \left(\frac{\partial y}{\partial t}\right)_0, \qquad w = \left(\frac{\partial z}{\partial t}\right)_0, \qquad (2\text{–}2)$$

$$a_x = \left(\frac{\partial^2 x}{\partial t^2}\right)_0, \qquad a_y = \left(\frac{\partial^2 y}{\partial t^2}\right)_0, \qquad a_z = \left(\frac{\partial^2 z}{\partial t^2}\right)_0. \qquad (2\text{–}3)$$

The subscript zero on the partial derivatives is a reminder to hold the original coordinates, a, b, and c, constant. This approach is the one commonly used in the dynamics of solids where it is convenient to identify a discrete particle, e.g., the center of mass of a spring-mass system, and to determine the subsequent history of its movement in time.

When dealing with a fluid as a continuum of particles, the Lagrangian approach becomes extremely cumbersome because the description of the flow field requires three times as many parameters in the form of Eqs. (2–1). Furthermore, due to the deformable nature of the fluid medium, we are not usually concerned with the detailed history of an individual particle, but rather with the interrelation of flow properties at individual points in the flow field.

The second method of describing fluid motion, the one which will be used in this book, enables us to fix attention at discrete points without regard to the identity of the particles occupying the points at a given instant. When using this approach, known as the Eulerian method, the observer notes the flow characteristics in the vicinity of a fixed point as particles pass by. The description of the entire flow field is essentially an instantaneous picture of the velocities and accelerations of every particle. The basic difference between the two methods lies in the fact that in the Lagrangian method the *coordinates* of particles are represented as functions of time, whereas in the Eulerian method the particle *velocities* at various points are given as functions of time. Hence x, y, and z are independent variables, whereas in the Lagrangian method, they are dependent variables. The Eulerian velocity field is given by

$$\mathbf{q} = \mathbf{i}u + \mathbf{j}v + \mathbf{k}w, \tag{2–4}$$

where

$$u = f_1(x, y, z, t),$$
$$v = f_2(x, y, z, t),$$
$$w = f_3(x, y, z, t),$$

and $\mathbf{i}$, $\mathbf{j}$, and $\mathbf{k}$ are vectors of magnitude equal to unity and pointing in the directions of the x-, y-, z-axes.

Using this method, we must express the change in velocity in the vicinity of a point in terms of the partial derivatives of the four independent variables, x, y, z, t; for example, the velocity change in the x-direction is

$$du = \frac{\partial u}{\partial t}\, dt + \frac{\partial u}{\partial x}\, dx + \frac{\partial u}{\partial y}\, dy + \frac{\partial u}{\partial z}\, dz,$$

as a given particle moves through a small distance in the vicinity of the point in the time dt. The components of the distance moved are not independent but are $dx = u\, dt$, $dy = v\, dt$, and $dz = w\, dt$. By substituting these values into the above equation, we obtain the rate of change

$$\frac{du}{dt} = \frac{\partial u}{\partial t} + u\frac{\partial u}{\partial x} + v\frac{\partial u}{\partial y} + w\frac{\partial u}{\partial z}.$$

This is the total or substantial derivative representing the rate of change for a particle occupying a particular point in space at a particular time. It is made up of a "local" change as a function of time $[\partial u/\partial t]$ and a "convective" change dependent on the motion of the particle in space $[u(\partial u/\partial x) + v(\partial u/\partial y) + w(\partial u/\partial z)]$. Any other property of the fluid or its motion can be treated in this way. Thus the total rate of density change of a compressible fluid would be

$$\frac{d\rho}{dt} = \frac{\partial \rho}{\partial t} + u\frac{\partial \rho}{\partial x} + v\frac{\partial \rho}{\partial y} + w\frac{\partial \rho}{\partial z}.$$

2–2 STEADY VERSUS UNIFORM MOTION

The rules for the total derivative give, for the three acceleration components,

$$a_x = \frac{du}{dt} = \frac{\partial u}{\partial t} + u\frac{\partial u}{\partial x} + v\frac{\partial u}{\partial y} + w\frac{\partial u}{\partial z},$$

$$a_y = \frac{dv}{dt} = \frac{\partial v}{\partial t} + u\frac{\partial v}{\partial x} + v\frac{\partial v}{\partial y} + w\frac{\partial v}{\partial z}, \qquad (2\text{–}5)$$

$$a_z = \frac{dw}{dt} = \frac{\partial w}{\partial t} + u\frac{\partial w}{\partial x} + v\frac{\partial w}{\partial y} + w\frac{\partial w}{\partial z}.$$

In vector notation, we have

$$\mathbf{a} = \mathbf{i}a_x + \mathbf{j}a_y + \mathbf{k}a_z,$$

or

$$\mathbf{a} = \frac{d\mathbf{q}}{dt} = \frac{\partial\mathbf{q}}{\partial t} + (\mathbf{q}\cdot\nabla)\mathbf{q}, \qquad (2\text{–}5\text{a})$$

where the vector differential operator $\nabla = \mathbf{i}(\partial/\partial x) + \mathbf{j}(\partial/\partial y) + \mathbf{k}(\partial/\partial z)$ in Cartesian coordinates.

If all the local accelerations are zero, the motion is *steady*. The velocity may change from point to point in space, but at a fixed point there are no changes with time.

If all the convective accelerations are zero, the motion is *uniform*. Zero convective acceleration implies parallel flow as can be seen by inspecting the preceding equations. Hence we state the definition that motion is uniform if the velocity vectors are everywhere parallel.

2–3 ROTATING AND ACCELERATING COORDINATE SYSTEMS

The preceding relations for acceleration pertain to a *fixed* coordinate system. By this we mean fixed with respect to the apparent fixed stars of the universe. This is known as an inertial frame of reference. Many problems involve a moving frame of reference, the most common one being a coordinate system fixed to the earth, since in many cases it is obviously more convenient to describe fluid motions with respect to the earth than to the fixed stars. On the other hand, it is desirable to use the inertial framework in describing astronomical events or the dynamics of space travel.

Consider a moving frame whose origin translates and rotates with vector velocities $d\mathbf{R}/dt$ and $\boldsymbol{\Omega}$, respectively, relative to the fixed reference [1]. Using the notation in Fig. 2–1, we find that the velocity and acceleration of a particle in the moving frame have the following values relative to the fixed reference:

$$\text{Velocity} = \mathbf{q}_1 = \mathbf{q}_2 + \frac{d\mathbf{R}}{dt} + \boldsymbol{\Omega}\times\mathbf{r}, \qquad (2\text{–}6)$$

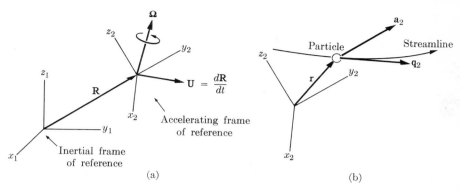

FIG. 2–1. Notation for inertial and accelerating frames of reference: (a) position, velocity and rotation vectors for accelerating frame; (b) particle location, velocity and acceleration vectors relative to accelerating frame.

where

$\mathbf{q}_1$ = vector velocity of fluid particle relative to fixed or inertial frame,

$\mathbf{q}_2$ = vector velocity of fluid particle relative to the moving frame,

$\mathbf{r}$ = vector position of a particular particle in the moving frame.

$$\text{Acceleration} = \mathbf{a}_1 = \frac{d\mathbf{q}_2}{dt} + \frac{d^2\mathbf{R}}{dt^2} + 2\mathbf{\Omega} \times \mathbf{q}_2 + \mathbf{\Omega} \times (\mathbf{\Omega} \times \mathbf{r}) + \frac{d\mathbf{\Omega}}{dt} \times \mathbf{r},$$

(2–7)

where

$\mathbf{a}_1$ = vector acceleration relative to the fixed frame,

$\dfrac{d\mathbf{q}_2}{dt} = \mathbf{a}_2$ = vector acceleration relative to the moving frame.

The term $\mathbf{\Omega} \times (\mathbf{\Omega} \times \mathbf{r})$ is called the *centrifugal acceleration* and is due to the angular velocity of the moving frame. Let z be the axis of rotation of $\mathbf{\Omega}$, then the magnitude of the centrifugal acceleration is $\Omega^2 r_{xy}$, where r_{xy} is the distance to the fluid particle measured in the xy-plane from the z-axis. The vector lies in the xy-plane and points toward the z-axis.

The term $2\mathbf{\Omega} \times \mathbf{q}_2$ is called the *Coriolis acceleration*. If z is the axis of rotation of $\mathbf{\Omega}$, the magnitude of the Coriolis acceleration is $2\Omega q_{2_{xy}}$, where $q_{2_{xy}}$ is the projection of $\mathbf{q}_2$ on the xy-plane. The vector lies in the xy-plane and points in a direction normal to $q_{2_{xy}}$.

The term $d\mathbf{\Omega}/dt \times \mathbf{r}$ is the tangential acceleration due to the angular acceleration of the moving frame.

The above interpretations for acceleration are directly applicable to a rotating coordinate system whose origin is at the center of the earth with the z-axis coincident with the axis of rotation and x and y in the equatorial plane, as shown in Fig. 2–2(a). Thus the vector $\mathbf{\Omega}$ is the angular velocity of rotation of the earth.

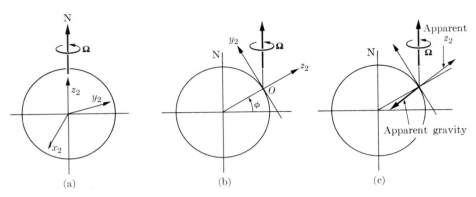

FIG. 2–2. Rotating coordinate systems fixed to the earth.

In treating fluid motion on the surface of the earth, it is usually permissible to neglect the curvature of the earth and to assume that within the region of interest its surface is a plane. Consider a right-handed coordinate system whose origin is fixed to the surface of the earth at some latitude ϕ, as shown in Fig. 2–2(b). The surface is represented by the xy-plane and z is directed normally outward which is the direction we call the vertical. In this case, the coordinate system still rotates with constant angular velocity $\mathbf{\Omega}$ as the point O translates in a circle at latitude ϕ. Further, let the xy-plane be oriented so that the x-axis is directed toward the East [into the plane of the paper in Fig. 2–2(b)] and the y-axis toward the North. Then, assuming that the origin O is not accelerating with respect to the fixed stars, the three acceleration components are [2]:

$$a_{1x} = \frac{du_2}{dt} - \Omega^2 x - 2\Omega(v_2 \sin \phi - w_2 \cos \phi),$$

$$a_{1y} = \frac{dv_2}{dt} - \Omega^2(R \cos \phi + y \sin \phi - z \cos \phi) \sin \phi + 2\Omega u_2 \sin \phi, \quad (2\text{–}8)$$

$$a_{1z} = \frac{dw_2}{dt} + \Omega^2(R \cos \phi + y \sin \phi - z \cos \phi) \cos \phi - 2\Omega u_2 \cos \phi,$$

where
$$\Omega = \text{magnitude of angular rotation vector of the earth,}$$
$$u_2, v_2, w_2 = x\text{-}, y\text{-}, z\text{-components of } \mathbf{q}_2, \quad R = \text{radius of earth.}$$

The terms containing Ω^2 are the centrifugal acceleration components.

Since the earth is not a true sphere, in practice the direction of z is adjusted slightly (by a few minutes of angle depending on the latitude) to coincide with the direction of "apparent gravity." Both gravity and centrifugal accelerations have vertical and horizontal components in this new system (see Fig. 2–2c). In the new system, the horizontal component of the "true" acceleration of gravity cancels the horizontal components of the centrifugal acceleration. The vertical component of the gravity force combines with the vertical component of the centrifugal acceleration to give an "apparent gravity." Only the "apparent gravity" can be determined by observation, and g as used throughout the book will refer to this quantity.

The terms in Eqs. (2–8) which contain Ω are the components of the Coriolis acceleration. Their effect upon fluid motion in a system using the apparent gravity can be seen if we consider that motion occurs only in the xy-plane. Therefore, $w_2 = 0$, and

$$a_{1x} = \frac{du_2}{dt} - (2\Omega \sin \phi)v_2, \qquad a_{1y} = \frac{dv_2}{dt} + (2\Omega \sin \phi)u_2. \qquad (2\text{–}9)$$

Since $\Omega = 0.73 \times 10^{-4}$ rad/sec, at latitude $\phi = 45°$ the value of $(2\Omega \sin \phi)$ is 1×10^{-4} rad/sec. Thus the magnitude of the acceleration is small and the Coriolis acceleration is usually neglected except when dealing with large masses such as the oceans and the atmosphere in which other accelerations (i.e., inertia forces per unit mass) are also small.

Finally, having accounted for centrifugal accelerations by the use of apparent gravity and by neglecting Coriolis accelerations, we have $\mathbf{a}_1 = \mathbf{a}_2$. This is the justification for applying Eqs. (2–5) to the rotating coordinate system fixed to the surface of the earth as used in most engineering problems, i.e., we treat the earth as though it were stationary.

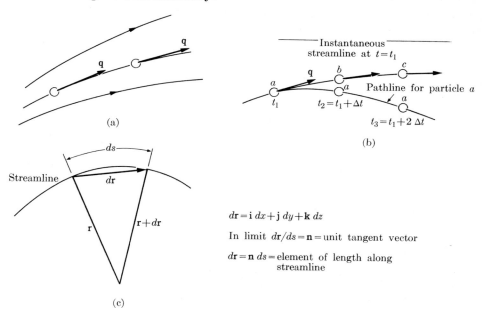

FIG. 2–3. Streamlines and path lines: (a) velocity vectors tangent to streamline; (b) streamline and path line; (c) element of arc length along streamline.

2–4 STREAMLINES VERSUS PATH LINES

A streamline is an imaginary line connecting a series of points in space at a given instant in such a manner that all particles falling on the line at that instant have velocities whose vectors are tangent to the line, as indicated in Fig. 2–3(a). Thus streamlines indicate the *direction of motion* at each point along the line at the given instant.

A stream tube or stream filament is a small imaginary tube or "conduit" bounded by streamlines. The streamlines are boundaries in the same sense that real conduit walls are boundaries. Conversely, the boundaries of a real conduit or of any immersed solid are streamlines. At solid boundaries, there is no velocity component normal to the boundary.

In steady motion the streamlines remain fixed with respect to the frame of reference. Furthermore, the steady flow streamlines represent the *path lines* for moving particles. In unsteady motion, a fluid particle will not, in general, remain on the same streamline, and hence path lines and streamlines do not coincide. Unsteady uniform flow is an exception to this rule. Figure 2–3(b) shows both a streamline and a path line for an unsteady nonuniform flow. At time t_1, the velocity vectors are shown for particles a, b, and c on the streamline; at times t_2 and t_3, the particle a is shown to occupy the successive positions shown on the path line.

For a streamline in a two-dimensional plane, xy, for example, the differential equation for the streamline is obtained by noting that

$$u = \frac{dx}{dt}, \qquad v = \frac{dy}{dt},$$

whence it follows that

$$\frac{dx}{u} = \frac{dy}{v}. \tag{2–10}$$

Alternatively, we note that since the velocity vector is tangent to a streamline, the equation of a streamline is

$$\mathbf{q} \times d\mathbf{r} = 0, \tag{2–10a}$$

where

$d\mathbf{r} =$ an element of length along the streamline as defined in Fig. 2–3(c).

Thus, in general for a three-dimensional field, we have in Cartesian coordinates

$$v\,dx = u\,dy,$$
$$w\,dx = u\,dz, \tag{2–11}$$
$$w\,dy = v\,dz.$$

We will have occasion to use these equations in later work involving integration along a streamline.

2–5 VELOCITY GRADIENTS AND SHEAR

The gradients $\partial u/\partial x$, $\partial u/\partial y$, etc., have appeared in previous descriptions. We have already seen in Chapter 1 how velocity gradients measure rate of deformation, and how shear can result in viscous fluids due to this deformation behavior. The deformation is a kinematic feature, while the shear stress is dynamic (involves a force). Thus the viscosity is a mechanism which, in addition to fluid mass (inertia), links the kinematics to the dynamic behavior of fluids in motion. Shear deformation and viscosity relations are treated more generally in Chapter 5.

PROBLEMS

2–1. Verify the components of acceleration on the surface of the earth with respect to a fixed coordinate system as given by Eqs. (2–8) by applying the necessary modifications and transformations to Eq. (2–7).

2–2. A water spreader is formed by a 20-ft long horizontal section of pipe of constant diameter which rotates in a horizontal plane about a vertical axis passing through one end of the pipe. The angular velocity is constant at 0.5 rev/sec. Find the components of the acceleration of a water particle relative to a fixed frame just before it emerges from the open end of the pipe; the magnitude of its exit velocity is 100 ft/sec relative to the pipe.

2–3. Imagine you are standing fixed to the center of a rotating turntable. A girl enters the room and walks straight toward the center of the turntable. What do *you* find for her velocity and acceleration if, relative to fixed coordinates, (a) her velocity $\mathbf{v}_t$ = const and the turntable rotates at $\boldsymbol{\Omega}$ = const; (b) her velocity $\mathbf{v}_t$ = const and the turntable is accelerating?

2–4. A rotating disk slings liquid particles outward. The particles leave the disk at an angle of 45° to the peripheral velocity direction. Obtain a formula for the acceleration of the particles relative to a fixed frame at the instant they clear the edge of the disk, assuming that their tangential velocity equals the disk peripheral velocity. Evaluate for a 1-ft disk rotating at 1200 rpm.

2–5. A rotating lawn sprinkler operating on the jet reaction principle discharges water from two diametrically opposite jet nozzles which are oriented at an angle of 60° to a radius vector and 30° to the horizontal. Obtain a relation for the acceleration of the water relative to a fixed frame at the instant of discharge from the nozzle when the sprinkler is accelerating up to speed. Express answer in terms of flow rate and radius of the sprinkler.

2–6. Compute the approximate deviation in magnitude and direction of the apparent gravity from the true gravity at a latitude of 45°.

2–7. Compute the approximate value of the Coriolis acceleration as a fraction of gravity acceleration at a latitude of 45° N for a stream of water flowing east at a velocity of 20 ft/sec.

2–8. Consider a particle of water on the surface of the ocean at latitude ϕ which is set into motion with a constant velocity $\mathbf{q}$ due to a rapidly moving atmospheric pressure disturbance. In the plane of the ocean surface the Coriolis acceleration acts at right angles to the direction of the velocity vector. In the Northern Hemisphere the deflecting force due to the Coriolis acceleration on a unit mass of water is to the right as one looks in the direction of the velocity vector. In the absence of friction the velocity remains constant in magnitude, and the particle will perform a series of approximately circular gyrations while drifting in the direction of the original motion. The equilibrium condition is obtained by a balance between the centrifugal acceleration due to the circular motion of the particle and the Coriolis acceleration due to the rotation of the earth. Derive expressions for the radius of the particle orbit and the time for one revolution of the particle, given that $\mathbf{q} = 1$ ft/sec at latitude 45°[3].

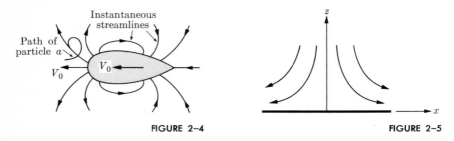

FIGURE 2–4   FIGURE 2–5

2–9. The instantaneous pattern of streamlines is shown in Fig. 2–4 for time $t = t_0$ for a strut moving with a constant velocity V_0 in an otherwise stationary fluid. The path line for a particle located at point a at $t = t_0$ is also shown for subsequent time intervals. This motion is unsteady relative to a fixed observer. It may be made steady by superimposing a flow field of uniform horizontal velocity V_0 from left to right, thereby making the velocity of the strut zero with respect to the observer. (a) Sketch the streamline pattern for this steady flow. (b) Show that the streamlines and path lines coincide for this condition.

2–10. A jet of incompressible fluid, symmetrical about the z-axis and moving in the negative z-direction, strikes a horizontal plate in the xy-plane as shown in Fig. 2–5. The velocity components are given as follows: $u = 10x, v = 10y, w = -20z$ (ft/sec). Find the equation of the streamlines in the zx-plane.

REFERENCES

1. INGARD, U., and W. L. KRAUSHAAR, *An Introduction to Mechanics, Matter, and Waves*, Addison-Wesley Publishing Co., Inc., Reading, Mass., 1960. Section 11–5 discusses the relations for rotating coordinate systems.

2. HAURWITZ, B., *Dynamic Meteorology*, McGraw-Hill Book Co., New York, 1941, p. 123.

3. VON ARX, W. S., *Introduction to Physical Oceanography*, Addison-Wesley Publishing Co., Inc., Reading, Mass., 1962, p. 102.

CHAPTER 3

Dynamic Features and Methods of Analysis

3–1 INTRODUCTION

When studying the dynamic behavior of fluids, we are usually concerned with some aspect of the fluid transport phenomena, namely the ability of fluids in motion to convey materials and properties from place to place and the mechanism by which these materials and properties are diffused and transmitted through the fluid medium. It is useful to categorize the available methods of analysis in terms of the various types of transport processes. In other words, the method of analysis should be chosen so as to apply physical laws which are pertinent to the problem to be solved. The fundamental transport phenomena which are associated with the motion of a fluid are mass, heat, and momentum transport. Each of these processes is, in turn, associated with a basic physical law which has been formulated as a result of observation and experience. The processes and laws may be summarized as follows:

Process	Observational law
Mass transport	Conservation of matter
Heat transport	Conservation of energy (first law of thermodynamics)
Momentum transport	Newton's second law (equation of motion).

The second law of thermodynamics and Maxwell's electrodynamic equations are other universal laws of comparable rank but of somewhat more restricted application in fluid dynamics. In addition, there are numerous subsidiary laws for properties of the continuum which enable us to describe molecular phenomena in terms of macroscopic quantities. The observational laws relating stress and strain (e.g., Hooke's elastic equation for solids and Newton's viscosity equation for fluids) and the equation of state for a perfect gas are examples of the subsidiary laws.

The scope and sophistication of technological problems in every branch of engineering has brought about the realization that basic knowledge of the transport process can no longer be compartmented with respect to the type of engineer being trained. In dealing with fluid flow, we should, from the outset, form the habit of thinking of methods of analysis in terms of the fundamental transport phenomena.

3–2 MASS TRANSPORT

With the exception of relativistic and quantum effects, all fluid motions must satisfy the principle of conservation of matter. If we are to analyze the motion of fluid, it is axiomatic that we must be concerned with mass transport. A more detailed discussion requires that we distinguish between homogeneous and nonhomogeneous fluids. A *homogeneous* fluid is one which exists throughout the region of interest as a single species. For example, air may undergo changes in density, velocity, and temperature, yet it may remain identifiable as the stable mixture of gases which we call air. Water, benzene, or mercury may likewise be compressed, heated, and accelerated, but unless a change of phase occurs, the liquids may be considered homogeneous.

A *nonhomogeneous* fluid is one in which two or more identifiable species exist within the region of interest. Nonhomogeneous fluids are characterized by variations in the amount of one substance relative to another from point to point in the system. The species may be of the same or different phases. For example, if a jet of carbon dioxide is discharged into the atmosphere, the concentration of CO_2 in the air varies from point to point, and the fluid is nonhomogeneous but of a single phase. A similar category occurs when a jet of fresh water is discharged into the ocean. A stream carrying solid sediment particles in suspension is an example of a nonhomogeneous, two-phase flow. A mixture of air bubbles and water is another nonhomogeneous, two-phase example.

In homogeneous fluids, the law of conservation of matter leads to an expression, known as the continuity equation, which relates the time and spatial variations in density and velocity. If the homogeneous fluid is also considered to be incompressible, the continuity equation reduces to a relation for the spatial velocity variation alone.

For a nonhomogeneous fluid, the conservation of matter principle must be satisfied for each component or species of the fluid mixture. In addition to the mass transport due to the local velocity of the flow of the mixture, there is an independent mass transport process due to the tendency of a component of the mixture to move in the direction of decreasing concentration of that component. Thus the individual species move with a velocity relative to the local velocity of the mixture. This is readily visualized by considering a beaker with the bottom half filled with a sodium chloride solution and the top half filled with fresh water. In this case, the local velocity of the system is zero, yet, after a period of time, sodium chloride will be detected in the region above the original interface. This process is known as molecular diffusion.

The many applications of diffusion and mass transport theory include the following: mixing and absorption processes of chemical technology; the pollution of air, surface and ground waters by contaminants; evaporation from oceans, lakes, and reservoirs; the aeration (oxygen transfer) of rivers with a consumption of oxygen by biological processes; the intrusion of saline ocean water in estuaries; and the separation of mixtures by distillation.

3–3 HEAT TRANSPORT

The fundamental concepts and definitions related to the macroscopic phenomenon known as heat form the basis of the science of *thermodynamics*. In *fluid dynamics*, we are primarily concerned with the transport of heat by fluid motion, whereas classical thermodynamics has largely been involved with relations between the equilibrium states of matter in bulk, the so-called nonflow processes.

By applying the principle of conservation of energy, also known as the first law of thermodynamics, to a flow process, we derive an equation which provides a useful relation between pressure, density, temperature, velocity, elevation, mechanical work, and heat input (or output). As we will see in Chapter 4, when analyzing liquid and low-speed gas flows, the thermodynamic conservation laws can be simplified since the heat capacity of the fluid is large compared to its kinetic energy. Hence the temperature and density remain essentially constant even though large amounts of the kinetic energy may be dissipated by friction. The energy conservation laws are used for nearly incompressible flow since it is a useful way of separating mechanically interchangeable energy from thermally degraded energy.

The most general formulations based upon the conservation of energy equations involve temperature variations from point to point in the flow. Hence, in addition to heat transfer by convection due to the velocity of the flow, we are concerned with the transfer of heat by conduction, which is the tendency for heat to move in the direction of decreasing temperature. This is analogous to the previously mentioned observation that mass is transported in the direction of decreasing concentration.

All forms of fluid machinery, such as compressors, pumps, and turbines, involve energy transfer in flow processes. Other applications are found in heat-exchange equipment and in technological processes accomplished by boiling and condensation. In the field of water resources, more attention is now paid to heat pollution and to the optimum use of cooling water for steam and nuclear power plants. In many instances, a large part of the flow of a river may be diverted through such a plant. Circulation and mixing processes in the oceans, lakes, and reservoirs are greatly influenced by small density variations caused by thermal stratification.

3–4 MOMENTUM TRANSPORT

Momentum is defined as the product of the mass of a particle and its velocity vector. Newton's second law provides the fundamental, nonrelativistic relation between the summation of forces acting on a particle and the time rate of change of its momentum. The resultant expressions are known as equations of motion. Momentum transport phenomena are of primary interest in fluid mechanics since they encompass the mechanisms of fluid resistance, boundary and internal shear stresses, and propulsion and forces on immersed bodies.

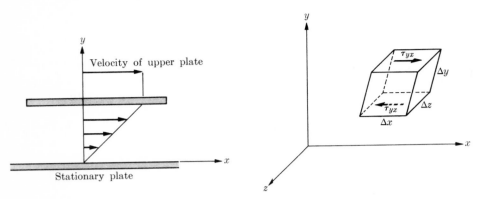

FIG. 3–1. Transverse momentum transport and shear stress in direction of momentum gradient for the case of laminar Couette flow.

As an example, consider the motion produced in the fluid occupying the space between two large parallel plates (Fig. 3–1). The upper plate is in motion and the lower plate is stationary. The fluid in immediate contact with either boundary takes on the velocity of the boundary in accordance with the no-slip condition discussed in Chapter 1. The fluid adjacent to the upper plate acquires a longitudinal momentum which in turn causes a longitudinal motion in the adjacent "layer." In order to satisfy the zero velocity condition required for the lowest layer, the velocity of each succeeding layer is smaller than that immediately above. Individual fluid masses thus acquire longitudinal momentum. Each layer obtains longitudinal momentum through momentum transport in the transverse direction. The transverse momentum transport is of the gradient type and is proportional to the transverse gradient of longitudinal momentum per unit volume of fluid. Note that the transverse momentum transport is in the direction of decreasing longitudinal momentum (toward the lower plate). This momentum transport process is therefore analogous to the transport of heat in the direction of decreasing temperature and to the transport of mass in the direction of decreasing concentration.

3–5 TRANSPORT ANALOGIES

In the previous paragraphs we have emphasized transport by two mechanisms. The first is convection, the direct process in which a fluid and any of its properties bodily move from place to place through a flow system. The second is conduction or diffusion, the process of movement of mass or heat or momentum in the direction of decreasing concentration of mass, temperature, or momentum. This common feature of transport due to a "driving force" arising from a gradient leads to analogous expressions relating rate of transport and magnitude of gradient, which we can state generally as

$$\frac{dA/dt}{\text{area}} = K\,\frac{dA/\mathrm{v}}{ds}\,, \tag{3–1}$$

where

$$\frac{dA/dt}{\text{area}} = \text{time rate of transport of } A \text{ per unit area}$$
$$\text{normal to transport direction,}$$

$$K = \text{a diffusivity constant,}$$

$$\frac{dA/\mathrm{v}}{ds} = \text{gradient of } A \text{ per unit volume of fluid in}$$
$$\text{the transport direction.}$$

For A we may substitute mass, heat, or momentum. In this formulation, the diffusivity constant depends on the mode of motion of the fluid. There are two modes, laminar flow and turbulent flow. We will discuss this distinction in more detail in Chapter 8. Here we merely note that if the flow moves in laminar streamlines without macroscopic mixing, then the transfer process is due to molecular diffusivity. If, on the other hand, turbulent eddying exists with consequent bodily mixing of fluid particles, then the transfer process is due to turbulent diffusivity. We will discuss transport under turbulent conditions in later chapters. Here we will summarize the analogies for the several molecular diffusion processes.

3–5.1 Momentum transport

In the parallel-plate problem previously discussed, flow is in the x-direction, and there is a gradient of momentum in the y-direction normal to the flow (Fig. 3–1). Then, in Eq. (3–1), we have

$$A = \text{momentum} = \Delta mu,$$

where

$$u = \text{local velocity,}$$
$$\Delta m = \text{the mass of a fluid particle,}$$

and

$$s = y = \text{direction of the gradient.}$$

From Eq. (3–1), we have

$$\frac{d(\Delta mu)}{dt} \cdot \frac{1}{\Delta x\,\Delta z} = K\,\frac{d}{dy}\left(\frac{\Delta mu}{\Delta \mathrm{v}}\right), \tag{3–2}$$

where $\Delta\mho$ is the volume of the particle. Newton's second law states that for any mass element the sum of the external forces acting on the element equals the time rate of change of momentum, or

$$\Delta F_x = \frac{d(\Delta mu)}{dt}.$$

In the parallel-flow problem with no effects except the relative motion of the two parallel boundaries, only shear stresses exist. Hence the left-hand term of Eq. (3–2) is

$$\frac{\Delta F_x}{\Delta x\,\Delta z} = \tau_{yx}, \tag{3–3}$$

where τ_{yx} is the shear stress parallel to the x-direction acting on a plane whose normal is parallel to the y-direction. A positive stress is a positively directed force acting on a positive face. The stresses on the two faces shown in Fig. 3–1 are both positive. Noting that by definition, $\Delta m/\Delta\mho = \rho$, the fluid density, we find that Eqs. (3–2) and (3–3) combine to give

$$\tau_{yx} = K\,\frac{d(\rho u)}{dy}. \tag{3–4}$$

If ρ is essentially constant, then

$$\tau_{yx} = \rho K\,\frac{du}{dy}. \tag{3–5}$$

By comparing Eq. (3–5) with Eq. (1–7) for the shear stress due to molecular mobility, we see that $\rho K = \mu$. Then it follows that the molecular diffusivity constant K for momentum transfer is the kinematic viscosity, so that

$$\tau_{yx} = \rho\nu\,\frac{du}{dy} = \mu\,\frac{du}{dy}. \tag{3–6}$$

The fundamental molecular process known as viscosity is therefore seen to result in a shear stress which is due to transverse molecular momentum transport. If a flow becomes turbulent, we shall see later that a much larger shear stress is developed due to the transverse momentum transport of macroscopic fluid particles. Then the diffusivity constant depends on mixing motions of the fluid particles and $K \ggg \nu$.

3–5.2   Heat transport

If the upper and lower parallel plates in our flow example are maintained at different temperatures, there will be a conduction of heat within the fluid. The quantity of heat, Q, in an element of mass is defined in the following manner:

$$A = \text{heat} = \mathrm{Q} = \Delta m c_p T \quad \text{(Btu)}.$$

Equation (3–1) becomes

$$\frac{dQ}{dt}\frac{1}{\Delta x\,\Delta z} = q_{H_y} = -K\frac{d}{dy}\left[\frac{\Delta m c_p T}{\Delta \mathcal{U}}\right], \tag{3–7}$$

where q_H is the time rate of heat transfer per unit area normal to the direction of transport (Btu/sec-ft^2). The negative sign indicates that a positive heat transfer is in the direction of the decreasing temperature. If ρ and c_p are essentially constant, then

$$q_{H_y} = -\rho c_p K \frac{dT}{dy}. \tag{3–8}$$

The product $\rho c_p K = k$, where k is the thermal conductivity expressed in Btu/(sec-ft-°R). Hence the molecular diffusivity constant becomes

$$K = \frac{k}{\rho c_p} = \alpha, \tag{3–9}$$

where α is the thermal diffusivity in ft^2/sec.

3–5.3 Mass transport

If one of the parallel plates is coated with a material which is slowly dissolving into the fluid, there will be a variation in the concentration of the material in the fluid in the direction normal to the flow. The quantity of dissolved material in an element of the fluid is given by

$$A = \text{dissolved mass} = \Delta m c_A,$$

where c_A is the concentration defined as the mass of dissolved substance per unit mass of fluid. Equation (3–1) becomes

$$\frac{d(\Delta m c_A)}{dt}\frac{1}{\Delta x\,\Delta z} = j_{A_y} = -K\frac{d}{dy}\left[\frac{\Delta m c_A}{\Delta \mathcal{U}}\right], \tag{3–10}$$

where j_A is the time rate of mass transfer per unit area normal to the direction of transport, in mass/(sec-ft^2). The negative sign indicates a positive transport in the direction of decreasing concentration. If the concentrations are small and the fluid density reasonably constant, then

$$j_{A_y} = -\rho K \frac{dc_A}{dy}. \tag{3–11}$$

For this case, the molecular constant K becomes the mass diffusivity, D, expressed in ft^2/sec, also known as the *molecular diffusion coefficient*.

In summary, each of the three molecular transport processes is described by a kinematic fluid property, ν, α, and D; each has the dimensions of a length squared divided by time.

3–6 PARTICLE AND CONTROL–VOLUME CONCEPTS

3–6.1 Infinitesimal elements and control volumes

Each of the observational laws for mass, heat, and momentum transport may be formulated in the Eulerian sense of focusing attention on a fixed point in space. There are two basic methods of arriving at the Eulerian equations of fluid mechanics in the general, three-dimensional framework. These will be referred to as the *material* method and the *control-volume* method.

In the *material* method we describe the flow characteristics at a fixed point x, y, z by observing the motion of a material particle of infinitesimal mass, dm, in the vicinity of the point. The rate at which any function $f(x, y, z, t)$ varies for this moving particle is given by the substantial derivative discussed in Section 2–1. For example, in an inertial frame of reference, the acceleration of the particle is given by Eqs. (2–5). The equations of motion for the material particle then follow from Newton's second law written in the form in which we consider the summation of forces acting on and accelerating the particle of mass dm,

$$d\mathbf{F} = dm\,\mathbf{a}. \tag{3–12}$$

If it is desired to describe the fluid motion relative to a noninertial frame of reference, the acceleration vector in Eq. (3–12) must be replaced by Eq. (2–7).

In the *control-volume* method, instead of looking at the incremental displacement of a particle, we are concerned with a fixed differential control volume of fluid. In the Cartesian framework this control volume has dimensions Δx, Δy, and Δz, which, in the limit, express conditions at a point x, y, z. In this approach the observational laws may be formulated in terms of flux through the control volume and the time rate of accumulation of either momentum, heat, or mass within the control volume. For example, the equations of motion may be obtained from Newton's second law in the form which states that the sum of the external forces on the differential control volume equals the time rate of change of momentum,

$$\Delta\mathbf{F} = \frac{d(\Delta m \mathbf{q})}{dt} = \frac{d(\rho\,\Delta x\,\Delta y\,\Delta z \mathbf{q})}{dt}. \tag{3–13}$$

For infinitesimal particles and differential control volumes the end result of either method is identical so long as the fluid may be considered a continuum. In later developments we will have occasion to use both methods to obtain the three-dimensional equations of fluid mechanics. Alternative formulations may be obtained for special frames of reference such as cylindrical or spherical coordinates. It should be apparent that the type of coordinate system chosen should be the one most readily adapted to the geometry of the problem. For instance, in studying the flow of fluids in a circular pipe, cylindrical coordinates would be more convenient than the Cartesian system. Provided that one can overcome the mathematical difficulties, it should be possible to treat all fluid-flow problems using the equations obtained from either the differential control volume or the material-particle method.

3–6.2　Finite control volume

Due to the nature of the problem under investigation, frequently the information sought is related to gross descriptions of the flow rather than to point-to-point variations of quantities such as velocity and internal shear stress. For example, consider the flow of a fluid in a tube with a constriction caused by the insertion of a circular plate with a hole whose diameter is less than that of the tube. We then ask ourselves whether we can find a relation between the pressures measured at two places, upstream and downstream of the plate, and the volume rate of the flow in the tube. We can, but the analytical formulation is most readily done by means of a *finite* control volume rather than with the differential control volume. The finite control volume would encompass the entire fluid-filled space within a longitudinal portion of the tube, including the section containing the plate. In accordance with the Eulerian viewpoint, the finite control volume is fixed in space and the observational laws of mass, heat, and momentum transfer can be applied to the instantaneous mass of fluid contained therein. This method is frequently used for a "one-dimensional" analysis, since the interest may be primarily in variations in flow characteristics in the *direction of flow* rather than in the variations at a cross section.

3–7　SCOPE OF ANALYTICAL TREATMENT

The method of finite control volumes is usually easier to treat analytically than the differential control-volume or particle approach. However, a thorough understanding of the internal flow mechanism can be gained only by one of the latter methods.

The emphasis in this book will be on the flow of real fluids as contrasted to the consideration of ideal fluids of zero viscosity. It is precisely in the realm of real fluid behavior that the analytical difficulties of the material or differential control-volume methods become most apparent. Chapter 4 will present the finite control-volume formulation of the conservation laws for matter, energy, and momentum. Chapter 5 will introduce the material-particle approach and will develop the necessary stress-strain relations for a fluid in the Cartesian coordinate system. Chapter 6 will treat the differential control-volume equations for conservation of matter in homogeneous fluids and the three-dimensional equations of motion for the material particle. The corresponding differential control-volume formulation of the energy equation will not be presented, since this form is primarily useful in more advanced work in heat transfer. Differential control-volume equations for conservation of matter in nonhomogeneous fluids will be developed in Chapter 16.

CHAPTER 4

Continuity, Energy, and Momentum Equations for Finite Control Volumes

4–1 CONSERVATION OF MATTER IN HOMOGENEOUS FLUIDS

For homogeneous (single species) fluids or uniform mixtures, the expression for the conservation of matter is known as the *continuity equation*. We will develop this equation using the finite control-volume analysis and, in order to gain some familiarity with the finite control-volume approach, we will use several formulations.

4–1.1 Arbitrary control volume

Consider an arbitrary control volume, fixed to the reference frame in the general velocity field $\mathbf{q}(x, y, z, t)$, as shown in Fig. 4–1. Although the control volume remains fixed, the mass of fluid originally enclosed occupies the volume within the dashed lines at a time dt later. We now distinguish three spatial regions (A), (B), and (C). Since the mass m is conserved, we have

$$(m_A)_t + (m_B)_t = (m_B)_{t+dt} + (m_C)_{t+dt}, \qquad (4\text{--}1)$$

or

$$\frac{(m_B)_{t+dt} - (m_B)_t}{dt} = \frac{(m_A)_t - (m_C)_{t+dt}}{dt}. \qquad (4\text{--}2)$$

In the limit, the term on the left-hand side of Eq. (4–2) becomes the time rate of change of mass in the original control volume,

$$\frac{\partial(m_B)}{\partial t} = \frac{\partial}{\partial t}\int_{\text{cv}} (\rho \, d\mathcal{v}), \qquad (4\text{--}3)$$

where $d\mathcal{v}$ is a volume element.

The term on the right-hand side of Eq. (4–2) represents the net flux of matter through the control surface. This flux is into the control volume at (1) and out at (2) and is due to the velocity of the flow. Thus Eq. (4–2) may be written

$$\frac{\partial}{\partial t}\int_{\text{cv}} (\rho \, d\mathcal{v}) = \int \rho q_n \, dA_1 - \int \rho q_n \, dA_2, \qquad (4\text{--}4)$$

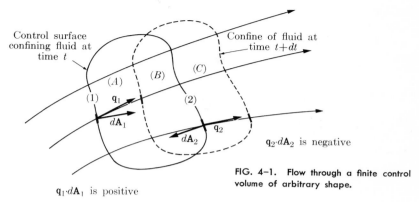

FIG. 4–1. Flow through a finite control volume of arbitrary shape.

where q_n is the component of the velocity vector normal to the surface of the control volume. In vector form we have

$$\frac{\partial}{\partial t} \int_{cv} (\rho \, d\mathcal{V}) = \oint_{cs} \rho \mathbf{q} \cdot d\mathbf{A}, \tag{4–5}$$

where $d\mathbf{A}$ = vector differential area pointing in the inward direction over an enclosed control surface. Hence $\mathbf{q} \cdot d\mathbf{A}$ is positive for an inflow into the control volume and negative for an outflow. This is the general equation of continuity for a fluid. It applies to a fluid of single species or to a uniform mixture.

It should be emphasized that the control volume defined at some initial time t remains fixed relative to the reference frame. If the fluid continues to occupy the entire control volume at subsequent times, the limits of integration of the first term of Eq. (4–5) are independent of time and the differentiation may be performed on the integrand rather than the integral. Under these conditions, Eq. (4–5) becomes

$$\int_{cv} \frac{\partial \rho}{\partial t} \, d\mathcal{V} = \oint_{cs} \rho \mathbf{q} \cdot d\mathbf{A}. \tag{4–5a}$$

To understand the difference between Eqs. (4–5) and (4–5a) we may visualize an open tank which is initially filled with a liquid. The control volume is coincident with the tank walls and the free surface of the liquid in its initial condition. If a port is opened at the bottom of the tank, the free surface elevation will decrease with time as the tank empties. Under these conditions the liquid no longer occupies the entire control volume and the time rate of change of the total mass of liquid in the tank should be evaluated in accordance with the left-hand side of Eq. (4–5).

As discussed in Section 3–2, for a nonhomogeneous fluid mixture we have, in addition, the conservation of mass equations for the individual species. In these we must also consider the mass flux across the boundaries of the control volume caused by concentration gradients of the individual species. Conservation-of-matter equations for the individual species, known as convective diffusion equations, are discussed in Chapter 16.

If the total mass within the control volume is constant in time, Eq. (4–5) becomes

$$\oint_{cs} \rho \mathbf{q} \cdot d\mathbf{A} = 0. \tag{4–6}$$

This equation applies to steady flow of a compressible fluid. With incompressible fluids the left side of Eq. (4–5) is zero for both steady and unsteady conditions, therefore

$$\oint_{cs} \mathbf{q} \cdot d\mathbf{A} = 0. \tag{4–6a}$$

In all of the preceding equations the control volume is fixed to an arbitrary xyz-reference system. Since the velocities are referred to the same coordinate system, it follows that the fluid velocities must be measured relative to the control volume. It is not necessary that the xyz-reference be an inertial system.

4–1.2   Stream-tube control volume

If a flow is steady, it is sometimes convenient to take a control volume with longitudinal boundaries coincident with a stream tube and transverse boundaries at right angles to the streamlines as shown in Fig. 4–2. Since, by definition, there is no flow across the longitudinal boundary, the control surface integral of Eq. (4–6) need be evaluated only at the transverse sections (1) and (2) shown in Fig. 4–2. Then the mass flow rate is

$$\rho_1 q_1 \, dA_1 = \rho_2 q_2 \, dA_2. \tag{4–7}$$

The positions of sections (1) and (2) are arbitrary; hence for steady flow the mass-flow rate along a stream tube is

$$\rho q \, dA = \text{const.} \tag{4–7a}$$

If the density is constant and uniform throughout, $\rho_1 = \rho_2$, and

$$q_1 \, dA_1 = q_2 \, dA_2 = dQ, \tag{4–8}$$

where dQ is the volume rate of flow.

The velocity ratio between any two points along a stream tube is

$$\frac{q_1}{q_2} = \frac{\rho_2 \, dA_2}{\rho_1 \, dA_1}. \tag{4–9}$$

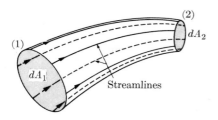

FIG. 4–2. Steady flow stream tube control volume.

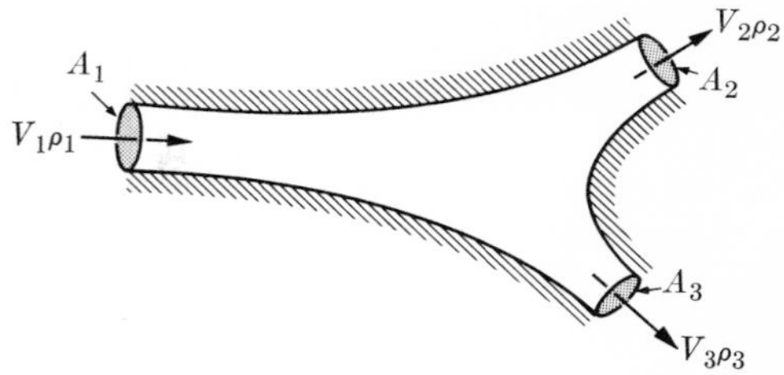

FIG. 4–3. Control volume coincident with boundaries of conduit.

4–1.3 Control volume coincident with conduit boundaries

The continuity concepts developed for the stream-tube control volume may be extended to include a control volume whose longitudinal boundaries coincide with the conduit boundaries. Then we may integrate Eq. (4–7) over A_1 and A_2,

$$\int_{A_1} \rho_1 q_1 \, dA_1 = \int_{A_2} \rho_2 q_2 \, dA_2 = \int \rho \, dQ. \tag{4–10}$$

If we define average velocities and densities as

$$V = \frac{\int q \, dA}{A}$$

and

$$\rho' = \frac{\int \rho \, dQ}{Q},$$

Eq. (4–10) becomes

$$\rho_1' V_1 A_1 = \rho_2' V_2 A_2, \tag{4–11}$$

which is the "one-dimensional" form of the continuity equation. Often ρ is essentially uniform normal to the flow direction, and we write

$$\rho_1 V_1 A_1 = \rho_2 V_2 A_2. \tag{4–12}$$

Again, if the density is constant and uniform throughout, $\rho_1 = \rho_2$, and

$$Q = V_1 A_1 = V_2 A_2. \tag{4–13}$$

We may also apply Eq. (4–6) to a branching conduit as shown in Fig. 4–3. Then we have

$$\int_{A_1} \rho_1 q_1 \, dA_1 = \int_{A_2} \rho_2 q_2 \, dA_2 + \int_{A_3} \rho_3 q_3 \, dA_3,$$

or

$$\rho_1' V_1 A_1 = \rho_2' V_2 A_2 + \rho_3' V_3 A_3. \tag{4–11a}$$

4–2   THE GENERAL ENERGY EQUATION

4–2.1   The first law of thermodynamics*

The concepts of continuity and energy will be combined with the first law of thermodynamics in a finite control-volume analysis to develop a general energy equation. First, let us restate two definitions already mentioned in Chapter 1 as they apply to systems of finite amounts of fluid.

(1) *A* **property** *of a system is* **any observable characteristic** *of the system. Properties include such physical quantities as location, velocity, pressure, temperature, mass, and volume.*

(2) *A* **state** *of a system is* **its condition as identified through properties** *of the system. Two states are identical if every property of the system is the same in both instances. The state of a system is fixed if a sufficient number of independent properties are established; e.g., for a gas at rest two of the three quantities, pressure, mass, and temperature, are sufficient to fix the state and establish all other properties.*

Now we will state the first law of thermodynamics as:

The difference between the heat added to a system of masses and the work done by the system depends only on the initial and final states of the system.

This difference is called the *change in energy, E*. By this definition, we see that energy is a "property"; that is, if for a given state of the system an arbitrary value of E is assigned, the corresponding value of E at any other state is unique. In equation form, the first law is

$$\delta Q - \delta W = dE, \tag{4–14}$$

where

$\delta Q =$ heat added *to* the system from its surroundings

$\delta W =$ work done *by* the system on its surroundings,

$dE =$ increase in energy of the system,

and

$\delta(\ )$ denotes an increment of a nonproperty,

$d(\ )$ denotes an increment of a property.

This equation is also known as the principle of *conservation of energy*. It may be expressed in terms of the rate of change of energy due to the rate of heat transfer and the rate of doing work:

$$\frac{\delta Q}{dt} - \frac{\delta W}{dt} = \frac{dE}{dt}. \tag{4–15}$$

---

* See references at end of chapter.

When applying this law to the flow of fluids, we note that both the work and energy terms may be composed of one or more components. The work done by the fluid system on its surroundings may include:

(a) $W_{pressure}$, *work of normal stresses* (pressure intensity) acting on the boundaries of the system;

(b) W_{shear}, *work of tangential stresses* (shear stresses) done at the boundaries of the system on adjacent external fluid which is in motion;

(c) W_{shaft}, *shaft work* done on a rotating element in the system and transmitted outside the system through a rotating shaft.

Thus the conservation-of-energy equation (4–15) may be written as

$$\frac{\delta Q}{dt} - \frac{\delta W_{pressure}}{dt} - \frac{\delta W_{shear}}{dt} - \frac{\delta W_{shaft}}{dt} = \frac{dE}{dt}. \tag{4–15a}$$

The energy E is stored energy and equals its specific value, energy per unit mass e, summed over the entire mass. The stored energy is composed of three types:

(a) $e_u = u$, *internal* energy per unit mass associated with the local temperature level of the fluid;

(b) e_p, *potential* energy per unit mass;

(c) $e_q = q^2/2$, *kinetic* energy per unit mass associated with the local velocity of the fluid, q relative to the reference frame.

Then the total stored energy per unit mass is

$$e = e_u + e_p + e_q = u + e_p + q^2/2. \tag{4–16}$$

For the gravity field of the earth, $e_p = gh$, where h is the local elevation of the fluid.

We stop here to emphasize that the system of units used is immaterial so long as it is a consistent single system represented by the simple form of Newton's equation $\mathbf{F} = m\mathbf{a}$. In particular, we will use force in pounds and mass in slugs. Equation (4–16) and the following equations will not hold for mixed units (e.g., force in pounds and mass in pounds). A suitable conversion must be used.

The internal energy of a substance is associated with (a) the activity of the molecules comprising the substance and of the elementary particles comprising the molecules, and (b) with forces existing between the molecules. The molecular activity represents a storage of kinetic energy: the intermolecular forces, a form of potential energy. The molecular kinetic energy is dependent primarily on temperature. The intermolecular potential energy is dependent primarily on the phase of the substance; changes in conditions which do not change the phase cause only relatively minor variations. Thus, for changes in conditions without a change in phase, the most significant variations in internal energy will be due to temperature variations. For changes in conditions leading to change in phase, large changes of internal energy may be obtained at constant temperature.

For systems in equilibrium with small or zero changes in kinetic or potential energy, dE/dt is due entirely to changes of internal energy. Thus, in classical thermodynamics, E is commonly called *internal energy*.

Example 4–1: Conversion of units for stored energy terms

Usually internal energy is given in units of Btu/lbm, elevation in ft, velocity in ft/sec, and acceleration in ft/sec². All terms in Eq. (4–16) can be expressed in any set of common units using the following table of conversion factors (Table 4–1).

TABLE 4–1

CONVERSION MULTIPLIERS FOR STORED ENERGY UNITS

Quantity	u	h	$q^2/2$
Multiply by value in table to obtain these units → Given in these units ↓	$\dfrac{\text{Btu}}{\text{lbm}}$	$\text{ft} = \dfrac{\text{ft-lb}}{\text{lb}}$	$\dfrac{\text{ft-lb}}{\text{slug}}$
$\dfrac{\text{Btu}}{\text{lbm}}$	1	$778.26\,(g_c/g)$ $\left(\dfrac{\text{ft-lb}}{\text{Btu}}\right)\left(\dfrac{\text{lbm}}{\text{lb}}\right)$	$(778.26)\,g_c$ $\left(\dfrac{\text{ft-lb}}{\text{Btu}}\right)\left(\dfrac{\text{lbm}}{\text{slug}}\right)$
ft	$\dfrac{g/g_c}{778.26}$ $\dfrac{\text{lb/lbm}}{(\text{ft-lb})/\text{Btu}}$	1	g $\dfrac{\text{lb}}{\text{slug}}$
$\left(\dfrac{\text{ft}}{\text{sec}}\right)^2 = \dfrac{\text{ft-lb}}{\text{slug}}$	$\dfrac{1/g_c}{778.26}$ $\dfrac{\text{slug/lbm}}{(\text{ft-lb})/\text{Btu}}$	$\dfrac{1}{g}$ $\dfrac{\text{sec}^2}{\text{ft}}$	1

Usually the ratio g/g_c is near unity and can be neglected.

4–2.2 The general energy equation

Let us apply the first law to flow through a generalized apparatus as represented in Fig. 4–4 by the control volume fixed relative to the reference frame. During a time interval dt, there will be a flux of mass and energy into the fixed control volume entering in region (1) and leaving at region (2). Mechanical work can be transmitted through the shaft, and heat can be transferred across the

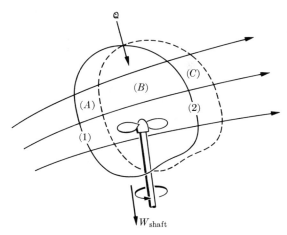

FIG. 4–4.   Control volume for energy balance with heat transfer and shaft work.

control-volume boundaries. As shown in Fig. 4–1, the fluid inside the control volume at time t will occupy the volume represented by the dashed outline at a time $t + dt$ later. As before, we are interested in the rate of happenings as dt becomes vanishingly small.

To apply Eq. (4-15a) to this general situation, we first note that the rate at which work of normal stresses will be done on a unit area equals the unit force due to local pressure times the velocity component in the force direction. This is $p(\mathbf{q} \cdot d\mathbf{A})$, where $(\mathbf{q} \cdot d\mathbf{A})$ is positive for flux into the control volume and negative for flux out. Thus the net rate at which work of pressure is done by the fluid on the surroundings outside the control volume is

$$\frac{\delta W_{\text{pressure}}}{dt} = -\oint_{\text{cs}} p(\mathbf{q} \cdot d\mathbf{A}).$$

The rate change of stored energy E will be the net rate of energy flux through the control volume plus the rate of change inside the volume,

$$\frac{dE}{dt} = -\oint_{\text{cs}} e\rho(\mathbf{q} \cdot d\mathbf{A}) + \frac{\partial}{\partial t} \int_{\text{cv}} (e\rho \, d\mathcal{V}).$$

We can now write the general energy equation as

$$\frac{\delta Q}{dt} - \frac{\delta W_{\text{shaft}}}{dt} - \frac{\delta W_{\text{shear}}}{dt} = -\oint_{\text{cs}} \left(\frac{p}{\rho} + e \right) \rho(\mathbf{q} \cdot d\mathbf{A}) + \frac{\partial}{\partial t} \int_{\text{cv}} (e\rho \, d\mathcal{V}),$$

$$(4\text{--}17)$$

where the work of normal stresses is combined with the stored energy integral.

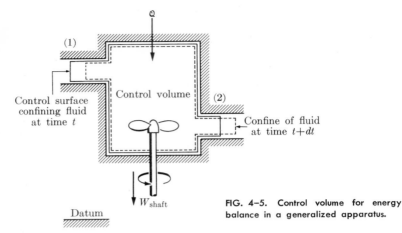

FIG. 4–5. Control volume for energy balance in a generalized apparatus.

Now assuming that e_p is due only to the gravitational field of the earth, we obtain the *general energy equation for the gravity field*,

$$\frac{\delta Q}{dt} - \frac{\delta W_{\text{shaft}}}{dt} - \frac{\delta W_{\text{shear}}}{dt} = -\oint_{\text{cs}} \left(\frac{p}{\rho} + u + gh + \frac{q^2}{2} \right) \rho (\mathbf{q} \cdot d\mathbf{A}) + \frac{\partial}{\partial t} \int_{\text{cv}} e\rho \, d\upsilon.$$

$$(4\text{–}18)$$

In application of Eq. (4–18), it is convenient to eliminate the term for work of shear stresses by a suitable choice of the control-volume boundaries. Thus in Fig. 4–5 the boundaries of the control volume coincide with the inner fixed boundary of the apparatus and are normal to the flow lines at (1) and (2). Because the velocity is zero at fixed boundaries and there are no tangential forces at (1) and (2), the shear stress work is zero. Thus Eq. (4–18) reduces to

$$\frac{\delta Q}{dt} - \frac{\delta W_{\text{shaft}}}{dt} = -\oint_{\text{cs}} \left(u + \frac{p}{\rho} + gh + \frac{q^2}{2} \right) \rho (\mathbf{q} \cdot d\mathbf{A}) + \frac{\partial}{\partial t} \int_{\text{cv}} (e\rho \, d\upsilon). \quad (4\text{–}19)$$

For steady motion, the last term is zero, so that

$$\frac{\delta Q}{dt} - \frac{\delta W_{\text{shaft}}}{dt} = -\oint_{\text{cs}} \left(u + \frac{p}{\rho} + gh + \frac{q^2}{2} \right) \rho (\mathbf{q} \cdot d\mathbf{A}). \quad (4\text{–}20)$$

For steady flow, the energy of the system within the control volume will remain constant, so that any net energy flux must be accounted for by the heat-transfer flux or shaft-work flux.

Equation (4–20) involves only the states at the boundaries of the control volume and the energy transfers across the control boundary to and from the system. Thus, without knowledge of interior local states, we see that the equation permits calculation of changes that occur between two separated points in a system.

Equation (4–20) is general within the restrictions of steady motion and absence of shear stress work. Allowances are made for energy transfers to and from the system, and the effects of friction are also included. Friction effects may not appear explicitly but will always be accounted for implicitly. Friction results in a degradation of mechanical energy into heat which in turn, (a) may be transferred away as Q without change in system temperature, (b) may cause a temperature change and consequent modification of internal energy, density, and other system properties and may also affect the magnitude of the shaft work. For the special case of a constant-density fluid in a system without shaft work, the total friction effect will be measured by the sum of the change in internal energy and heat transfer.

Equation (4–20) can also be applied to ideal nonviscous fluids for which shear stresses do not exist. The heat, work, and energy transfer relations will be the same as those for ideal frictionless processes.

Use of the previous equations requires knowledge of the properties of substances, and the reader is referred to Chapter 1 for tables and curves which give typical values for common fluids. We should also note that the sum $(u + p/\rho)$ which appears in the preceding equations is enthalpy as defined in Chapter 1. Let us caution again that consistent units must be used in each of the above equations. In particular, the units of u and p/ρ usually must be converted to the slug-density basis.

4–2.3   One-dimensional steady-flow equations

Equation (4–20) does not deal with the detailed variation of properties within the control volume. However, its evaluation does depend on variations in velocity, pressure, and density, and on all other properties over each of the cross sections (1) and (2) in Fig. 4–5. In many cases when Eq. (4–20) is applied to flow through conduits, the variations are small, and thus it is convenient to approximate the true conditions by assuming that properties are uniform normal to the flow direction. If this were true, the flow would be one-dimensional, i.e., the only significant changes in properties would occur with the dimension of distance along the conduit.

If we make this assumption and denote the average kinetic energy per unit mass as $V^2/2$, the integrated form of Eq. (4–20) is

$$\frac{\delta Q}{dt} - \frac{\delta W_{\text{shaft}}}{dt} = \left[u + \frac{p}{\rho} + gh + \frac{V^2}{2} \right]_2 \rho Q - \left[u + \frac{p}{\rho} + gh + \frac{V^2}{2} \right]_1 \rho Q,$$

where we have used the substitutions

$$\int_1 \rho\,(\mathbf{q} \cdot d\mathbf{A}) = \rho Q = \text{mass flow rate into control volume},$$

$$\int_2 \rho\,(\mathbf{q} \cdot d\mathbf{A}) = -\rho Q = \text{mass flow rate from control volume}.$$

Dividing by (ρQ),

$$\frac{\text{heat transfer}}{\text{slug}} - \frac{\text{shaft work}}{\text{slug}} = \left[u + \frac{p}{\rho} + gh + \frac{V^2}{2} \right]_2 - \left[u + \frac{p}{\rho} + gh + \frac{V^2}{2} \right]_1,$$

and dividing again by the gravity acceleration g yields

$$\frac{\text{heat transfer}}{\text{lb}} - \frac{\text{shaft work}}{\text{lb}} = \left[\frac{u}{g} + \frac{p}{\gamma} + h + \frac{V^2}{2g} \right]_2 - \left[\frac{u}{g} + \frac{p}{\gamma} + h + \frac{V^2}{2g} \right]_1.$$

$$(4\text{–}21)$$

Equation (4–21) pertains to one-dimensional steady flow of any fluid. In applying Eq. (4–21), we use average values for p, γ, h, u, and V at each flow section. This is a good approximation for the first four of these items so long as the flow is parallel at the observation section. The velocity, however, is known to vary from zero at the conduit wall to a maximum at the center. The true average for the flux of kinetic energy is

or

$$\left. \begin{array}{c} \displaystyle \int \left(\frac{\text{KE}}{\text{mass}} \right) \left(\frac{\text{mass}}{\text{sec}} \right) = \int \frac{q^2}{2} \, \rho (\mathbf{q} \cdot d\mathbf{A}) = K_e \frac{\rho}{2} V^3 A \\[3mm] \displaystyle \int \frac{\rho}{2} q^2 \, dQ = K_e \frac{\rho}{2} V^2 Q, \end{array} \right\} \qquad (4\text{–}22)$$

where

V = magnitude of average velocity over cross section = Q/A,

$K_e \geq 1$.

Equation (4–21) would then be written

$$\frac{\text{heat transfer}}{\text{lb}} - \frac{\text{shaft work}}{\text{lb}} = \left[\frac{p}{\gamma} + h + K_e \frac{V^2}{2g} \right]_2$$

$$- \left[\frac{p}{\gamma} + h + K_e \frac{V^2}{2g} \right]_1 + \frac{(u_2 - u_1)}{g}. \qquad (4\text{–}23)$$

Practical values of K_e for circular ducts are:

laminar flow (parabolic velocity distribution) $K_e = 2$,

turbulent flow (smooth pipe) $K_e \approx 1.06$.

The kinetic-energy flux correction K_e is a percentage correction on two subtractive terms. It can be important for cases where the difference in kinetic energy flux is large relative to other terms. Moreover, the effects tend to be more pronounced if the difference is between two large numbers, or if the shape of the velocity profile (and K_e) changes markedly from section to section.

For a fluid of uniform density, γ is constant and we customarily write

$$\left[\frac{p_1}{\gamma} + h_1 + K_{e_1}\frac{V_1^2}{2g}\right] = \left[\frac{p_2}{\gamma} + h_2 + K_{e_2}\frac{V_2^2}{2g}\right] + \frac{\text{shaft work}}{\text{lb}}$$

$$- \frac{\text{heat transfer}}{\text{lb}} + \frac{(u_2 - u_1)}{g}. \qquad (4\text{–}24)$$

Note that each term in Eq. (4–24) has the units of foot-pounds per pound of fluid flowing, or simply feet. Also note that the algebraic difference between internal energy gain and heat transfer from the fluid represents a net "decrease" in the mechanical energy of the system. For liquids this decrease is not conveniently reconverted to mechanical energy and is considered a "loss." We can write Eq. (4–24) as

$$H_1 = H_2 + \Delta H_M + H_{L_{1-2}}, \qquad (4\text{–}24a)$$

where

$H_1, H_2 =$ weight flow rate average values of energy, (ft-lb)/lb;

$\quad\quad\, =$ weight flow rate average values of total head, ft, for all streamlines passing through the section;

$\Delta H_M =$ shaft work transmitted from the system to the outside, (ft-lb)/lb or ft;

$H_{L_{1-2}} =$ loss of mechanical energy, (ft-lb)/lb or ft.

Here K_e is taken as unity unless the circumstances of the problem indicate that errors in the net flux will cause pronounced errors in the quantity sought.

For compressible fluids where density variations are important, losses due to heat of friction cannot be isolated using the general energy equation, as we mentioned in the preceding section.

When friction losses are negligible and there is no shaft work and no heat transfer, the internal energy is constant and Eq. (4–24) reduces to

$$\frac{p_1}{\gamma} + h_1 + K_{e_1}\frac{V_1^2}{2g} = \frac{p_2}{\gamma} + h_2 + K_{e_2}\frac{V_2^2}{2g}. \qquad (4\text{–}25)$$

If $K_{e_1} = K_{e_2} = 1$, we have the *one-dimensional Bernoulli equation*, namely,

$$H = \frac{p_1}{\gamma} + h_1 + \frac{V_1^2}{2g}$$

$$= \frac{p_2}{\gamma} + h_2 + \frac{V_2^2}{2g}. \qquad (4\text{–}26)$$

Equation (4–26) states that the total head along a conduit is constant. It represents a special form of a Bernoulli equation which we will derive and discuss in Chapter 6.

Example 4–2: Application of general energy equation

Consider the steady flow of air through a heat exchanger and into a turbine as shown in Fig. 4–6. The mass rate of air through-flow $= 0.621$ slug/sec, and the turbine develops 500 hp to power an external load. For the following given conditions find the rate of heat transfer and determine whether heat is added to or rejected from the air.

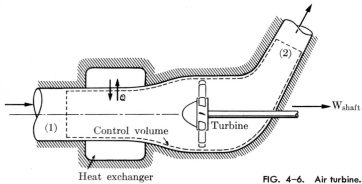

FIG. 4–6. Air turbine.

Given:

At inlet, section (1)	At outlet, section (2)
$p_1 = 15.3$ psig	$p_2 = 0$ psig
$T_1 = 200°F$	$\rho_2 = 0.00118$ slugs/ft^3
$A_1 = 2$ ft^2	$A_2 = 0.5$ ft^2.

Solution: (1) We complete the definition of flow at sections (1) and (2). Using Eq. (1–17) with the gas constant, $R_{slug} = g_c R_{lbm} = (32.2)(53.3)$ (see Eqs. (1–18) and Table 1–6), we have

$$p_1 = 15.3 + 14.7 = 30 \text{ psia} = 4320 \text{ psfa},$$

$$T_1 = 200°F + 460 = 660°R,$$

$$\rho_1 = \frac{p_1}{RT_1} = \frac{4320}{(32.2)(53.3)(660)} = 0.00382 \text{ slug/ft}^3,$$

$$V_1 = \frac{\text{mass flow rate}}{\rho_1 A_1} = \frac{0.621}{(0.00382)(2)} = 81.4 \text{ ft/sec},$$

$$p_2 = 14.7 \text{ psia} = 2118 \text{ psfa},$$

$$T_2 = \frac{p_2}{R\rho_2} = \frac{2118}{(32.2)(53.3)(0.00118)} = 1040°R = 580°F,$$

$$V_2 = \frac{0.621}{(0.00118)(0.5)} = 1050 \text{ ft/sec}.$$

(2) We apply the general energy equation (4–18). In this example

$$\frac{\delta W_{\text{shear}}}{dt} = 0$$

because the control-volume boundaries coincide with the fixed boundaries of the apparatus, and

$$\frac{\partial}{\partial t} \int_{\text{cv}} e\rho \, d\mathcal{U} = 0$$

because the flow is steady. Hence the general energy equation reduces to Eq. (4–20), namely,

$$\frac{\delta Q}{dt} - \frac{\delta W_{\text{shaft}}}{dt} = - \oint_{\text{cs}} \left(u + \frac{p}{\rho} + gh + \frac{q^2}{2} \right) \rho (\mathbf{q} \cdot d\mathbf{A}).$$

Now, assuming one-dimensional flow at (1) and (2) and the average kinetic energy per unit mass, $V^2/2$ [that is, $(K_e)_1 = (K_e)_2 = 1$], we can integrate Eq. (4–20) to give

$$\frac{\delta Q}{dt} = \frac{\delta W_{\text{shaft}}}{dt} + \left(u + \frac{p}{\rho} + gh + \frac{V^2}{2} \right)_2 \rho Q - \left(u + \frac{p}{\rho} + gh + \frac{V^2}{2} \right)_1 \rho Q.$$

Next, we introduce

$$\frac{\delta W_{\text{shaft}}}{dt} = (500 \text{ hp}) \, [550 \text{ (ft-lb)/(sec-hp)}]$$

$$= 2.8 \times 10^5 \text{ (ft-lb)/sec},$$

$$\left(u + \frac{p}{\rho} \right)_2 - \left(u + \frac{p}{\rho} \right)_1 = c_p(T_2 - T_1) \qquad \text{by Eq. (1–11b)}$$

$$= \frac{k}{k-1} R(T_2 - T_1) \quad \text{by Eq. (1–19) with mass in slugs}$$

$$= \frac{1.4}{1.4 - 1} g_c 53.3 \, (1040 - 660)$$

$$= 22.8 \times 10^5 \text{ (ft-lb)/slug},$$

and

$$\frac{V_2^2 - V_1^2}{2} = \tfrac{1}{2}[(1050)^2 - (81.4)^2]$$

$$= 5.5 \times 10^5 \text{ ft}^2/\text{sec}^2 = 5.5 \times 10^5 \text{ (ft-lb)/slug}.$$

Then, neglecting $(gh_2 - gh_1)$ which is small compared to other terms, we compute the rate of heat transfer to be

$$\frac{\delta Q}{dt} = (2.8 \times 10^5) + 0.621(5.5 + 22.8) \times 10^5$$

$$= + 20.4 \times 10^5 \text{ (ft-lb)/sec}.$$

Since $\delta Q/dt$ is positive, the heat is added to the air.

The several heat, work, and energy quantities are tabulated below. Note that the heat added to the air in the system goes to (a) increase the internal energy, (b) increase the potential for doing work by normal pressures, (c) increase the kinetic energy, (d) do work on the surroundings.

Quantity	Transferred into system	Transferred out of system
Internal energy	28.1×10^5 (ft-lb)/slug	44.3×10^5 (ft-lb)/slug
Work of normal pressures	11.3×10^5	17.9×10^5
Potential energy	0	0
Kinetic energy	0	5.5×10^5
Heat transfer	32.8×10^5	0
Shaft work	0	4.5×10^5
Total	72.2×10^5 (ft-lb)/slug	72.2×10^5 (ft-lb)/slug

4–2.4   Grade lines—energy (total head) and hydraulic (piezometric head)

When working with liquids, we noted that all terms in Eq. (4–24) have the units of energy per unit weight of fluid flowing. Thus each term is represented by a length, or "head." It is convenient and helpful to plot the various terms of the equation as vertical distances above an arbitrary horizontal datum. Figure 4–7 shows an example for flow through a pipe with a constant diameter. This is a case of zero shaft work. For this diagram we make the following definitions:

(1) *The* **energy** *or* **total head** *line is that line which lies everywhere a distance H vertically above the datum.*

(2) *The* **hydraulic grade** *or* **piezometric head** *line is that line which lies everywhere a distance $(p/\gamma + h)$ vertically above the datum.*

Therefore, by definition, the hydraulic grade line lies everywhere at a distance $V^2/2g$ below the energy line.

The following generalizations can be made:

(1) *The energy line may never be horizontal or slope upward in the direction of flow if the fluid is real and if no energy is being added. The vertical drop in total head line represents the head loss or energy dissipation per unit weight of fluid flowing.*

(2) *The energy line and hydraulic grade line are coincident and lie in the free surface for a body of liquid at rest (e.g., large reservoir).*

(3) *Whenever the hydraulic grade line falls below the point in the system for which it is being plotted, local pressure intensities are less than the reference pressure being used.*

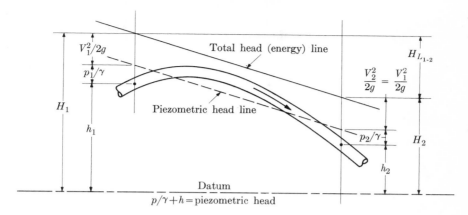

FIG. 4–7. Piezometric and total head lines.

Example 4–3: Power input and energy losses in a water line

A pump is used to supply a nozzle stream with water taken from a reservoir as shown in Fig. 4–8. The pump has an efficiency of 85% and a brake horsepower of 50 when delivering 2 ft³/sec. Under these conditions the pressure at (2) is —5 psig and the total and piezometric head lines are as shown. (a) What is the energy loss between the free surface at (1) and the pump inlet at (2)? (b) What is the pressure intensity at the pump discharge (3)? (c) What is the energy loss between the pump discharge (3) and the nozzle discharge (4)?

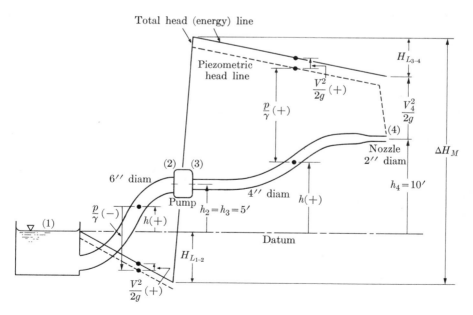

FIG. 4–8. Pumped flow through conduit with nozzle.

Solution: We proceed by using the constant density relations for continuity and energy. Applying Eq. (4–13) to the 6-in., 4-in., and 2-in. passages, respectively, we obtain

$$V = \frac{Q}{A} = 10.2 \text{ ft/sec}, \quad 22.9 \text{ ft/sec}, \quad \text{and} \quad 91.5 \text{ ft/sec}.$$

Now using Eq. (4–24a) with $K_e = 1.0$, we compute the following.
(a) Between (1) and (2),

$$H_1 = H_2 + H_{L_{1-2}},$$

or

$$0 = \frac{p_2}{\gamma} + h_2 + \frac{V_2^2}{2g} + H_{L_{1-2}} = \frac{(-5)(144)}{62.4} + 5 + \frac{(10.2)^2}{2g} + H_{L_{1-2}},$$

$$H_{L_{1-2}} = 5.0 \text{ (ft-lb)/lb}.$$

(b) Between (2) and (3). From (2) to (3) the pump transmits power to the water to cause a net increase in energy. Then $H_2 - \Delta H_M = H_3$, with

$$\Delta H_M = \frac{-(50)(0.85)(550)}{\gamma Q} = \frac{-23,400}{(62.4)(2)} = -188 \text{ (ft-lb)/lb}$$

(where the negative sign signifies the energy transmission *to* the water) and $H_2 = -H_{L_{1-2}}$. Then

$$-5 + 188 = \frac{p_3}{\gamma} + 5 + \frac{(22.9)^2}{2g},$$

$$p_3 = 170\gamma = 10,600 \text{ lb/ft}^2$$

$$p_3 = 73.5 \text{ lb/in}^2.$$

(c) Between (3) and (4); $H_3 = H_4 + H_{L_{3-4}}$,

$$170 + 5 + \frac{(22.9)^2}{2g} = 0 + 10 + \frac{(91.5)^2}{2g} + H_{L_{3-4}},$$

$$H_{L_{3-4}} = 43 \text{ (ft-lb)/lb}.$$

4–3   THE LINEAR MOMENTUM EQUATION FOR FINITE CONTROL VOLUMES

4–3.1   The momentum principle

The momentum principle is derived from Newton's second law as stated in the form

*The vector sum **F** of all the external forces acting on a fluid mass equals the time rate of change of the linear momentum vector **M** of the fluid mass.*

or

$$\mathbf{F} = \frac{d\mathbf{M}}{dt}. \tag{4–27}$$

External forces are of two types:

(a) *boundary forces,* which include
  (i) those acting normal to the control boundaries and which can be measured in terms of pressure intensities in fluid systems, $\mathbf{F}_p$;
  (ii) those acting parallel to the control boundaries and which can be measured in terms of tangential or shear stresses, $\mathbf{F}_s$.

(b) *body or field forces* such as those due to gravitational or magnetic fields, $\mathbf{F}_b$. If we use $\mathbf{F}_p$ and $\mathbf{F}_s$, and $\mathbf{F}_b$ and denote the body force per unit mass by $\mathbf{f}_b$, Eq. (4–27) becomes

$$\mathbf{F}_p + \mathbf{F}_s + \mathbf{F}_b = \frac{d\mathbf{M}}{dt},$$

or

$$\mathbf{F}_p + \mathbf{F}_s + \int_{\mathrm{cv}} \mathbf{f}_b(\rho\,d\mathcal{V}) = \frac{d\mathbf{M}}{dt}, \tag{4–28}$$

where $d\mathcal{V}$ is the volume element.

Equation (4–28) is valid for a control volume fixed in an inertial frame of reference or for a control volume translating with a uniform velocity relative to an inertial system. If the control volume is accelerating or rotating, additional or "apparent" body forces may be included in the left-hand side of Eq. (4–28) in accordance with the developments in Section 2–3. Simultaneously, then, $d\mathbf{M}_2/dt$ is taken as the time rate of change of momentum relative to the noninertial control volume. The additional body forces per unit mass are the four right-hand terms in Eq. (2–7). Using these terms, we may write Eq. (4–28) as

$$\mathbf{F}_p + \mathbf{F}_s + \int_{\mathrm{cv}} \mathbf{f}_b(\rho\,d\mathcal{V})$$

$$- \int_{\mathrm{cv}} \left[\frac{d^2\mathbf{R}}{dt^2} + 2\boldsymbol{\Omega} \times \mathbf{q}_2 + \boldsymbol{\Omega} \times (\boldsymbol{\Omega} \times \mathbf{r}) + \frac{d\boldsymbol{\Omega}}{dt} \times \mathbf{r} \right] (\rho\,d\mathcal{V}) = \frac{d\mathbf{M}_2}{dt}, \tag{4–29}$$

where

$\dfrac{d^2\mathbf{R}}{dt^2}$ = acceleration of the origin of a moving reference frame relative to the inertial frame,

$\boldsymbol{\Omega}$ = rotational velocity of a moving frame relative to the inertial frame,

$\mathbf{q}_2$ = fluid particle velocity relative to noninertial control volume fixed to moving frame.

Note that $d^2\mathbf{R}/dt^2$ and $\boldsymbol{\Omega}$ are independent of the coordinates of either reference frame. For the gravity field the body force per unit mass is $\mathbf{f}_b = \mathbf{g}$, where g is the acceleration due to gravity.

4–3.2 The general linear momentum equation

Inertial control volume. We will apply the momentum principle to flow through the generalized system represented in Fig. 4–1. The mass of the fluid within the control volume at time t will occupy the volume within the dashed outline at time $t + dt$. Consequently there will be a flux of momentum as well as mass, in at (1) and out at (2). The flux of momentum through a unit of area dA will equal the local velocity $\mathbf{q}$ times the mass per second $\rho(\mathbf{q} \cdot d\mathbf{A})$, or

$$\mathbf{q}\rho(\mathbf{q} \cdot d\mathbf{A}).$$

The total rate of change of momentum will be the net flux across the control-volume boundaries plus the rate of increase of momentum within the volume. Then Eq. (4–28) for the inertial reference frame can be written as

$$\mathbf{F}_p + \mathbf{F}_s + \mathbf{F}_b = -\oint_{cs} \mathbf{q}\rho(\mathbf{q} \cdot d\mathbf{A}) + \frac{\partial}{\partial t}\int_{cv}(\mathbf{q}\rho \, d\mathcal{V}), \qquad (4\text{–}30)$$

where we recall that $d\mathbf{A}$ is the vector unit area pointing inward over the control surface.

For *steady flow* and *negligible body forces,*

$$\mathbf{F}_p + \mathbf{F}_s = -\oint_{cs} \mathbf{q}\rho(\mathbf{q} \cdot d\mathbf{A}). \qquad (4\text{–}30a)$$

Noninertial control volume. For a noninertial control volume fixed to a moving reference frame the rate of change of momentum is evaluated relative to the moving frame, and the apparent body force of Eq. (4–29) is added to the external forces to give

$$\mathbf{F}_p + \mathbf{F}_s + \mathbf{F}_b - \int_{cv}\left[\frac{d^2\mathbf{R}}{dt^2} + 2\mathbf{\Omega} \times \mathbf{q}_2 + \mathbf{\Omega} \times (\mathbf{\Omega} \times \mathbf{r}) + \frac{d\mathbf{\Omega}}{dt} \times \mathbf{r}\right]\rho \, d\mathcal{V}$$

$$= -\oint_{cs} \mathbf{q}_2\rho(\mathbf{q}_2 \cdot d\mathbf{A}) + \frac{\partial}{\partial t}\int_{cv}(\mathbf{q}_2\rho \, d\mathcal{V}), \qquad (4\text{–}31)$$

where $\mathbf{q}_2$ is relative to the noninertial control volume. If the moving frame is in pure translation at constant velocity, this reduces to Eq. (4–30).

Equations (4–30) and (4–31) are general. They are applicable to ideal fluid systems and also to systems involving friction and energy dissipation. They hold whether or not there is heat transfer and whether or not the fluid is compressible. The combined effects of friction, energy loss, and heat transfer will appear implicitly in the magnitudes of the external forces, with corresponding effects on the local flow velocities, and hence the momenta of the fluid particles. The effects of density variations throughout the system likewise will appear implicitly in the velocity and force terms. Remember that these momentum equations are expressed as the sum of forces acting *on* the fluid. The forces of the fluid on the boundaries are, of course, equal and opposite.

Equation (4-30a) for steady flow is particularly powerful because in its application we need consider only the external conditions, i.e., the external forces and the flow conditions entering and leaving the system. Knowledge of the internal conditions is not necessary.

One of the many applications of Eqs. (4-30) and (4-31) is to the propulsion of objects through a fluid. In such cases the real and apparent body forces (and the rate of momentum change) for the system must be based on the total mass of the object plus the fluid. Often this is done conveniently by enclosing the object within the control volume as will be illustrated in Example 4-5.

4-3.3 Inertial control volume for a generalized apparatus

General equations. The components of the forces in Eq. (4-30) are

$$F_{p_x} + F_{s_x} + F_{b_x} = -\oint_{cs} u\rho(\mathbf{q} \cdot d\mathbf{A}) + \frac{\partial}{\partial t} \int_{cv} (u\rho \, d\mathcal{V}),$$

$$F_{p_y} + F_{s_y} + F_{b_y} = -\oint_{cs} v\rho(\mathbf{q} \cdot d\mathbf{A}) + \frac{\partial}{\partial t} \int_{cv} (v\rho \, d\mathcal{V}), \qquad (4\text{-}32)$$

$$F_{p_z} + F_{s_z} + F_{b_z} = -\oint_{cs} w\rho(\mathbf{q} \cdot d\mathbf{A}) + \frac{\partial}{\partial t} \int_{cv} (w\rho \, d\mathcal{V}),$$

where u, v, and w are the components of $\mathbf{q}$ in the x-, y-, and z-directions.

Equations (4-32) can be further reduced for application as we illustrate by considering flow through the generalized apparatus shown in Fig. 4-9. The boundaries of the control volume are taken to coincide with the inner solid boundary of the apparatus and are normal to the flow lines at (1) and (2). For this case, the net flux of momentum is the difference between that leaving at (2)

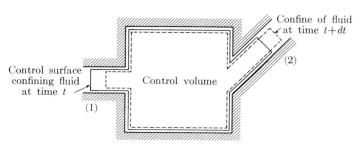

FIG. 4-9. Control volume for generalized apparatus.

and entering at (1). Then, for Eqs. (4–32), we have

$$F_{p_x} + F_{s_x} + F_{b_x} = \int_{(2)} u\rho \, dQ - \int_{(1)} u\rho \, dQ + \frac{\partial}{\partial t} \int_{cv} (u\rho \, d\mathcal{V}),$$

$$F_{p_y} + F_{s_y} + F_{b_y} = \int_{(2)} v\rho \, dQ - \int_{(1)} v\rho \, dQ + \frac{\partial}{\partial t} \int_{cv} (v\rho \, d\mathcal{V}), \qquad (4\text{–}32a)$$

$$F_{p_z} + F_{s_z} + F_{b_z} = \int_{(2)} w\rho \, dQ - \int_{(1)} w\rho \, dQ + \frac{\partial}{\partial t} \int_{cv} (w\rho \, d\mathcal{V}).$$

One-dimensional steady-flow equations. Although the steady-flow momentum equations, (Eq. 4–30a), or the steady-flow form of Eqs. (4–32a) does not deal with the detailed variation of properties and flow within the control volume, their evaluation does depend on variations in velocity and density over the cross sections (1) and (2). As we previously discussed in reference to the energy equation, there are many cases of application to flow through conduits where the variations are small. For these cases, it is convenient to approximate the true conditions by assuming that the velocity and density are constant normal to the flow direction. Thus we assume the flow to be *one-dimensional* with the only significant changes occurring with the dimension of distance along the conduit. If we make this assumption and denote the average momentum per unit mass as the average velocity V, we write the steady-flow form of Eqs. (4–32a) as

$$F_{p_x} + F_{s_x} + F_{b_x} = (V_x \rho Q)_2 - (V_x \rho Q)_1,$$
$$F_{p_y} + F_{s_y} + F_{b_y} = (V_y \rho Q)_2 - (V_y \rho Q)_1, \qquad (4\text{–}33)$$
$$F_{p_z} + F_{s_z} + F_{b_z} = (V_z \rho Q)_2 - (V_z \rho Q)_1.$$

When applying Eqs. (4–33), we use average values for ρ and V at each flow section. This is a good approximation for ρ so long as the flow is parallel at the observation section. The velocity, however, varies over the section, and the true average flux of momentum is

$$\int \mathbf{q}\rho(\mathbf{q} \cdot d\mathbf{A}) = K_m \mathbf{V}(\rho V A),$$

or

$$\int \mathbf{q}\rho \, dQ = K_m \mathbf{V} \rho Q,$$

where

V = magnitude of average velocity over cross section = Q/A,

$\mathbf{V}$ = average velocity vector,

$K_m \geq 1$.

Equations (4–33) would then be written

$$F_{p_x} + F_{s_x} + F_{b_x} = (K_m V_x \rho Q)_2 - (K_m V_x \rho Q)_1, \qquad (4\text{–}34)$$

and in a similar manner for the y- and z-directions. Practical values of K_m for circular ducts are:

laminar flow (parabolic velocity distribution) $K_m = 1.33$;

turbulent flow (smooth pipe) $K_m = 1.03 - 1.04$.

The momentum-flux coefficient K_m, as well as the kinetic-energy flux coefficient, is a percentage correction on two terms which are subtractive. It can be seen that for small differences between large numbers this may prove important even if K_m is constant or changes only slightly from section to section. If the shape of the velocity profile and K_m changes markedly from section to section, the effect can be pronounced even for small differences in the momentum terms.

Since these momentum equations involve vector quantities, the sense as well as the magnitude of each term is important. In Eq. (4–33), for example, while ρ and Q are scalar quantities, V_{x_2} and V_{x_1} are vector components which are positive if acting in the plus x-direction or negative if acting in the negative x-direction. With this sign convention, if the difference $(V_{x_2} - V_{x_1})$ is positive, the quantity F_x represents a resultant acting in the plus x-direction; if $(V_{x_2} - V_{x_1})$ is negative, F_x acts in the negative x-direction. Similarly the several terms comprising F_x may not all be of the same sense, and care must be taken with their individual signs.

Equations (4–5), (4–19), and (4–29) for continuity, energy, and linear momentum apply to unsteady- as well as steady-flow cases. However, it is not always convenient or possible to evaluate the integrals, especially the volume integrals. Sometimes it is possible to introduce approximations and obtain reliable results with a simplified solution. The following is an example of unsteady flow in which the time rate of change of the kinetic energy is neglected relative to the rate of change of potential energy. It is also an example in which the amount of liquid, and hence the volume of liquid, enclosed by the constant control volume varies.

Example 4–4: Continuity, energy, and linear momentum with unsteady flow

A large cylindrical tank mounted on rollers is filled with water to a depth of 16 ft above a discharge port near the base of the tank as shown in Fig. 4–10. At time $t = 0$, a fast-acting valve on the discharge nozzle is opened. With the tank held stationary, determine the instantaneous values of the following quantities at $t = 50$ sec: (a) the depth of water above the nozzle centerline h; (b) the discharge rate Q; (c) the force F necessary to keep the tank stationary.

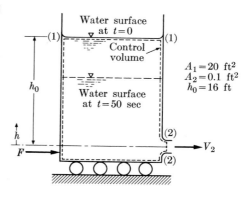

FIG. 4–10. Draining tank.

$A_1 = 20$ ft^2
$A_2 = 0.1$ ft^2
$h_0 = 16$ ft

Consider the fixed control volume coinciding with the initial water surface as shown in the figure. Then, from the continuity equation (4–4), we have

$$\frac{\partial}{\partial t} \int_{cv} (\rho \, d\mathcal{V}) = \int \rho q_n \, dA_1 - \int \rho q_n \, dA_2,$$

where $d\mathcal{V} = A_1 \, dh$ and $\rho q_n \, dA_1 = 0$, since there is no inflow across section (1). Therefore,

$$\rho A_1 \frac{\partial}{\partial t} \int_0^h dh = -\rho V_2 A_2,$$

or

$$A_1 \frac{dh}{dt} = -V_2 A_2. \tag{A}$$

Now consider the energy equation (4–19). There is no shaft work, and it will be assumed that heat transfer and temperature changes due to friction are negligible. Therefore,

$$0 = - \oint_{cs} \left(u + \frac{p}{\rho} + gh + \frac{q^2}{2} \right) \rho(\mathbf{q} \cdot d\mathbf{A}) + \frac{\partial}{\partial t} \int_{cv} (e\rho \, d\mathcal{V}),$$

where $e = u + gh + q^2/2$. In the last term of the time-dependent volume integral, namely,

$$\frac{\partial}{\partial t} \int_{cv} \rho \frac{q^2}{2} \, d\mathcal{V},$$

only the velocities in the tank near the nozzle will be significant. The volume integral of the kinetic energy in the tank will be nearly constant with respect to time provided the elevation of the free surface in the tank does not become too small. Therefore, Eq. (4–19) becomes

$$0 = \left(u + \frac{p}{\rho} + gh + \frac{V^2}{2} \right)_2 \rho V_2 A_2 + A_1 \rho \frac{\partial}{\partial t} \int_0^h (u + gh) \, dh$$

or

$$0 = u V_2 A_2 + \frac{V_2^2}{2} V_2 A_2 + u A_1 \frac{dh}{dt} + A_1 gh \frac{dh}{dt},$$

since $p_2 = h_2 = 0$ and $\rho = \text{const}$. From the continuity condition, Eq. (A), the terms containing the internal energy cancel, which they must if u is independent of time and volume. The remaining terms of the energy equation can be written as

$$V_{2(t)} = \sqrt{2gh_{(t)}}.$$

Substituting the above result into Eq. (A) gives

$$A_2 \sqrt{2gh} = -A_1 \frac{dh}{dt},$$

and integrating yields

$$\int_{h_0}^{h} \frac{dh}{\sqrt{h}} = -\int_0^t \frac{A_2}{A_1} \sqrt{2g}\, dt,$$

or

$$h = \left(h_0^{1/2} - \frac{A_2}{A_1} \frac{\sqrt{2g}}{2} t \right)^2$$

with $A_1 = 20\ \text{ft}^2$, $A_2 = 0.1\ \text{ft}^2$, $h_0 = 16\ \text{ft}$,

$$h = (4 - 0.020t)^2.$$

Therefore:

   (a) $t = 50\ \text{sec}$, $h = 9\ \text{ft}$.
   (b) $V_{2(t)} = \sqrt{2gh_{(t)}} = 24\ \text{ft/sec}$, $Q_{(t)} = (VA)_2 = 2.4\ \text{cfs}$.
   (c)   The momentum equation (4–29) is used to find the instantaneous force, F,

$$\mathbf{F}_p = -\oint_{\text{cs}} \mathbf{q}\rho(\mathbf{q} \cdot d\mathbf{A}) + \frac{\partial}{\partial t}\int_{\text{cv}} (\mathbf{q}\rho\, d\mathcal{V}).$$

Using the previous argument about the velocities within the tank, we may neglect the time rate of change of momentum inside if the tank area is large compared to the nozzle area. Then

$$F_{px} = V_2\rho V_2 A_2 = \rho V_2 Q_2$$
$$= (1.94)(24)(2.4) = 112\ \text{lb}.$$

Example 4–5: Linear momentum equation applied to a rocket

A rocket engine carries its own fuel and oxidizer. These are mixed in the combustion chamber and burned to develop gases at a high pressure. The hot gases expand through the nozzle to a lower pressure and a very high velocity.

Consider the following case of a rocket ascending radially from the surface of the earth:

   fuel burning rate $= 4\ \text{lb/sec}$,
   oxidizer burning rate $= 11\ \text{lb/sec}$,
   velocity of exiting gases relative to the rocket $=$
      $6000\ \text{ft/sec}$ and constant,
   nozzle discharge area $= 15\ \text{in}^2$,
   pressure at nozzle exit $= 13.7\ \text{psia}$,
   atmospheric pressure $= 13.7\ \text{psia}$,
   initial weight of rocket and its fuel $= 900\ \text{lb}$.

Compute the acceleration of the rocket after 10 sec if air resistance is neglected.

In the general case the rocket motion may be unsteady. Let the control volume be fixed relative to the rocket as shown in Fig. 4–11. Assume that the ascent is with only translation motion.

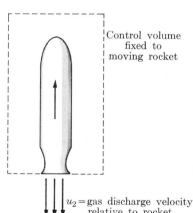

Control volume fixed to moving rocket

$u_2 =$ gas discharge velocity relative to rocket

FIG. 4–11. Rocket in noninertial control volume.

Then Eq. (4–31), for a noninertial control volume, reduces to

$$\mathbf{F}_p + \mathbf{F}_s + \mathbf{F}_b - \int_{cv} \frac{d^2\mathbf{R}}{dt^2}\, dm = -\oint_{cs} \mathbf{q}\rho(\mathbf{q} \cdot d\mathbf{A}) + \frac{\partial}{\partial t} \int_{cv} \mathbf{q}\rho\, d\mathcal{V},$$

where the volume integrals are taken over the total mass enclosed by the control volume. In this equation the right-hand terms represent the time rate of momentum change relative to the rocket. The quantity $d^2\mathbf{R}/dt^2$ is the rocket acceleration.

In the flight direction the known terms are:

$$\mathbf{i} \cdot \mathbf{F}_p = \mathbf{i} \cdot \left[\oint_{cs} p\, d\mathbf{A} \right] = 0$$

since the pressure over the control surface is everywhere the same in this example,

$$\mathbf{i} \cdot \mathbf{F}_s = 0,$$

$$\mathbf{i} \cdot \mathbf{F}_b = \text{rocket weight} = -(900 - 15t)\ \text{lb},$$

$$\mathbf{i} \cdot \left[\int_{cv} \frac{d^2\mathbf{R}}{dt^2}\, dm \right] = \frac{d^2\mathbf{R}}{dt^2}\, m = \frac{d^2\mathbf{R}}{dt^2} \frac{(900 - 15t)}{g}\ \text{lb},$$

$$\mathbf{i} \cdot \left[\oint_{cs} \mathbf{q}\rho(\mathbf{q} \cdot d\mathbf{A}) \right] = u_2\rho\, dQ = 6000\, \frac{15}{g} = 2790\ \text{lb},$$

$$\mathbf{i} \cdot \left[\frac{\partial}{\partial t} \int_{cv} \mathbf{q}\rho\, d\mathcal{V} \right] = \frac{\partial}{\partial t} \int_{cv} u\rho\, d\mathcal{V} \approx 0$$

since u is essentially constant (with time) relative to the rocket. The acceleration is computed as follows:

$$(900 - 15t) + \frac{d^2\mathbf{R}}{dt^2} \frac{(900 - 15t)}{g} = 2790,$$

and at $t = 10\,\text{sec}$,

$$\frac{d^2\mathbf{R}}{dt^2} = \frac{(2790 - 750)}{750}\, g = \frac{2040}{750}\, g = 87.7\ \text{ft/sec}^2.$$

4–4   THE MOMENT OF MOMENTUM EQUATION FOR FINITE CONTROL VOLUMES

4–4.1   The moment of momentum principle for inertial reference systems

The moment of momentum principle is Newton's second law applied to rotating fluid masses. In the following we will develop this principle only for inertial control volumes.

The vector sum ($\mathbf{r} \times \mathbf{F}$) *of all the external moments acting on a fluid mass equals the time rate of change of the moment of momentum vector* ($\mathbf{r} \times \mathbf{M}$) *of the fluid mass,*

or

$$\mathbf{r} \times \mathbf{F} = \frac{d}{dt} (\mathbf{r} \times \mathbf{M}), \qquad (4\text{–}35)$$

where $\mathbf{r}$ is the position vector of a mass in an arbitrary curvilinear motion (with instantaneous velocity $\mathbf{q}$) and linear momentum $\mathbf{M}$ as indicated in Fig. 4–12.

Equation (4–35) is obtained by taking the vector cross product of $\mathbf{r}$ with both sides of Eq. (4–27) and noting that

$$\mathbf{r} \times \frac{d\mathbf{M}}{dt} = \frac{d}{dt} (\mathbf{r} \times \mathbf{M}), \qquad \text{because} \qquad \frac{d\mathbf{r}}{dt} \times \mathbf{M} = \mathbf{q} \times (\text{mass}) \, \mathbf{q} \equiv 0.$$

The external moments arise from the external forces and are of two types: (a) *boundary forces;* (b) *body or field forces.* If we denote the external forces by $\mathbf{F}_p$, $\mathbf{F}_s$ and $\mathbf{F}_b$ as before, and their respective moments by $\mathbf{T}_p$, $\mathbf{T}_s$ and $\mathbf{T}_b$, Eq. (4–35) becomes

$$(\mathbf{r} \times \mathbf{F}_p) + (\mathbf{r} \times \mathbf{F}_s) + (\mathbf{r} \times \mathbf{F}_b) = \frac{d}{dt} (\mathbf{r} \times \mathbf{M}),$$

or

$$\mathbf{T}_p + \mathbf{T}_s + \mathbf{T}_b = \frac{d}{dt} (\mathbf{r} \times \mathbf{M}). \qquad (4\text{–}36)$$

The quantity ($\mathbf{r} \times \mathbf{M}$) is commonly called the angular momentum.

4–4.2   The general moment of momentum equation

We can apply the principle of Eq. (4–36) to a generalized system by considering flow through a fixed control volume (Fig. 4–13). The sum of the moments (about a fixed point) of the flux of momentum through the control

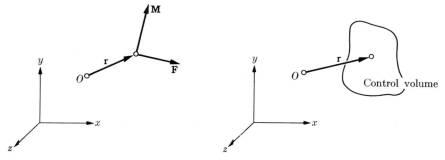

FIG. 4–12.   Position, force and momentum vectors.   FIG. 4–13.   Control volume for moment of momentum analysis.

surface plus the rate of increase of moment of momentum within the control volume will equal the sum of the external torque. Thus we have the general moment of momentum equation as

$$\mathbf{T}_p + \mathbf{T}_s + \mathbf{T}_b = -\oint_{cs} (\mathbf{r} \times \mathbf{q})\rho(\mathbf{q} \cdot d\mathbf{A}) + \frac{\partial}{\partial t} \int_{cv} (\mathbf{r} \times \mathbf{q})\rho \, d\mathcal{V}, \quad (4\text{–}37)$$

where we recall that $d\mathbf{A}$ is the vector unit area pointing inward over the control surface. Equation (4–37) for moment of momentum is analogous to Eq. (4–30) for linear momentum.

The components of the moments in Eq. (4–37) about the x-, y-, and z-axes are obtained by using the following definitions: r_{yz}, q_{yz} are the components lying in yz-plane; α_{yz} is the angle between q_{yz} and normal to r_{yz}; and $(\pi/2 - \alpha_{yz})$ is the angle between q_{yz} and r_{yz}. Thus we have $(\mathbf{r} \times \mathbf{q})_{yz}$ as the component of $(\mathbf{r} \times \mathbf{q})$ in the x-direction, and

$$|(\mathbf{r} \times \mathbf{q})_{yz}| = r_{yz}q_{yz} \sin (\pi/2 - \alpha_{yz}) = (rq \cos \alpha)_{yz}.$$

Then about the x-axis, we have

$$T_{p_x} + T_{s_x} + T_{b_x} = -\oint_{cs} (rq \cos \alpha)_{yz}\rho(\mathbf{q} \cdot d\mathbf{A}) + \frac{\partial}{\partial t} \int_{cv} (rq \cos \alpha)_{yz}\rho \, d\mathcal{V},$$
$$(4\text{–}38a)$$

and similarly for the y- and z-axes,

$$T_{p_y} + T_{s_y} + T_{b_y} = -\oint_{cs} (rq \cos \alpha)_{zx}\rho(\mathbf{q} \cdot d\mathbf{A}) + \frac{\partial}{\partial t} \int_{cv} (rq \cos \alpha)_{zx}\rho \, d\mathcal{V},$$
$$(4\text{–}38b)$$

$$T_{p_z} + T_{s_z} + T_{b_z} = -\oint_{cs} (rq \cos \alpha)_{xy}\rho(\mathbf{q} \cdot d\mathbf{A}) + \frac{\partial}{\partial t} \int_{cv} (rq \cos \alpha)_{xy}\rho \, d\mathcal{V}.$$
$$(4\text{–}38c)$$

Equations (4–38), as well as Eqs. (4–32), are general, applying alike to systems with and without friction, heat transfer, and compressibility.

4–4.3 Steady-flow equations for turbomachinery

Turbomachines are pumps, turbines, fans, blowers, or compressors in which a dynamic reaction occurs between a rotating vaned element (known as a runner) and fluid passing through the element. When the rotating speed is constant, the moment of momentum relations reduce to the following form, known as Euler's equation:

$$T_r = \int_{A_2} r_2 V_2 \cos \alpha_2 \rho \, dQ - \int_{A_1} r_1 V_1 \cos \alpha_1 \rho \, dQ, \quad (4\text{–}39)$$

where with the aid of Fig. 4–14 we see that

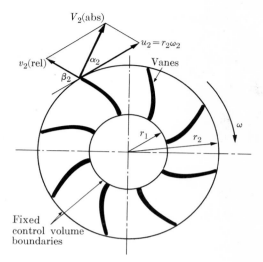

r_1, r_2 = radii of fluid elements at the entrance and exit of the runner,

V_1, V_2 = velocities relative to the fixed reference frame (absolute velocities),

α_1, α_2 = angles of absolute velocities with tangential direction,

A_1, A_2 = entrance and exit areas of a control volume which encloses the runner,

T_r = torque exerted on fluid by the runner (known as runner torque).

FIG. 4–14. Notation and control volume for a turbomachine. (Example shown is for a radial flow pump or compressor.)

The runner torque T_r will be positive in the above equation for a pump or compressor and negative for a turbine. The torque at the shaft of a turbomachine will differ from the runner torque by the magnitude of any frictional resistances, mechanical or hydrodynamical, occurring externally to the runner flow passages. For a pump or blower, shaft torque is greater than runner torque, and for a turbine, shaft torque is less than runner torque.

As are Eqs. (4–38), Eq. (4–39) is general (so long as the rotating speed is constant) and applies to systems with and without friction, heat transfer, and compressibility.

If $r_2 V_2 \cos \alpha_2$ and $r_1 V_1 \cos \alpha_1$ are constant, the change in moment of momentum (or angular momentum) is the same for all streamlines. If, in addition, the density is constant, we obtain the specialized form

$$T_r = \rho Q(r_2 V_2 \cos \alpha_2 - r_1 V_1 \cos \alpha_1). \tag{4–40}$$

In general, real machines will not exactly satisfy the conditions of $rV \cos \alpha$ being constant over inlet or exit. However, when we use average values of $rV \cos \alpha$, Eq. (4–40) will approximate the runner torque for liquid machines.

If $r_2 V_2 \cos \alpha_2 = r_1 V_1 \cos \alpha_1$, we have

$$\rho Q r V \cos \alpha = \rho Q r v_\theta = \text{const}, \tag{4–41}$$

which is the special condition of a fluid moving with spiral streamlines with zero torque. This is the condition of constant angular momentum. The limiting case of $\alpha = 0$ gives a flow in circles with a constant angular momentum which is called free vortex flow.

PROBLEMS

4–1. Water flows out from under a gate as shown in Fig. 4–15. Velocity measurements give the magnitudes and directions shown. The gate is 10 ft wide. Compute the total discharge rate in ft^3/sec.

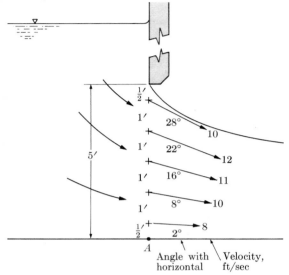

FIGURE 4–15

4–2. Suppose that the water depth in the channel to the left of the gate in the previous problem is very large. Estimate the velocity with which water particles on a cylindrical arc at a 50-ft radius from point A approach the outlet.

4–3. A method of flow control in a circular converging-diverging nozzle involves suction along the expanding wall. Consider the nozzle in Fig. 4–16 in which the area of the diverging section expands linearly from 0.1 ft^2 at its throat to 0.5 ft^2. Wall suction velocity is arranged to increase parabolically from zero at the throat to a maximum of 10 ft/sec at the exit edge. Given a mean velocity of the outflow from the nozzle of 30 ft/sec and a nozzle length of 1 ft, compute the inflow rate in cfs. The fluid is water.

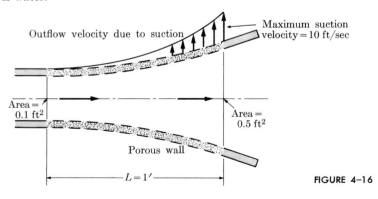

FIGURE 4–16

4–4. Consider a jet pump operated by water as illustrated in Fig. 4–17. The water flow rate is 1 cfs and the pumped material is oil having a specific gravity of 0.9. Determine the amount pumped when the specific gravity of the mixture is 0.95.

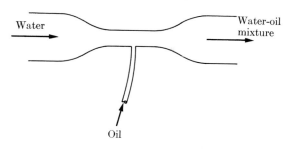

Water

Water-oil mixture

Oil

FIGURE 4–17

4–5. A steam turbine uses 15,000 lbm/hr of steam when it delivers 1500 hp to a generator coupled to its shaft. The operating conditions are:

	Inlet	Outlet
Temperature	$T = 1000°F$	500°F
Enthalpy	$(p/\rho + u) = 1505$ Btu/lbm	1202 Btu/lbm
Pressure	$p = 1000$ psia	680 psia
Velocity	$V = 200$ ft/sec	900 ft/sec

Determine the rate at which heat transfer occurs between the turbine casing and the surrounding air. Which direction does the heat flow? Neglect elevation differences.

4–6. A vapor flows through a nozzle expanding from a pressure of 500 psia down to 100 psia. The initial and final enthalpies $(p/\rho + u)$ are 1470 and 1270 Btu/lbm, respectively. The velocity of flow into the nozzle is 100 ft/sec. Compute the exit velocity, assuming adiabatic conditions. Neglect elevation differences.

4–7. Compressed air flows through a horizontal converging-diverging nozzle. The conditions are:

	Inlet	Throat	Exit
Nozzle area, in^2	1	0.113	0.153
Velocity, ft/sec	100		
Pressure, psia	500		100
Temperature, °F	500	340	

Assuming that the air expands isentropically, compute the exit velocity by two methods.

4–8. Air enters a horizontal pipe 14 in. in diameter at a pressure of 14.8 psia and a temperature of 60°F. At the exit the pressure is 14.5 psia. Entrance velocity is 35 ft/sec. Exit velocity is 50 ft/sec. Also $c_p = 0.24$, $c_v = 0.17$, Btu/(lbm-°R). What amount of heat is added or subtracted?

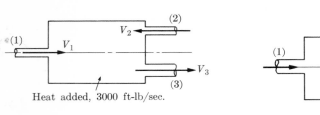

FIGURE 4-18

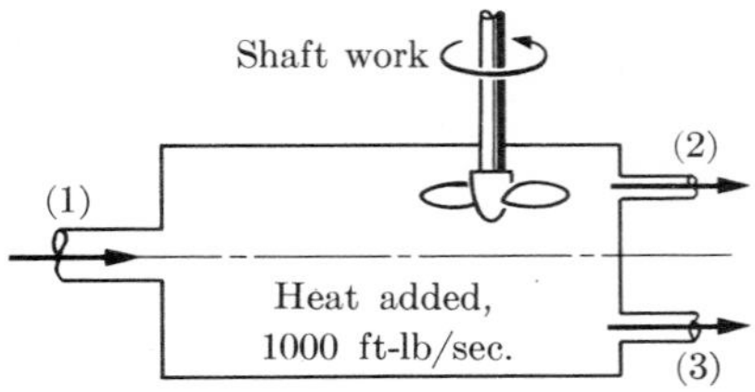

FIGURE 4-19

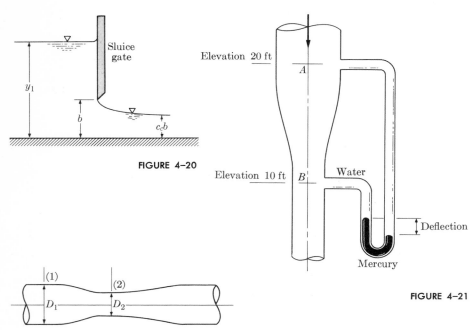

FIGURE 4-20

FIGURE 4-21

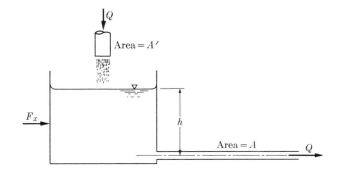

FIGURE 4-22

FIGURE 4-23

4–9. A gas that is neither a pure substance nor a perfect gas flows steadily through a heat exchanger. Heat from an external source is transmitted to the gas in the amount of 3000 ft-lb/sec. No work is done by the gas. Properties at the three cross sections (Fig. 4–18) where fluid flows are:

$$\rho_1 = 0.1 \text{ slug/ft}^3, \qquad \rho_2 = 0.2 \text{ slug/ft}^3, \qquad \rho_3 = 0.15 \text{ slug/ft}^3$$
$$A_1 = 0.1 \text{ ft}^2, \qquad A_2 = 0.1 \text{ ft}^2, \qquad A_3 = 0.2 \text{ ft}^2.$$
$$V_1 = 70 \text{ fps}, \qquad V_2 = 40 \text{ fps},$$
$$p_1 = 15 \text{ psia}, \qquad p_2 = 20 \text{ psia},$$

The internal energy is constant. Compute V_3 and p_3.

4–10. A perfect gas flows steadily through the machine shown in Fig. 4–19. The gas constant for this gas is 40(ft-lb)/(lbm-°R). One thousand foot-pounds of heat are added per second. Calculate the shaft work of the machine.

$$T_1 = 60°F \qquad T_2 = 60°F \qquad T_3 = 60°F$$
$$p_1 = 40 \text{ psia} \qquad p_2 = 80 \text{ psia} \qquad p_3 = 120 \text{ psia}$$
$$V_1 = 30 \text{ fps} \qquad V_2 = 30 \text{ fps}$$
$$A_1 = 2 \text{ ft}^2 \qquad A_2 = \tfrac{1}{2} \text{ ft}^2 \qquad A_3 = \tfrac{2}{3} \text{ ft}^2$$

4–11. Derive the equation for the volume rate of flow per unit width for the sluice gate shown in Fig. 4–20 in terms of the geometric variables b, y_1, and c_c. Assume the pressure is hydrostatic at y_1 and $c_c b$ and the velocity is constant over the depth at each of these sections.

4–12. Derive the expression for the total force per unit width exerted by the above sluice gate on the fluid in terms of vertical distances shown in Fig. 4–20.

4–13. Given that $D_A = 6$ in., $D_B = 3$ in., and water is flowing at a rate $Q = 1.20 \text{ ft}^3/\text{sec}$, find the magnitude of the deflection of the mercury manometer shown in Fig. 4–21.

4–14. Consider the flow of an incompressible fluid through the Venturi meter shown in Fig. 4–22. Assuming uniform flow at sections (1) and (2) and neglecting all losses, find the pressure difference between these sections as a function of the flow rate Q, the diameters of the sections, and the density of the fluid, ρ. Note that for a given configuration, Q is a function of only the pressure drop and fluid density. The meter is named for Venturi, who investigated its principle in about 1791. However, in 1886 Clemens Herschel first used the meter to measure discharge, and he is usually credited with its invention.

4–15. Water flows into a tank from a supply line and out of the tank through a horizontal pipe as shown in Fig. 4–23. The rates of inflow and outflow are the same, and the water surface in the tank remains a distance h above the discharge pipe center-line. All velocities in the tank are negligible compared to those in the pipe. The head loss between the tank and the pipe exit is H_L. (a) Find the discharge Q in terms of h, A, and H_L. (b) What is the horizontal force, F_x, required to keep the tank from moving? (c) If the supply line has an area A', what is the vertical force exerted on the water in the tank by the vertical jet?

4–16. A large closed tank is filled with ammonia gas under a pressure of 30.0 psia, at a temperature of 65°F, and $c_p/c_v = 1.29$. The gas discharges into the atmosphere through a small opening in the side of the tank. Calculate the velocity of the gas assuming isentropic flow conditions.

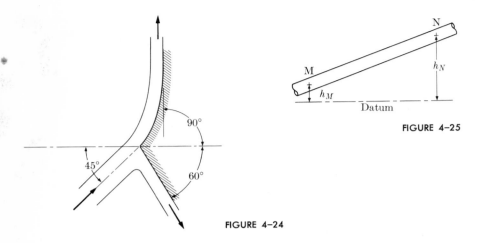

FIGURE 4-25

FIGURE 4-24

4-17. A pump draws water from a sump through a vertical pipe 6 in. in diameter. The pump has a horizontal discharge pipe 4 in. in diameter which is 10.6 ft above the sump level. While pumping 1.25 cfs, gauges near the pump at suction and discharge read −4.6 psig and +25.6 psig, respectively. The discharge pressure gauge is 3.0 ft above the suction pressure gauge. Compute the horsepower output of the pump.

4-18. The fixed surface shown in Fig. 4–24 divides the water jet so that 1.0 cfs flows in both directions. For an initial jet velocity of 48 ft/sec, find the x- and y-components of the force required to keep the surface in equilibrium. Neglect frictional resistance.

4-19. A jet propulsion system is to develop a thrust of 2500 lb. If it takes in air at 200 mph (293 ft/sec) and discharges it at 5000 ft/sec, what weight of air per second must it handle? Neglect fuel consumption.

4-20. A stationary nozzle discharges a 200-ft/sec water jet which pushes a vane at 50 ft/sec in a straight line. The vane absorbs energy from the jet at the rate of 50 hp. The jet diameter is 3 in. Neglecting friction, compute the angle through which the jet water is deflected as it passes over the vane.

4-21. A 6-in. horizontal pipe line bends through 90° and while bending changes its diameter to 3 in. The pressure in the 6-in. pipe is 30 psig. Calculate the magnitude and direction of the total force on the bend when 2.0 cfs of water flows through the pipe.

4-22. Using the continuity condition and momentum equation applied to a control volume which encloses the traveling density discontinuity of a sound-wave front, derive Eq. (1–15a).

4-23. Oil is flowing through the 8-in. diameter circular pipe shown in Fig. 4–25. The flow is steady and the discharge is 4.0 ft³/sec. The specific weight of the oil is 48 lb/ft³. The pressure and elevation conditions are:

$$p_M = 8 \text{ psig,} \qquad h_M = 5 \text{ ft;} \qquad p_N = 5 \text{ psig} \qquad h_N = 20 \text{ ft.}$$

Determine the direction of flow and the rate of energy dissipation between the two points M and N.

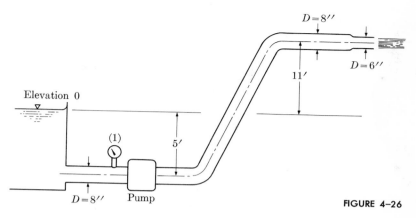

FIGURE 4-26

4-24. The pump in the system shown in Fig. 4-26 delivers 4.25 hp to the water flowing at a rate of 2 ft³/sec. The flow discharges into the atmosphere at the nozzle. The pressure gauge at point (1) reads $p_1 = 0.7$ psig. Sketch the total head line and the piezometric head line and indicate numerical values for the elevations of the two lines at appropriate points.

4-25. A pipeline has been designed to deliver water from Northern California to the Los Angeles metropolitan area. The original design called for a tunnel through the coastal mountain range between points (2) and (4). The original design had no pumps or turbines in the region shown in the Fig. 4-27. The pressure at point (1) in this design was 100 psig and at point (5) the pressure was 50 psig due to pipe friction losses. The flow rate is 1000 ft³/sec and the pipe is 10 ft in diameter.

(a) Make a sketch and draw the energy and hydraulic grade lines between points (1) and (5). Assume that the pipe is horizontal.

Later investigations showed that a fault passes through the tunnel, and it was decided to route the pipeline over the coastal range following the ground surface of the mountain (assume that the 4000-ft mountain may be represented by an isosceles triangle) to facilitate repair in case of an earthquake.

(b) Explain why a pump is necessary for this alternate route and compute the power the pump will transmit to the water when the flow rate is 1000 ft³/sec. The gauge pressure in the pipeline at the summit of the mountain should not go below atmospheric

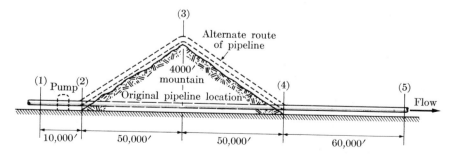

FIGURE 4-27

pressure. Draw the energy and hydraulic grade lines between points (1) and (5) for the alternate route, assuming that the pressure at point (1) remains at 100 psig.

4–26. It is decided to adopt the alternate route for the pipeline described in Problem 4–25. In this design the pump is located at point (1) and the pipeline follows the ground surface of the mountain. It is possible to recover a portion of the power supplied to the water at the pump by installing a turbine at point (4). The pipe diameter, flow rate, and pressures at points (1) and (5) are to be as specified in the original design. Draw the hydraulic and energy-grade lines for the system using both the pump and turbine. Compute the power removed by the turbine.

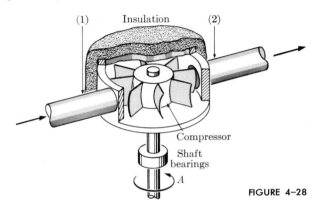

FIGURE 4–28

4–27. Consider an air compressor in a pipeline (Fig. 4–28). The pipe cross-sectional area is 0.2 ft² at both the inlet and discharge of the compressor. A motor of 400-hp capacity is specified to drive the compressor shaft at A. The compressor is rated to deliver a mass rate of flow of 5.76 lbm/sec.

A test is made during which the compressor is insulated as shown and the following results are obtained for the rated flow rate:

$$p_1 = 20 \text{ psia}, \qquad V_1 = 400 \text{ ft/sec},$$
$$\gamma_1 = 0.072 \text{ lb/ft}^3, \qquad p_2 = 69.5 \text{ psia},$$
$$T_1 = 290°\text{F}, \qquad T_2 = 480°\text{F}.$$

Determine the power actually delivered to the air by the compressor.

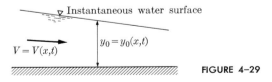

FIGURE 4–29

4–28. Derive the one-dimensional continuity equation for the unsteady, nonuniform flow of an incompressible liquid in a horizontal open channel as shown in Fig. 4–29. The channel has a rectangular cross section of constant width, b. Both the depth, y_0, and the mean velocity, V, are functions of x and t.

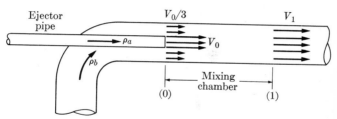

FIGURE 4-30

4-29. We wish to analyze the performance of a liquid-liquid ejector in which ejector liquid a entrains liquid b and pumps the mixture into a delivery line (Fig. 4–30). The ejector liquid of density ρ_a and average velocity V_0 and the entrained liquid of density ρ_b and average velocity $V_0/3$ enter the cylindrical mixing chamber at section (0). The cross-sectional area of the mixing chamber is A. The cross-sectional area of the ejector pipe is $A/3$. At section (1) the two liquids are completely mixed and the mean velocity of the mixture is V_1.

Assume that the pressures across sections (0) and (1) are constant, the flow is steady, and both liquids are incompressible. Neglect friction on the mixing chamber walls. Derive an expression for the pressure change between (0) and (1) if $\rho_b = 3\rho_a$.

4-30. A horizontal circular jet of water (density $= 1.94$ slug/ft^3) 6-in. in diameter strikes a conical deflector whose included angle at the vertex is 60°. A horizontal force of 5 lb is required to hold the deflector stationary in the jet. (a) Determine the flow rate from the nozzle in cfs. (b) Explain how you would solve a similar problem if the jet striking the cone were discharging vertically upward. Would any additional information have to be specified?

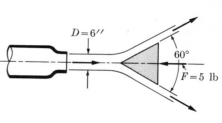

FIGURE 4-31

4-31. The direction and magnitude of the water velocity at the discharge from a centrifugal pump runner are measured with a special velocity probe. The results are indicated in Fig. 4–32. At the inlet to the runner the absolute water velocity has no tangential component. (a) Calculate the runner torque T_r. (b) The discharge vane angle β_2 theoretically necessary to produce this torque equals the angle of the relative velocity. Find this angle.

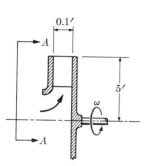

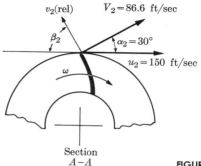

Section
$A\text{--}A$

FIGURE 4-32

4–32. Determine the functional relationship between torque T_r and flow rate Q for a turbomachine having the form of Fig. 4–14 with the vanes set between parallel disk walls. Assume that the angle β_2 at which the fluid leaves the runner is constant and the speed ω is constant.

4–33. The runner pictured in Fig. 4–14 is used for an inward flow turbine. Assume the following conditions: $r_{\text{outer}} = 10$ ft, $r_{\text{inner}} = 5$ ft, speed $= 50$ rpm, radial velocity into runner $= 10$ ft/sec, absolute velocity into runner $= 48.3$ ft/sec, flow passages, of constant width in the axial direction, $= 1$ ft. (a) If the flow angle β_2 must equal the vane angle at the outer radius for optimum entry flow conditions, determine the vane angle. (b) If the exiting flow at the inner radius is tangent to the vane surface, determine the exit vane angle for minimum exiting kinetic energy.

4–34. For the turbine of Problem 4–33, (a) Compute the runner torque T_r. (b) If friction losses are neglected, the head converted to power can be computed using $\gamma Q H_r = T_r \omega$. Assume no friction and compute the pressure drop from the outer to the inner radius.

REFERENCE

1. KEENAN, J. H., and G. H. HATSOPOULOS, *General Principles of Thermodynamics*, John Wiley and Sons, Inc., New York, 1965; LEE, J. F., and F. W. SEARS, *Thermodynamics*, 2nd ed., Addison-Wesley Publishing Co., 1963; and A. H. SHAPIRO, *Compressible Fluid Flow*, Vol. 1, Ronald Press Co., New York, 1953 are useful references suggested for supplementary reading.

CHAPTER 5

Stress-Strain Relations

A certain familiarity with the basic concepts of stress and strain is assumed on the basis of previous work in physics and the mechanics of solids [1]. The aim of this chapter is to draw upon the students' background in solid mechanics insofar as it is useful in deducing the stress-rate of strain relations for a fluid. These relations are used in Chapter 6 to obtain the three-dimensional equations of motion for a material particle.

5–1 GENERAL STRESS-STRAIN SYSTEM

5–1.1 Surface stresses

When a system of external forces acts upon a body, the stress components acting on the six surfaces of a small element within the body, having sides Δx, Δy, and Δz, may be represented as shown in Fig. 5–1. The stress on the face perpendicular to the x-axis at point O may be expressed in terms of a normal stress and two orthogonal shear stresses.

$$\sigma_x = \lim_{\Delta A_x \to 0} \frac{\Delta F_x}{\Delta A_x}, \qquad \tau_{xy} = \lim_{\Delta A_x \to 0} \frac{\Delta F_y}{\Delta A_x}, \qquad \tau_{xz} = \lim_{\Delta A_x \to 0} \frac{\Delta F_z}{\Delta A_x}. \qquad (5\text{–}1)$$

In the above definitions, ΔF_x, ΔF_y, and ΔF_z are the components of the force vector, $\Delta \mathbf{F}$, acting on the face, and ΔA_x is the area of the x-face of the element. The subscript for the normal stress indicates the direction of the stress component. The subscript convention for the shear stress components is that the first subscript indicates the direction of the normal to the face on which it acts, and the second indicates the direction in which it acts. In all cases, a stress component is positive when a positively directed force component acts on a positive face. In the same manner, a positive stress results when a negatively directed force component acts on a negative face.

The general stress system requires nine scalar components (one normal and two shear stresses for each surface). However, it can be shown that the pairs of shearing stresses with subscripts differing only in their order are equal. In elementary mechanics of solids, it is usual to consider the stressed element to be in a state of static equilibrium, which is equivalent to specifying that the sum of all moments and the sum of all forces each equal zero for the element. However, we wish to treat the more general case in which the element is in an arbitrary state of motion. Considering the summation of moments, for example, about a

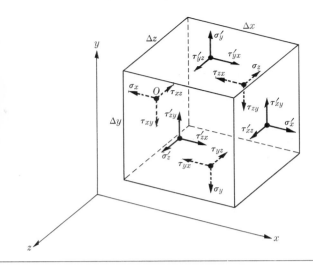

$$\sigma'_x = \sigma_x + \frac{\partial \sigma_x}{\partial x}\Delta x, \qquad \sigma'_y = \sigma_y + \frac{\partial \sigma_y}{\partial y}\Delta y, \qquad \sigma'_z = \sigma_z + \frac{\partial \sigma_z}{\partial z}\Delta z;$$

$$\tau'_{xy} = \tau_{xy} + \frac{\partial \tau_{xy}}{\partial x}\Delta x, \qquad \tau'_{yx} = \tau_{yx} + \frac{\partial \tau_{yx}}{\partial y}\Delta y, \qquad \tau'_{zx} = \tau_{zx} + \frac{\partial \tau_{zx}}{\partial z}\Delta z;$$

$$\tau'_{xz} = \tau_{xz} + \frac{\partial \tau_{xz}}{\partial x}\Delta x, \qquad \tau'_{yz} = \tau_{yz} + \frac{\partial \tau_{yz}}{\partial y}\Delta y, \qquad \tau'_{zy} = \tau_{zy} + \frac{\partial \tau_{zy}}{\partial z}\Delta z.$$

FIG. 5–1. Surface stresses on a material element.

centroidal axis in the z-direction, we have

$$\sum M_z = \text{(mass of element)}(\text{radius of gyration})^2(\text{angular acceleration}).$$

This results in

$$\tau_{xy} - \tau_{yx} = \lim_{\substack{\Delta x \to 0 \\ \Delta y \to 0 \\ \Delta z \to 0}} [\rho(\text{radius of gyration})^2(\text{angular acceleration})] = 0.$$

In the limit as Δx, Δy, and Δz approach zero, the right-hand side approaches zero because the radius of gyration is a higher-order quantity. By applying this procedure to all three directions, we obtain

$$\tau_{xy} = \tau_{yx}, \qquad \tau_{yz} = \tau_{zy}, \qquad \tau_{zx} = \tau_{xz}, \tag{5–2}$$

which is the same conclusion as that reached for static equilibrium. Thus the general stress system is reduced to six scalar components. The next objective is to consider the strain components of the element as it deforms under the applied stresses.

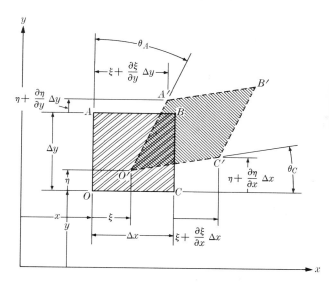

FIG. 5–2.   Deformation and strain in a plane.

5–1.2   Strain components

The strain of any continuous medium, solid, liquid, or gas, can be described in terms of normal and shear strain components. These components, in turn, can be related to the rates of linear and angular deformation. For small strains, relations which hold with good accuracy can be defined by using the principle of infinitesimal deformations of the material element in the following manner.

Consider a small element $OABC$ of an undeformed body lying in the xy-plane as shown in Fig. 5–2. If the body is subjected to a system of external forces, the element may be deformed into $O'A'B'C'$ as shown. The coordinates of point O before deformation are x, y, z, and after deformation they become $x + \xi$, $y + \eta, z + \zeta$. The normal strain component is by definition equal to the change in length of a side of the element divided by the original length. The symbol ϵ is used to designate the normal strain with a subscript which indicates the direction in which the strain occurs. Hence, at point O, for the x-direction, we have

$$\epsilon_x = \lim_{\Delta x \to 0} \frac{O'C' - OC}{OC} = \lim_{\Delta x \to 0} \frac{\{[\Delta x - \xi] + [\xi + (\partial \xi/\partial x)\, \Delta x]\} - \Delta x}{\Delta x} = \frac{\partial \xi}{\partial x}.$$

Repeating for the other two directions gives the three normal strains

$$\epsilon_x = \frac{\partial \xi}{\partial x}, \qquad \epsilon_y = \frac{\partial \eta}{\partial y}, \qquad \epsilon_z = \frac{\partial \zeta}{\partial z}. \tag{5–3}$$

The normal strain is positive when the element elongates under deformation.

The *shear strain* is defined as the change in the angle between two originally perpendicular elements as deformation occurs. The shear strain is designated by the symbol γ with two subscripts indicating the direction of the perpendicular axes in the plane of deformation. Hence, for the xy-plane in Fig. 5–2, we have

$$\gamma_{xy} = \lim_{\substack{\Delta x \to 0 \\ \Delta y \to 0}} (\theta_C + \theta_A),$$

$$\gamma_{xy} = \lim_{\substack{\Delta x \to 0 \\ \Delta y \to 0}} \left[\frac{(\partial \eta / \partial x)\, \Delta x}{\Delta x} + \frac{(\partial \xi / \partial y)\, \Delta y}{\Delta y} \right] = \frac{\partial \eta}{\partial x} + \frac{\partial \xi}{\partial y}.$$

Repeating for the other two directions, we obtain the three shear strains

$$\gamma_{xy} = \frac{\partial \eta}{\partial x} + \frac{\partial \xi}{\partial y}, \qquad \gamma_{yz} = \frac{\partial \zeta}{\partial y} + \frac{\partial \eta}{\partial z}, \qquad \gamma_{zx} = \frac{\partial \xi}{\partial z} + \frac{\partial \zeta}{\partial x}. \qquad (5\text{–}4)$$

With the above notation, the displacement of a point (such as O) under deformation can be written as the displacement vector, $\boldsymbol{\delta}$, where

$$\boldsymbol{\delta} = \xi \mathbf{i} + \eta \mathbf{j} + \zeta \mathbf{k}, \qquad (5\text{–}5)$$

and $\mathbf{i}$, $\mathbf{j}$, $\mathbf{k}$ are the unit vectors in the x- y- and z-directions.

The change in volume of the deformed element divided by the original volume is known as the *volume dilation*, e, and hence

$$e = \text{volume dilation} = \frac{d(\Delta \mathcal{V})}{\Delta \mathcal{V}} = \epsilon_x + \epsilon_y + \epsilon_z, \qquad (5\text{–}6)$$

or, using Eqs. (5–3), we have

$$e = \frac{\partial \xi}{\partial x} + \frac{\partial \eta}{\partial y} + \frac{\partial \zeta}{\partial z} = \nabla \cdot \boldsymbol{\delta}. \qquad (5\text{–}7)$$

5–2 RELATIONS BETWEEN STRESS AND STRAIN FOR ELASTIC SOLIDS

The solid matter composing the deformable element is assumed to have elastic properties which are independent of the orientation of the coordinate axes (i.e., *isotropic*). If it is further assumed to be an ideal elastic solid, Hooke's law expresses the linear proportionality between stress and strain as

$$\epsilon_x^0 = \frac{\sigma_x}{E}, \qquad \epsilon_y^0 = \frac{\sigma_y}{E}, \qquad \epsilon_z^0 = \frac{\sigma_z}{E}, \qquad (5\text{–}8)$$

where E is Young's modulus of elasticity of the solid and ϵ_x^0 is the elongation in the x-direction due to the normal stress, σ_x. Hooke's law, which states that stress is linear with strain, is an empirical approximation of the behavior of many real solids under small deformations. Because of the lateral contraction of matter under tension, the normal stresses, σ_y and σ_z, will also cause strains in the x-direction. For example, the strain ϵ_x' in the x-direction due to σ_y is given in terms of Poisson's ratio n, as

$$\epsilon_x' = -n\epsilon_y^0 = -\frac{n\sigma_y}{E},\tag{5–9}$$

Similarly ϵ_x'' due to σ_z is

$$\epsilon_x'' = -n\epsilon_z^0 = -\frac{n\sigma_z}{E}.\tag{5–10}$$

Hence

$$\epsilon_x = \epsilon_x^0 + \epsilon_x' + \epsilon_x'' = \frac{\sigma_x}{E} - \frac{n}{E}(\sigma_y + \sigma_z).\tag{5–11}$$

Therefore, for all three directions,

$$\epsilon_x = \frac{1}{E}[\sigma_x - n(\sigma_y + \sigma_z)],\tag{5–12a}$$

$$\epsilon_y = \frac{1}{E}[\sigma_y - n(\sigma_z + \sigma_x)],\tag{5–12b}$$

$$\epsilon_z = \frac{1}{E}[\sigma_z - n(\sigma_x + \sigma_y)].\tag{5–12c}$$

The shear stresses are related to the shear strains by the shear modulus of elasticity, G, where we again assume a linear proportionality in accordance with Hooke's law,

$$\gamma_{xy} = \frac{\tau_{xy}}{G}, \qquad \gamma_{yz} = \frac{\tau_{yz}}{G}, \qquad \gamma_{zx} = \frac{\tau_{zx}}{G}.\tag{5–13}$$

Equations (5–12) and (5–13) are the generalized form of Hooke's law for an elastic solid. They contain both the shear and Young's moduli of elasticity. Since these quantities are related, it is desirable to algebraically rearrange Eq. (5–12) to express the relation between the normal stresses and strains in terms of the shear modulus.

Young's modulus, the shear modulus, and Poisson's ratio are related in the following manner [2]:

$$G = \frac{E}{2(1 + n)}.\tag{5–14}$$

By means of Eq. (5–12), the volume dilation expressed by Eq. (5–6) can be written

$$e = \frac{1 - 2n}{E}(\sigma_x + \sigma_y + \sigma_z).\tag{5–15}$$

We now define $\bar{\sigma}$ as the arithmetic mean of the three normal stresses,

$$\bar{\sigma} = \tfrac{1}{3}(\sigma_x + \sigma_y + \sigma_z).\tag{5–16}$$

From Eqs. (5–12a), (5–14), and (5–15), we have

$$\sigma_x = 2G\left[\epsilon_x + \frac{ne}{1 - 2n}\right].\tag{5–17}$$

Poisson's ratio can be eliminated by adding the zero quantity

$$\bar{\sigma} - \tfrac{1}{3}(\sigma_x + \sigma_y + \sigma_z) = 0$$

to Eq. (5–17) and simplifying. Performing these steps for all three directions gives

$$\sigma_x - \bar{\sigma} = 2G\left(\epsilon_x - \frac{e}{3}\right), \qquad \sigma_y - \bar{\sigma} = 2G\left(\epsilon_y - \frac{e}{3}\right),$$

$$\sigma_z - \bar{\sigma} = 2G\left(\epsilon_z - \frac{e}{3}\right). \tag{5–18}$$

To complete the listing of the nine stress terms for an elastic solid, we have from Eqs. (5–4) and (5–13):

$$\tau_{yx} = \tau_{xy} = G\left(\frac{\partial \eta}{\partial x} + \frac{\partial \xi}{\partial y}\right), \qquad \tau_{zy} = \tau_{yz} = G\left(\frac{\partial \zeta}{\partial y} + \frac{\partial \eta}{\partial z}\right),$$

$$\tau_{xz} = \tau_{zx} = G\left(\frac{\partial \xi}{\partial z} + \frac{\partial \zeta}{\partial x}\right). \tag{5–19}$$

5–3 RELATIONS BETWEEN STRESS AND RATE OF STRAIN FOR NEWTONIAN FLUIDS

Experimental evidence suggests that stresses in a fluid are related to the time rate of strain rather than to the strain itself, as are the stresses in the solid state. Hence, instead of assuming stress to be linearly proportional to strain [as in Eqs. (5–12) and (5–13)], we assume a linear relation between *stress* and *time rate of strain*. As pointed out in the first chapter, fluids which satisfy this assumption are called Newtonian fluids, and the constant of proportionality is the dynamic viscosity μ.

The appropriate stress-strain relations for a Newtonian fluid can be obtained by analogy from Eqs. (5–18) and (5–19). For example, considering the x-component from Eq. (5–18) and replacing the shear modulus by a quantity which expresses its dimensions, we have

Hookeian elastic solid: $\sigma_x - \bar{\sigma} = 2\left(\dfrac{F}{L^2}\right)\left(\epsilon_x - \dfrac{e}{3}\right).$ (5–18a)

By analogy, we can write a similar equation for fluids, relating the stresses on the left-hand side to the time rate of strains on the right-hand side as

Newtonian fluid: $\sigma_x - \bar{\sigma} = 2\left(\dfrac{FT}{L^2}\right)\dfrac{\partial}{\partial t}\left(\epsilon_x - \dfrac{e}{3}\right).$ (5–20)

Note that in order to preserve the dimensions, the new proportionality constant must contain time. Thus the dynamic viscosity μ for a fluid is analogous to the shear modulus of elasticity for a solid and, as shown in Table 1–1, μ has the dimensions FT/L^2. The fluid stress-strain relation may then be written

$$\sigma_x - \bar{\sigma} = 2\mu \frac{\partial \epsilon_x}{\partial t} - \frac{2}{3} \mu \frac{\partial e}{\partial t}. \tag{5-21}$$

The term $\partial \epsilon_x / \partial t$ is the time rate of change of the normal strain component as derived from Eq. (5–3),

$$\frac{\partial \epsilon_x}{\partial t} = \frac{\partial}{\partial t} \left(\frac{\partial \xi}{\partial x} \right) = \frac{\partial}{\partial x} \left(\frac{\partial \xi}{\partial t} \right) = \frac{\partial u}{\partial x}. \tag{5-22}$$

The quantity $\partial e / \partial t$ expresses the time rate of volume dilation of the fluid element. Hence from Eq. (5–7),

$$\frac{\partial e}{\partial t} = \nabla \cdot \frac{\partial \boldsymbol{\delta}}{\partial t} = \nabla \cdot \mathbf{q}. \tag{5-23}$$

By definition, an incompressible fluid is one in which the time rate of volume expansion of a fluid element is zero. Hence the mathematical statement of incompressibility is

$$\frac{\partial e}{\partial t} = \nabla \cdot \mathbf{q} = 0. \tag{5-23a}$$

As we have pointed out in Chapter 1, liquids are nearly incompressible and hence they will obey Eq. (5–23a) closely. For gases and vapors which undergo only small pressure changes, compressibility is small, and Eq. (5–23a) holds approximately.

After making the substitutions in Eq. (5–21), we see that the normal stress becomes

$$\sigma_x = \bar{\sigma} + 2\mu \frac{\partial u}{\partial x} - \frac{2}{3} \mu (\nabla \cdot \mathbf{q}). \tag{5-24}$$

Following the same analogy, the shear stresses of Eqs. (5–19) are transformed for the fluid into the form

$$\tau_{yx} = \tau_{xy} = \mu \left(\frac{\partial v}{\partial x} + \frac{\partial u}{\partial y} \right). \tag{5-25}$$

It remains now to introduce the pressure p which exists in the fluid. It is customary to identify a mean fluid pressure with the thermodynamic pressure [such as $p = \rho RT$ from Eq. (1–17)]. The question then arises as to how the thermodynamic pressure p and the mean normal stress $\bar{\sigma}$ are related. There are two possibilities.

One possibility is to use the assumption, which is supported by experiments with incompressible fluids, that the viscous effects are represented completely by the viscosity μ relating shear stress and rate of strain. This is a case of complete analogy with the elastic solid equations based on Hooke's law, and we use the substitution

$$\bar{\sigma} = -p = \tfrac{1}{3}(\sigma_x + \sigma_y + \sigma_z). \tag{5-26}$$

The negative sign accounts for the convention of pressure, a compression, being taken as positive in treating fluids. Equation (5-26) would be expected to apply to an incompressible fluid or to any other zero-volume dilation case.

The second possibility is to use the experimental evidence of dilation effects with compressible liquids and gases. To account for these effects, the $\bar{\sigma}$ in Eq. (5-24) can be represented as the sum of the thermodynamic pressure and an amount dependent on a *second coefficient* of viscosity. For an isotropic fluid, the relation can be stated as

$$\bar{\sigma} = -p + \mu'(\nabla \cdot \mathbf{q}), \tag{5-27}$$

where

$$\mu' = \text{second coefficient of viscosity associated solely} \\ \text{with dilation,}$$

so that

$$-p = \bar{\sigma} - \mu'(\nabla \cdot \mathbf{q}) = \tfrac{1}{3}(\sigma_x + \sigma_y + \sigma_z) - \mu'(\nabla \cdot \mathbf{q}). \tag{5-28}$$

The last term is important only when the rate of volume change $\nabla \cdot \mathbf{q}$ becomes very large. For most cases it is small and we equate $\bar{\sigma}$ to $-p$.

Therefore, making the assumption of zero-dilation viscosity effects, and by introducing Eq. (5-26) into Eq. (5-24) to obtain the stress-rate of strain relations as usually given for fluids, we have

$$\sigma_x = -p + 2\mu \frac{\partial u}{\partial x} - (\tfrac{2}{3})\mu(\nabla \cdot \mathbf{q}),$$

$$\sigma_y = -p + 2\mu \frac{\partial v}{\partial y} - (\tfrac{2}{3})\mu(\nabla \cdot \mathbf{q}), \tag{5-29}$$

$$\sigma_z = -p + 2\mu \frac{\partial w}{\partial z} - (\tfrac{2}{3})\mu(\nabla \cdot \mathbf{q}).$$

The complete set of shear stresses becomes

$$\tau_{yx} = \tau_{xy} = \mu \left(\frac{\partial v}{\partial x} + \frac{\partial u}{\partial y}\right), \qquad \tau_{yz} = \tau_{zy} = \mu \left(\frac{\partial w}{\partial y} + \frac{\partial v}{\partial z}\right),$$

$$\tau_{zx} = \tau_{xz} = \mu \left(\frac{\partial u}{\partial z} + \frac{\partial w}{\partial x}\right). \tag{5-30}$$

Equations (5-29) show that, in general, the normal stress in a given coordinate direction is not equal to the mean of the normal stresses unless the viscous effects are zero or the fluid is at rest. Furthermore, for zero viscous effects, the three normal stresses are all equal and the shear stresses are all zero. This leads to the statement

$$\sigma_x = \sigma_y = \sigma_z = \bar{\sigma}, \quad \text{and} \quad \tau_{xy} = \tau_{yz} = \tau_{zx} = 0, \quad (5\text{-}31)$$

for inviscid fluids in motion and all fluids at rest. The fluid at rest corresponds to the state of hydrostatic equilibrium discussed in Chapter 1. Thus we conclude that if shear stresses are present in a real fluid, the fluid must be in motion.

PROBLEMS

5-1. Verify eq. (5–14).

5-2. Verify eq. (5–18).

5-3. Consider a fluid element under a general state of stress as illustrated in Fig. 5–1. Given that the element is in a gravity field, show that the equilibrium requirement between surface, body and inertial forces leads to the equations

$$\frac{\partial \sigma_x}{\partial x} + \frac{\partial \tau_{yx}}{\partial y} + \frac{\partial \tau_{zx}}{\partial z} + \rho g_x = \rho a_x,$$

$$\frac{\partial \tau_{xy}}{\partial x} + \frac{\partial \sigma_y}{\partial y} + \frac{\partial \tau_{zy}}{\partial z} + \rho g_y = \rho a_y,$$

$$\frac{\partial \tau_{xz}}{\partial x} + \frac{\partial \tau_{yz}}{\partial y} + \frac{\partial \sigma_z}{\partial z} + \rho g_z = \rho a_z.$$

5-4. Consider a fluid in two-dimensional motion. Using plane polar coordinates r, θ, and z, show that the rate of strain components are

$$\frac{\partial \epsilon_r}{\partial t} = \frac{\partial v_r}{\partial r}, \quad \frac{\partial \epsilon_\theta}{\partial t} = \frac{1}{r}\frac{\partial v_\theta}{\partial \theta} + \frac{v_r}{r}, \quad \frac{\partial \gamma_{r\theta}}{\partial t} = \frac{\partial v_\theta}{\partial r} + \frac{1}{r}\frac{\partial v_r}{\partial \theta} - \frac{v_\theta}{r}.$$

REFERENCES

1. CRANDALL, S. H., and N. C. DAHL, *An Introduction to the Mechanics of Solids,* McGraw-Hill Book Co., New York, 1959.
2. CRANDALL, S. H., and N. C. DAHL, op. cit. supra., Section 5.4.

CHAPTER 6

Equations of Continuity and Motion

The three-dimensional expressions for conservation of matter and the momentum equations of motion are developed in this chapter. In a sense, this will complete the formulation of the analytical methods of fundamental fluid dynamics. The subsequent chapters are concerned primarily with their application to particular types of problems.

6–1 CONTINUITY EQUATION

Consider a differential control volume $\Delta x\,\Delta y\,\Delta z$ in a region where density and velocity are functions of position in space and time. Referring to Fig. 6–1, we compute the flux of mass per second through each face of the cube to get, for the three directions,

$$-\left[\frac{\partial(\rho u)}{\partial x}\,\Delta x\right]\Delta y\,\Delta z,\qquad -\left[\frac{\partial(\rho v)}{\partial y}\,\Delta y\right]\Delta z\,\Delta x,\qquad -\left[\frac{\partial(\rho w)}{\partial z}\,\Delta z\right]\Delta x\,\Delta y.$$

From the principle of conservation of matter, the sum of these must equal the time rate of change of mass,

$$\frac{\partial}{\partial t}\left(\rho\,\Delta x\,\Delta y\,\Delta z\right).$$

Here $\Delta x\,\Delta y\,\Delta z$ is independent of time since the control volume is fixed. After combining and factoring out $\Delta x\,\Delta y\,\Delta z$, we get

$$\frac{\partial\rho}{\partial t}+\frac{\partial(\rho u)}{\partial x}+\frac{\partial(\rho v)}{\partial y}+\frac{\partial(\rho w)}{\partial z}=0,\tag{6–1a}$$

or

$$\frac{\partial\rho}{\partial t}+\nabla\cdot\rho\mathbf{q}=0.\tag{6–1b}$$

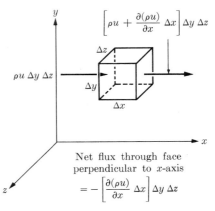

$$\left[\rho u+\frac{\partial(\rho u)}{\partial x}\,\Delta x\right]\Delta y\,\Delta z$$

Net flux through face perpendicular to x-axis

$$=-\left[\frac{\partial(\rho u)}{\partial x}\,\Delta x\right]\Delta y\,\Delta z$$

FIG. 6–1. Differential control volume for conservation of matter.

106

This can be expressed alternatively as

$$\frac{d\rho}{dt} + \rho \left(\frac{\partial u}{\partial x} + \frac{\partial v}{\partial y} + \frac{\partial w}{\partial z} \right) = 0, \tag{6–2a}$$

or

$$\frac{d\rho}{dt} + \rho (\nabla \cdot \mathbf{q}) = 0. \tag{6–2b}$$

Equation (6–1) or (6–2) is the general *continuity equation* for a fluid in unsteady flow. For steady-state conditions, where $\partial \rho / \partial t = 0$, we have

$$\frac{\partial (\rho u)}{\partial x} + \frac{\partial (\rho v)}{\partial y} + \frac{\partial (\rho w)}{\partial z} = \nabla \cdot \rho \, \mathbf{q} = 0. \tag{6–3}$$

We can note at this point that if the right-hand term of Eq. (4–5a) is transformed to a volume integral by Gauss' theorem,* then

$$\oint_{\text{cs}} \rho \, \mathbf{q} \cdot d\mathbf{A} = - \int_{\text{cv}} (\nabla \cdot \rho \, \mathbf{q}) \, d\upsilon. \tag{6–4}$$

Thus Eq. (4–5a) becomes

$$\int_{\text{cv}} \frac{\partial \rho}{\partial t} \, d\upsilon = - \int_{\text{cv}} (\nabla \cdot \rho \, \mathbf{q}) \, d\upsilon.$$

Since the integrands must be equal, Eq. (4–5a) may be written in flux per unit volume terms as

$$\frac{\partial \rho}{\partial t} + \nabla \cdot \rho \, \mathbf{q} = 0,$$

which is Eq. (6–1b).

The condition for an incompressible fluid is given by Eq. (5–23a) as

$$\nabla \cdot \mathbf{q} = 0 = \frac{\partial u}{\partial x} + \frac{\partial v}{\partial y} + \frac{\partial w}{\partial z} . \tag{6–5}$$

Equation (6–5) is the common form of the continuity equation for steady and unsteady flows of incompressible fluids. Using Eq. (6–5) in Eq. (6–2), we see for an incompressible fluid

$$\frac{d\rho}{dt} = 0 = \frac{\partial \rho}{\partial t} + u \frac{\partial \rho}{\partial x} + v \frac{\partial \rho}{\partial y} + w \frac{\partial \rho}{\partial z} . \tag{6–6}$$

---

* If $\mathbf{X}$ is a vector function of position and $d\mathbf{A}$ is the vector differential area pointing into a region, the following relates an integral over the volume of the region to an integral over the enclosing surface [1],

$$\int_{\upsilon} (\nabla \cdot \mathbf{X}) \, d\upsilon = - \int_{A} \mathbf{X} \cdot d\mathbf{A}.$$

For a uniform-density incompressible fluid, each individual term in Eq. (6–6) is independently zero. For a nonuniform density fluid, Eq. (6–6) expresses the relations among density gradients that must exist if the fluid particles are to be incompressible. The conservation of mass equation for a single species in a non-homogeneous fluid mixture is discussed in Chapter 16.

6–2 STREAM FUNCTION IN TWO-DIMENSIONAL, INCOMPRESSIBLE FLOWS

A useful relation between the streamlines and the continuity equation may be formulated for the special case of two-dimensional incompressible fluid flow. Equation (6–5) can be written for two-dimensional flow as

$$\frac{\partial u}{\partial x} + \frac{\partial v}{\partial y} = 0. \tag{6–7}$$

If a continuous function $\psi\,(x, y)$, called the stream function, is defined such that

$$u = -\frac{\partial \psi}{\partial y}, \qquad v = \frac{\partial \psi}{\partial x}, \tag{6–8}$$

the continuity equation (6–7) is exactly satisfied, since

$$-\frac{\partial^2 \psi}{\partial x\, \partial y} + \frac{\partial^2 \psi}{\partial y\, \partial x} \equiv 0. \tag{6–9}$$

The analogous relations for velocities in cylindrical coordinates are

$$v_r = -\frac{\partial \psi}{r\, \partial \theta}, \qquad v_\theta = \frac{\partial \psi}{\partial r}. \tag{6–10}$$

From Eq. (2–10) the equation for a streamline in two-dimensional flow is

$$v\, dx - u\, dy = 0. \tag{6–11}$$

Using Eqs. (6–8), we find that along a streamline

$$\frac{\partial \psi}{\partial x}\, dx + \frac{\partial \psi}{\partial y}\, dy = d\psi = 0. \tag{6–12}$$

Since the differential $d\psi$ is zero, then

$$\psi = \text{const along a streamline.} \tag{6–13}$$

As a consequence of Eqs. (6–7), (6–8), (6–9), the differential of ψ is exact and its integral gives $(\psi_2 - \psi_1)$, which is dependent only on the end points of integration. This

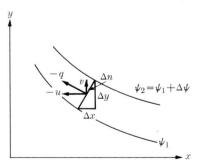

FIG. 6–2. Stream functions and volume flow rate continuity.

is the basis for a useful property of the stream function which may be deduced by considering two adjacent streamlines separated by a distance Δn, as shown in Fig. 6–2. From the law of conservation of mass, we know that the mass flux across Δn must equal the combined flux across both Δx and Δy, and hence in the limit for an incompressible fluid,

$$-q \, dn = -u \, dy + v \, dx, \tag{6–14}$$

and again using Eq. (6–8), we have

$$-q \, dn = \frac{\partial \psi}{\partial y} \, dy + \frac{\partial \psi}{\partial x} \, dx = d\psi. \tag{6–15}$$

Therefore, between adjacent streamlines the change in ψ is dimensionally and numerically equal to the volume rate of flow per unit width of the two-dimensional flow. The sign convention is shown by Eqs. (6–14) and (6–15). When ψ is increasing in the $+y$-direction, as in Fig. 6–2, the direction of flow is in the negative x-direction, from right to left.

For axially symmetric motions, the streamlines in any radial plane containing the symmetry axis lie in stream surfaces concentric about the axis. As in two-dimensional motions, these streamlines can be described by two coordinates leading also to a single stream function (known as Stokes' stream function). For general three-dimensional motion, three coordinates are necessary. This leads to the situation in which a pair of functions describes two sets of intersecting stream surfaces. The lines of intersection are the streamlines [2].

6–3   ROTATIONAL AND IRROTATIONAL MOTION

6–3.1   Rotation and vorticity

The rate of rotation of a fluid element, Δx, Δy, about the z-axis may be expressed in terms of the velocity components u and v and their changes in the x- and y-directions, as shown in Fig. 6–3. Defining counterclockwise as positive, we find that the rate of rotation of the Δx-face of the element is

$$\frac{v + (\partial v/\partial x) \, \Delta x - v}{\Delta x} = \frac{\partial v}{\partial x}.$$

For the Δy-face,

$$\frac{-[u + (\partial u/\partial y) \, \Delta y - u]}{\Delta y} = -\frac{\partial u}{\partial y},$$

where the negative sign is introduced to preserve the convention that the counterclockwise rota-

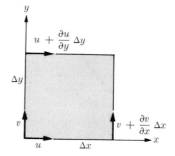

FIG. 6–3. Definition diagram for computing rotation of a fluid element.

tion is positive. The net rate of rotation of the fluid element about the z-axis is the average of the sum of the rotation of the Δx- and Δy-faces,

$$\omega_z = \frac{1}{2}\left(\frac{\partial v}{\partial x} - \frac{\partial u}{\partial y}\right).$$

In a similar manner,

$$\omega_x = \frac{1}{2}\left(\frac{\partial w}{\partial y} - \frac{\partial v}{\partial z}\right),$$

and

$$\omega_y = \frac{1}{2}\left(\frac{\partial u}{\partial z} - \frac{\partial w}{\partial x}\right). \tag{6-16a}$$

The resultant rate-of-rotation vector is

$$\boldsymbol{\omega} = \tfrac{1}{2}(\nabla \times \mathbf{q}), \tag{6-16b}$$

with the magnitude

$$|\boldsymbol{\omega}| = \sqrt{\omega_x^2 + \omega_y^2 + \omega_z^2}.$$

The vector $\nabla \times \mathbf{q}$ is the curl of the velocity vector and is known as the *vorticity*, ζ. The components of the vorticity are ξ, η, and ζ, each twice the corresponding rate of rotation component.

An *irrotational flow* is defined as one in which the rate of rotation components $\omega_x = \omega_y = \omega_z = 0$. Hence $\nabla \times \mathbf{q} = 0$, and

$$\frac{\partial w}{\partial y} = \frac{\partial v}{\partial z}, \qquad \frac{\partial u}{\partial z} = \frac{\partial w}{\partial x}, \qquad \frac{\partial v}{\partial x} = \frac{\partial u}{\partial y}. \tag{6-17}$$

The rotation components in cylindrical coordinates become

$$2\omega_r = \frac{1}{r}\frac{\partial v_z}{\partial \theta} - \frac{\partial v_\theta}{\partial z}, \qquad 2\omega_\theta = \frac{\partial v_r}{\partial z} - \frac{\partial v_z}{\partial r}, \qquad 2\omega_z = -\frac{1}{r}\frac{\partial v_r}{\partial \theta} + \frac{v_\theta}{r} + \frac{\partial v_\theta}{\partial r}, \tag{6-18}$$

and the corresponding condition for irrotational flow is $\omega_r = \omega_\theta = \omega_z = 0$.

6-3.2 Circulation

In later developments, we will use the concept of circulation designated by the symbol Γ. *Circulation* is defined as the line integral of the tangential velocity component about any closed contour s,

$$\Gamma = \oint \mathbf{q} \cdot d\mathbf{s}, \tag{6-19}$$

where a positive sense corresponds to a counterclockwise integrating path around the surface enclosed by s. Applying this to the infinitesimal fluid element shown in Fig. 6–3, we have

$$d\Gamma = \left(\frac{\partial v}{\partial x} - \frac{\partial u}{\partial y}\right) dx\, dy.$$

When we use Eq. (6–16), the circulation is

$$\Gamma = \iint_A 2\omega_z\, dA = \iint_A (\nabla \times \mathbf{q})_z\, dA. \tag{6–20}$$

If the flow in the region bounded by the contour is entirely irrotational, the circulation $\Gamma = 0$, since $\omega_z = 0$. Equations (6–19) and (6–20) are a statement of Stokes' theorem in two dimensions.

6–4 EQUATIONS OF MOTION

We now apply Newton's second law in the form of Eq. (3–12) and sum forces on the material particle of fixed mass Δm shown in Fig. 5–1. The material method, described in Section 3–6, leads to a simpler formulation of the equations of motion than does the control-volume method which was used to obtain the three-dimensional continuity equation. In writing the summation of forces on the fluid particle, it is necessary to consider body forces acting on the particle in addition to the surface forces discussed in Chapter 5. Body forces may arise, for example, through the action of gravitational or electromagnetic fields. In either case, the forces act at the centroid of the element. Other centroidal forces having the nature of body forces may arise due to the choice of an accelerating or rotating frame of reference with respect to inertial space as discussed in Chapter 2. The Coriolis force, for example, is of this type. For the present, only apparent gravitational force fields will be considered in the production of body forces. The gravitational force per unit mass is the acceleration of gravity $\mathbf{g}$ with the components given by

$$\mathbf{g} = \mathbf{i}g_x + \mathbf{j}g_y + \mathbf{k}g_z.$$

The summation of the x-components of force on the element (Fig. 5–1) in accordance with Newton's second law is

$$\Delta F_x = \Delta m a_x = (\rho\, \Delta x\, \Delta y\, \Delta z)a_x,$$

with

$$\Delta F_x = (\rho\, \Delta x\, \Delta y\, \Delta z)g_x - \sigma_x\, \Delta y\, \Delta z + \left(\sigma_x + \frac{\partial \sigma_x}{\partial x}\, \Delta x\right) \Delta y\, \Delta z$$

$$- \tau_{yx}\, \Delta x\, \Delta z + \left(\tau_{yx} + \frac{\partial \tau_{yx}}{\partial y}\, \Delta y\right) \Delta x\, \Delta z$$

$$- \tau_{zx}\, \Delta x\, \Delta y + \left(\tau_{zx} + \frac{\partial \tau_{zx}}{\partial z}\, \Delta z\right) \Delta x\, \Delta y.$$

Dividing by the volume of the element gives

$$\rho g_x + \frac{\partial \sigma_x}{\partial x} + \frac{\partial \tau_{yx}}{\partial y} + \frac{\partial \tau_{zx}}{\partial z} = \rho a_x. \tag{6–21}$$

Similarly, for the y- and z-directions, we have

$$\rho g_y + \frac{\partial \tau_{xy}}{\partial x} + \frac{\partial \sigma_y}{\partial y} + \frac{\partial \tau_{zy}}{\partial z} = \rho a_y, \qquad \rho g_z + \frac{\partial \tau_{xz}}{\partial x} + \frac{\partial \tau_{yz}}{\partial y} + \frac{\partial \sigma_z}{\partial z} = \rho a_z. \tag{6–21a, b}$$

6–4.1 The Navier-Stokes equations

Equations (6–21) are general and apply to any fluid with gravitational body forces. For Newtonian fluids with a single viscosity coefficient, we make use of the stress-strain relationships of Chapter 5 as given by Eqs. (5–29) and (5–30). By substituting these normal and tangential shear stresses in Eqs. (6–21), we obtain for the x-component of the equation of motion,

$$\rho g_x - \frac{\partial p}{\partial x} + \frac{\partial}{\partial x} \left[2\mu \frac{\partial u}{\partial x} - \tfrac{2}{3}\mu (\nabla \cdot \mathbf{q}) \right]$$
$$+ \frac{\partial}{\partial y} \left[\mu \left(\frac{\partial v}{\partial x} + \frac{\partial u}{\partial y} \right) \right] + \frac{\partial}{\partial z} \left[\mu \left(\frac{\partial u}{\partial z} + \frac{\partial w}{\partial x} \right) \right] = \rho a_x. \tag{6–22}$$

This equation is valid for a Newtonian fluid of varying density and viscosity in a gravitational field. Viscosity measurements indicate that μ is a function of temperature and very slightly a function of pressure. The latter effect is almost always negligible, and if temperature changes are not large, the assumption of a constant viscosity corresponding to the mean temperature of the fluid is justified. Therefore, Eq. (6–22) becomes

$$\rho g_x - \frac{\partial p}{\partial x} + \mu \frac{\partial}{\partial x} \left[2 \frac{\partial u}{\partial x} - \tfrac{2}{3}(\nabla \cdot \mathbf{q}) \right]$$
$$+ \mu \frac{\partial}{\partial y} \left(\frac{\partial v}{\partial x} + \frac{\partial u}{\partial y} \right) + \mu \frac{\partial}{\partial z} \left(\frac{\partial u}{\partial z} + \frac{\partial w}{\partial x} \right) = \rho a_x. \tag{6–23}$$

After expanding and simplifying, we have

$$\rho g_x - \frac{\partial p}{\partial x} + \mu \left[\frac{\partial^2 u}{\partial x^2} + \frac{\partial^2 u}{\partial y^2} + \frac{\partial^2 u}{\partial z^2} \right] + \frac{\mu}{3} \frac{\partial}{\partial x} (\nabla \cdot \mathbf{q}) = \rho a_x,$$

and similarly for the y- and z-directions,

$$\rho g_y - \frac{\partial p}{\partial y} + \mu \left[\frac{\partial^2 v}{\partial x^2} + \frac{\partial^2 v}{\partial y^2} + \frac{\partial^2 v}{\partial z^2} \right] + \frac{\mu}{3} \frac{\partial}{\partial y} (\nabla \cdot \mathbf{q}) = \rho a_y,$$

$$\rho g_z - \frac{\partial p}{\partial z} + \mu \left[\frac{\partial^2 w}{\partial x^2} + \frac{\partial^2 w}{\partial y^2} + \frac{\partial^2 w}{\partial z^2} \right] + \frac{\mu}{3} \frac{\partial}{\partial z} (\nabla \cdot \mathbf{q}) = \rho a_z.$$

$$\left. \right\} \tag{6–24}$$

Equations (6–24) are the Cartesian form of the *Navier-Stokes* equation for compressible fluids with constant viscosity. In vector notation and using Eq. (2–5a) for the acceleration vector, we find that Eqs. (6–24) become

$$\rho \mathbf{g} - \nabla p + \mu \nabla^2 \mathbf{q} + \frac{\mu}{3} \nabla(\nabla \cdot \mathbf{q}) = \rho \frac{\partial \mathbf{q}}{\partial t} + \rho(\mathbf{q} \cdot \nabla)\mathbf{q}. \tag{6–24a}$$

For incompressible fluids, $\nabla \cdot \mathbf{q} = 0$ and the Navier-Stokes equations become

$$\rho \mathbf{g} - \nabla p + \mu \nabla^2 \mathbf{q} = \rho \frac{\partial \mathbf{q}}{\partial t} + \rho(\mathbf{q} \cdot \nabla)\mathbf{q}. \tag{6–25}$$

Dense fluids, like liquids, have a large heat capacity. Therefore, temperature changes due to internal friction are small. The density and viscosity in these cases are very little affected and may be assumed to be constant. Thus Eq. (6–25) may be used even though isothermal conditions are not preserved completely.

If h is taken as the vertical direction (measured positive upward), the components of the acceleration due to gravity are

$$g_x = -g \frac{\partial h}{\partial x}, \qquad g_y = -g \frac{\partial h}{\partial y}, \qquad g_z = -g \frac{\partial h}{\partial z}, \tag{6–26}$$

or

$$\mathbf{g} = -g\nabla h. \tag{6–26a}$$

For example, if the Cartesian axes are oriented so that h and z coincide, then $g_x = g_y = 0$, and

$$g_z = -g. \tag{6–27}$$

The minus sign appears because the acceleration due to gravity is in the negative h-direction.

For convenience in later use, the incompressible Navier-Stokes equations of motion Eq. (6–25), are written in Cartesian components, with the accelerations given by Eqs. (2–5):

$$\frac{\partial u}{\partial t} + u \frac{\partial u}{\partial x} + v \frac{\partial u}{\partial y} + w \frac{\partial u}{\partial z} = -g \frac{\partial h}{\partial x} - \frac{1}{\rho} \frac{\partial p}{\partial x} + \frac{\mu}{\rho} \left[\frac{\partial^2 u}{\partial x^2} + \frac{\partial^2 u}{\partial y^2} + \frac{\partial^2 u}{\partial z^2} \right],$$

$$\frac{\partial v}{\partial t} + u \frac{\partial v}{\partial x} + v \frac{\partial v}{\partial y} + w \frac{\partial v}{\partial z} = -g \frac{\partial h}{\partial y} - \frac{1}{\rho} \frac{\partial p}{\partial y} + \frac{\mu}{\rho} \left[\frac{\partial^2 v}{\partial x^2} + \frac{\partial^2 v}{\partial y^2} + \frac{\partial^2 v}{\partial z^2} \right],$$

$$\frac{\partial w}{\partial t} + u \frac{\partial w}{\partial x} + v \frac{\partial w}{\partial y} + w \frac{\partial w}{\partial z} = -g \frac{\partial h}{\partial z} - \frac{1}{\rho} \frac{\partial p}{\partial z} + \frac{\mu}{\rho} \left[\frac{\partial^2 w}{\partial x^2} + \frac{\partial^2 w}{\partial y^2} + \frac{\partial^2 w}{\partial z^2} \right].$$

$$\tag{6–28}$$

For incompressible fluids and isothermal flows in a gravitational field, there are four flow variables, u, v, w, and p, which appear in the equations of motion.

Therefore, in principle, the three equations of motion plus the incompressible conservation of mass (Eq. 6–5),

$$\frac{\partial u}{\partial x} + \frac{\partial v}{\partial y} + \frac{\partial w}{\partial z} = 0,$$

are sufficient to obtain a solution when the boundary conditions are specified. The entire system of equations (6–28) and (6–5) must satisfy both kinematic and physical boundary conditions. The kinematic conditions are that the velocities normal to any rigid boundary or wall shall equal the boundary velocity (e.g., zero, for stationary boundaries). The physical condition is a consequence of the property of any real fluid which can be treated as a continuum to "stick" to a rigid boundary. The result is a no-slip condition for the tangential velocity component at the wall. Consequently, both the normal and tangential components of velocity relative to the wall vanish at the wall surface.

General solutions for the Navier-Stokes equations have not as yet been found because of the nonlinear, second-order nature of the partial differential equations. However, many particular solutions can be obtained by introducing various simplifications. One of the primary objectives of a first course in fluid mechanics will be to develop a feeling for the approximations which are appropriate for the solution of various engineering problems.

In many cases, it is convenient to work in a coordinate system other than the Cartesian. Thus the equations of motion and continuity may be transformed into cylindrical or spherical coordinates. The equations for an incompressible fluid in cylindrical coordinates (r, θ, z) are given below in terms of the velocity components v_r, v_θ, and v_z.

The Navier-Stokes equations in cylindrical coordinates for constant density and viscosity are given below.

r-component:

$$\rho\left(\frac{\partial v_r}{\partial t} + v_r\frac{\partial v_r}{\partial r} + \frac{v_\theta}{r}\frac{\partial v_r}{\partial \theta} - \frac{v_\theta^2}{r} + v_z\frac{\partial v_r}{\partial z}\right)$$

$$= -\frac{\partial p}{\partial r} + \mu\left\{\frac{\partial}{\partial r}\left(\frac{1}{r}\frac{\partial}{\partial r}[rv_r]\right) + \frac{1}{r^2}\frac{\partial^2 v_r}{\partial \theta^2} - \frac{2}{r^2}\frac{\partial v_\theta}{\partial \theta} + \frac{\partial^2 v_r}{\partial z^2}\right\} + \rho g_r;$$

$$(6\text{–}29a)$$

θ-component:

$$\rho\left(\frac{\partial v_\theta}{\partial t} + v_r\frac{\partial v_\theta}{\partial r} + \frac{v_\theta}{r}\frac{\partial v_\theta}{\partial \theta} + \frac{v_r v_\theta}{r} + v_z\frac{\partial v_\theta}{\partial z}\right)$$

$$= -\frac{1}{r}\frac{\partial p}{\partial \theta} + \mu\left\{\frac{\partial}{\partial r}\left(\frac{1}{r}\frac{\partial}{\partial r}[rv_\theta]\right) + \frac{1}{r^2}\frac{\partial^2 v_\theta}{\partial \theta^2} + \frac{2}{r^2}\frac{\partial v_r}{\partial \theta} + \frac{\partial^2 v_\theta}{\partial z^2}\right\} + \rho g_\theta;$$

$$(6\text{–}29b)$$

z-component:

$$\rho \left(\frac{\partial v_z}{\partial t} + v_r \frac{\partial v_z}{\partial r} + \frac{v_\theta}{r} \frac{\partial v_z}{\partial \theta} + v_z \frac{\partial v_z}{\partial z} \right)$$

$$= -\frac{\partial p}{\partial z} + \mu \left\{ \frac{1}{r} \frac{\partial}{\partial r} \left(r \frac{\partial v_z}{\partial r} \right) + \frac{1}{r^2} \frac{\partial^2 v_z}{\partial \theta^2} + \frac{\partial^2 v_z}{\partial z^2} \right\} + \rho g_z.$$

$$(6\text{–}29\text{c})$$

The continuity equation in cylindrical coordinates for constant density is

$$\frac{1}{r} \frac{\partial}{\partial r} (rv_r) + \frac{1}{r} \frac{\partial}{\partial \theta} (v_\theta) + \frac{\partial}{\partial z} (v_z) = 0. \qquad (6\text{–}30)$$

Normal and shear stresses in cylindrical coordinates for constant density and viscosity are

$$\sigma_r = -p + 2\mu \frac{\partial v_r}{\partial r}, \qquad\qquad \tau_{r\theta} = \mu \left[r \frac{\partial}{\partial r} \left(\frac{v_\theta}{r} \right) + \frac{1}{r} \frac{\partial v_r}{\partial \theta} \right],$$

$$\sigma_\theta = -p + 2\mu \left(\frac{1}{r} \frac{\partial v_\theta}{\partial \theta} + \frac{v_r}{r} \right), \qquad \tau_{\theta z} = \mu \left(\frac{\partial v_\theta}{\partial z} + \frac{1}{r} \frac{\partial v_z}{\partial \theta} \right),$$

$$\sigma_z = -p + 2\mu \frac{\partial v_z}{\partial z}, \qquad\qquad \tau_{zr} = \mu \left(\frac{\partial v_r}{\partial z} + \frac{\partial v_z}{\partial r} \right). \qquad (6\text{–}30\text{a})$$

6–5 EXAMPLES OF LAMINAR MOTION

Although the Navier-Stokes equations cannot readily be solved in their general form, they are nevertheless of great value in many viscous-flow problems in which certain terms are zero or negligible. The term *laminar motion* is used to describe an orderly state of flow in which macroscopic fluid particles move in layers. Thus, in laminar flow through a tube of constant diameter, the instantaneous velocity at any point would always be unidirectional (along the axis of the tube). The moving fluid which is bounded by the tube wall is characterized by a velocity gradient as we move away from the wall. The fluid in contact with the solid boundary is assumed to satisfy the "no-slip" condition. In accordance with the concept of the Newtonian viscosity, the velocity gradient gives rise to viscous forces within the fluid. The viscous forces, being dissipative, have a stabilizing or damping effect on the motion when subjected to disturbances. The capability of the viscous forces to stabilize the flow depends upon their relative magnitude compared with the inertial (destabilizing) forces. When the flow becomes unstable, the instantaneous velocity vector is no longer unidirectional even though the time-average velocity remains in the axial direction. This type of motion is known as turbulent flow and will be discussed in detail in a later section.

6–5.1 Laminar flow between parallel plates

We consider the two-dimensional, steady, laminar flow of an incompressible fluid between parallel plates in which the upper plate is in motion with a velocity U in the x-direction relative to the lower plate, as shown in Fig. 6–4.

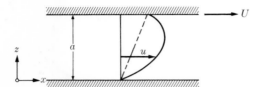

FIG. 6–4. Flow between parallel plates.

Let z be the coordinate perpendicular to the plates. For two-dimensional conditions, v and all derivatives with respect to y are zero. The z-axis is taken to coincide with the vertical direction h. Hence, in Eqs. (6–28), $\partial h/\partial x = 0$ and $\partial h/\partial z = 1$. The flow is steady, thereby making the time derivatives of the velocity zero. We assume parallel flow in the x-direction so that w and all its derivatives are zero. By continuity, $\partial u/\partial x = 0$ at every x, and therefore, $\partial^2 u/\partial x^2 = 0$ and $u = f(z)$ only. Under these conditions, the equations of motion for the x- and z-directions may be written

$$0 = -\frac{1}{\rho}\frac{\partial p}{\partial x} + \frac{\mu}{\rho}\left(\frac{\partial^2 u}{\partial z^2}\right), \tag{6–31a}$$

$$0 = -g - \frac{1}{\rho}\frac{\partial p}{\partial z}. \tag{6–31b}$$

Note that all accelerations are zero and all inertia terms vanish from the equations of motion. From Eq. (6–31b), for the z-direction, we have

$$p = -\gamma z + f(x), \tag{6–32}$$

which indicates that the pressure distribution is hydrostatic normal to the flow direction. While in this case the z-direction coincides with the vertical, it can be shown that for any orientation of a parallel flow the pressure is distributed hydrostatically in a direction normal to the flow.

Because of the hydrostatic condition of Eq. (6–32), $\partial p/\partial x$ is independent of z and may be written as dp/dx. Equation (6–31a) may then be integrated twice to give

$$\frac{dp}{dx}\frac{z^2}{2} = \mu u + C_1 z + C_2. \tag{6–33}$$

Using the boundary conditions,

$$z = 0, \quad u = 0,$$
$$z = a, \quad u = U,$$

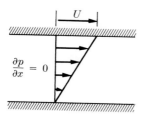

FIG. 6–5.   Couette flow.

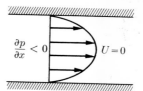

FIG. 6–6.   Poiseuille flow.

we obtain

$$u = \frac{Uz}{a} - \frac{az}{2\mu}\frac{dp}{dx}\left(1 - \frac{z}{a}\right). \tag{6-34}$$

If $dp/dx = 0$, the flow is known as a *Couette flow* (Fig. 6–5), and the velocity distribution is given by

$$u = \frac{Uz}{a}. \tag{6-35}$$

If $U = 0$, we have parallel flow between stationary plates (Fig. 6–6) known as two-dimensional Poiseuille flow. The velocity distribution is parabolic with the maximum velocity at $z = a/2$. Hence

$$u_{\max} = \frac{-a^2}{8\mu}\frac{dp}{dx}, \tag{6-36}$$

and the average velocity Q/A is given by

$$V = \frac{Q}{A} = \frac{2}{3}u_{\max} = \frac{-a^2}{12\mu}\frac{dp}{dx}. \tag{6-37}$$

6–5.2   Laminar flow in a circular tube of constant diameter

Steady laminar flow due to pressure drop along a tube is called Poiseuille flow. The velocity distribution for such a flow in a tube of radius r_0 may be obtained by simplifying the equations of motion in cylindrical coordinates. If we take the z-axis as coincident with the axis of the tube, v_θ and v_r are everywhere equal to zero. The velocity v_z and its derivatives are independent of z (continuity for parallel flow) and of θ (symmetry). In this example, z will be taken arbitrarily and different from the vertical h. From Eqs. (6–29), the z-component for steady motion becomes

$$0 = -\frac{\partial(p + \gamma h)}{\partial z} + \mu\frac{1}{r}\frac{\partial}{\partial r}\left(r\frac{\partial v_z}{\partial r}\right). \tag{6-38}$$

Using the condition of hydrostatic pressure distribution normal to the direction of parallel flows, we note that $\partial(p + \gamma h)/\partial z$ is independent of r. Now integrating

Eq. (6–38) twice with respect to r and using the boundary conditions

$$r = 0, \qquad v_z = v_{z,\,\max},$$

$$r = r_0, \qquad v_z = 0,$$

we have

$$v_z = - \frac{d(p + \gamma h)}{dz} \frac{r_0^2}{4\mu} \left[1 - \left(\frac{r}{r_0} \right)^2 \right], \tag{6–39}$$

which is the equation of a paraboloid of revolution. At $r = 0$,

$$v_z = v_{z,\,\max} = - \frac{d(p + \gamma h)}{dz} \frac{r_0^2}{4\mu}. \tag{6–40}$$

The average velocity is

$$V_z = \frac{v_{z,\,\max}}{2} = - \frac{d(p + \gamma h)}{dz} \frac{r_0^2}{8\mu}, \tag{6–41}$$

and the head loss per unit length of pipe of diameter D is

$$\frac{h_f}{L} = - \frac{1}{\gamma} \frac{d(p + \gamma h)}{dz} = \frac{32\mu V_z}{\gamma D^2}. \tag{6–42}$$

We also note that in polar coordinates the shear stress acting in the z-direction is

$$\tau_{zr} = \mu \left(\frac{\partial v_r}{\partial z} + \frac{\partial v_z}{\partial r} \right), \tag{6–43}$$

and this reduces to

$$\tau = \mu \frac{\partial v_z}{\partial r} \tag{6–44}$$

for uniform flow in a tube. Equation (6–38) then becomes

$$0 = - \frac{d(p + \gamma h)}{dz} + \frac{1}{r} \frac{d}{dr} (r\tau),$$

and

$$\tau = \frac{r}{2} \frac{d(p + \gamma h)}{dz}. \tag{6–45}$$

Finally, note that this example of steady parallel flow involves no inertia force terms.

6–6 EQUATIONS FOR IRROTATIONAL MOTION

The equations of motion developed in this chapter are derived from Newton's second law and are therefore momentum equations. In Chapter 4 it was shown that the one-dimensional energy equation which is derived from the first law of thermodynamics reduces to the so-called Bernoulli equation for the steady flow of an incompressible fluid with zero friction and no shaft work. Inasmuch as the Bernoulli equation is an important first approximation in the analysis of many flow problems, it is pertinent to show that it may also be obtained from the momentum equations within the three-dimensional frame of reference. In this section we will show this for the case of irrotational motion. In a subsequent section we will show the derivation when only frictionless motion is assumed. We will begin by introducing the velocity potential function, which is useful in treating irrotational motion.

6–6.1   Velocity potential and stream function

If $\phi\ (x, y, z, t)$ is any scalar quantity having continuous first and second derivatives, then, by a fundamental vector identity,

$$\text{curl (grad } \phi) \equiv \nabla \times (\nabla \phi) \equiv 0. \tag{6–46}$$

Thus, in irrotational flow, where $\nabla \times \mathbf{q} = 0$, there must exist a scalar function ϕ whose gradient is equal to the velocity vector $\mathbf{q}$. It is convenient to define the positive direction of the flow in the direction in which the scalar quantity is decreasing, and therefore

$$\mathbf{q} = -\nabla \phi(x, y, z, t). \tag{6–47}$$

Since the negative gradient of ϕ is equal to the velocity vector, ϕ is known as the *velocity potential* and irrotational flow is frequently called *potential flow*. Either compressible or incompressible fluids may be in irrotational motion, and a potential function will exist in either case.

In incompressible flow, Eq. (6–5) requires

$$\nabla \cdot \mathbf{q} = 0.$$

Therefore

$$\nabla \cdot (-\nabla \phi) = -\nabla^2 \phi = 0. \tag{6–48}$$

This is Laplace's equation, and the operator ∇^2 is known as the Laplacian. In Cartesian coordinates,

$$\nabla^2 \phi = \frac{\partial^2 \phi}{\partial x^2} + \frac{\partial^2 \phi}{\partial y^2} + \frac{\partial^2 \phi}{\partial z^2} = 0, \tag{6–49}$$

and in cylindrical coordinates,

$$\nabla^2 \phi = \frac{1}{r} \frac{\partial}{\partial r} \left(r \frac{\partial \phi}{\partial r} \right) + \frac{1}{r^2} \frac{\partial^2 \phi}{\partial \theta^2} + \frac{\partial^2 \phi}{\partial z^2} = 0. \tag{6–50}$$

For two-dimensional incompressible irrotational motion, the velocity potential ϕ and the stream function ψ are related. Writing Eq. (6–47) in Cartesian coordinates, we have

$$u = -\frac{\partial \phi}{\partial x}, \qquad v = -\frac{\partial \phi}{\partial y}. \tag{6–47a}$$

Equating to Eqs. (6–8), we obtain

$$\frac{\partial \psi}{\partial y} = \frac{\partial \phi}{\partial x}, \qquad \frac{\partial \psi}{\partial x} = -\frac{\partial \phi}{\partial y}. \tag{6–51}$$

These are known as the Cauchy-Riemann equations.

In irrotational flow, the stream function ψ also satisfies the two-dimensional Laplace equation. This may be shown by substituting Eqs. (6–8) into the two-dimensional condition for irrotational motion from Eqs. (6–17),

$$\frac{\partial v}{\partial x} = \frac{\partial u}{\partial y},$$

and hence

$$\frac{\partial^2 \psi}{\partial x^2} + \frac{\partial^2 \psi}{\partial y^2} = 0. \tag{6–52}$$

Also, for two-dimensional flow, Eq. (6–49) becomes

$$\frac{\partial^2 \phi}{\partial x^2} + \frac{\partial^2 \phi}{\partial y^2} = 0. \tag{6–53}$$

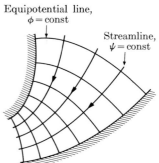

FIG. 6–7. Flow net for irrotational converging flow illustrating orthogonality of lines of constant ψ and constant ϕ.

Since both ϕ and ψ satisfy the Laplace equation, stream functions and velocity potentials may be interchanged. Furthermore, the Laplace equation is linear, and flows may be constructed by superposition of various stream functions. Finally, it can be shown that lines of constant ψ and ϕ must form an orthogonal mesh, known as a *flow net*, of the type shown in Fig. 6–7. From Eq. (2–10), which is the equation for a streamline, we get

$$\left.\frac{dy}{dx}\right|_{\psi=\text{const}} = \frac{v}{u}. \tag{6–54}$$

Along lines of constant velocity potential, $d\phi = 0$, and hence

$$d\phi = 0 = \frac{\partial \phi}{\partial x}\,dx + \frac{\partial \phi}{\partial y}\,dy.$$

Therefore,

$$\left.\frac{dy}{dx}\right|_{\phi=\text{const}} = -\frac{\partial \phi/\partial x}{\partial \phi/\partial y} = -\frac{u}{v}, \tag{6–55}$$

and

$$\left.\frac{dy}{dx}\right|_{\phi=\text{const}} = -\left.\frac{dx}{dy}\right|_{\psi=\text{const}}. \tag{6–56}$$

Hence the slopes are the negative reciprocal of each other. Any streamline may represent a solid boundary since there can be no flow across it. The analytical solution of the Laplace equation is difficult for complex boundaries, and in these cases a graphical solution may be obtained by sketching a network of curvilinear squares. Some examples of this procedure will be given in a later chapter. It should be noted that the velocity potential exists only for irrotational flows; however, the stream function is not subject to this restriction.

6–6.2 The Bernoulli equation

If the irrotational motion conditions of Eqs. (6–17) are substituted in the incompressible Navier-Stokes equations (6–28), we obtain the following set of equations (using the x-component as an example):

$$\frac{\partial u}{\partial t} + u\frac{\partial u}{\partial x} + v\frac{\partial v}{\partial x} + w\frac{\partial w}{\partial x} = -g\frac{\partial h}{\partial x} - \frac{1}{\rho}\frac{\partial p}{\partial x} + \frac{\mu}{\rho}\left[\frac{\partial^2 u}{\partial x^2} + \frac{\partial^2 v}{\partial y\,\partial x} + \frac{\partial^2 w}{\partial z\,\partial x}\right],$$

$$\frac{\partial u}{\partial t} + \frac{\partial}{\partial x}\left(\frac{u^2}{2} + \frac{v^2}{2} + \frac{w^2}{2}\right) = -g\frac{\partial h}{\partial x} - \frac{1}{\rho}\frac{\partial p}{\partial x} + \frac{\mu}{\rho}\frac{\partial}{\partial x}\left(\frac{\partial u}{\partial x} + \frac{\partial v}{\partial y} + \frac{\partial w}{\partial z}\right).$$

$$(6\text{–}57)$$

Then, since

$$q^2 = u^2 + v^2 + w^2,$$

and from the continuity equation

$$\nabla \cdot \mathbf{q} = \frac{\partial u}{\partial x} + \frac{\partial v}{\partial y} + \frac{\partial w}{\partial z} = 0,$$

we obtain

$$\frac{\partial u}{\partial t} + \frac{\partial}{\partial x}\left(\frac{q^2}{2}\right) = -g\frac{\partial h}{\partial x} - \frac{1}{\rho}\frac{\partial p}{\partial x}. \tag{6–58}$$

Grouping terms under the operator $\partial/\partial x$, we have

$$\frac{\partial u}{\partial t} + \frac{\partial}{\partial x}\left[\frac{q^2}{2} + gh + \frac{p}{\rho}\right] = 0, \tag{6–59a}$$

and the corresponding y- and z-equations become

$$\frac{\partial v}{\partial t} + \frac{\partial}{\partial y}\left[\frac{q^2}{2} + gh + \frac{p}{\rho}\right] = 0 \quad \text{and} \quad \frac{\partial w}{\partial t} + \frac{\partial}{\partial z}\left[\frac{q^2}{2} + gh + \frac{p}{\rho}\right] = 0.$$

$$(6\text{–}59\text{b, c})$$

Using the velocity potential, we can substitute the equalities

$$\frac{\partial u}{\partial t} = -\frac{\partial^2 \phi}{\partial t\,\partial x}, \qquad \frac{\partial v}{\partial t} = -\frac{\partial^2 \phi}{\partial t\,\partial y}, \qquad \frac{\partial w}{\partial t} = -\frac{\partial^2 \phi}{\partial t\,\partial z}$$

into Eqs. (6–59). This allows integration to give

$$-\frac{\partial \phi}{\partial t} + \frac{q^2}{2} + gh + \frac{p}{\rho} = F(t). \tag{6–60}$$

This relation holds throughout the entire field of irrotational motion. Thus for a steady irrotational flow of an incompressible fluid,

$$\frac{q^2}{2} + gh + \frac{p}{\rho} = \text{const} \tag{6-61}$$

throughout the entire flow field (both along and normal to any streamline). This is known as the *Bernoulli* equation for an incompressible fluid.

Dividing each term of Eq. (6–61) by the acceleration of gravity gives the "head" terms familiar from the energy equation. Hence, between any two points in the fluid field,

$$H = \frac{p_1}{\gamma} + h_1 + \frac{q_1^2}{2g} = \frac{p_2}{\gamma} + h_2 + \frac{q_2^2}{2g}, \tag{6-62}$$

where

$H =$ total head at a point in the flow.

Equation (6–62) is the three-dimensional form of the one-dimensional Bernoulli equation for negligible friction as previously derived in Chapter 4, Eq. (4–26). Remember that in Eq. (4–26), H is the weight-flow-rate average of the total head over a cross section of a conduit, and each of the quantities p, h and V is its average value over the flow section. In Eq. (6–62), on the other hand, all the quantities are the values at a point in the flow.

Note that in the derivation, the assumptions made were incompressibility, steady flow, and irrotational motion. The only condition placed on viscosity was that it be a constant. In fact, no additional assumptions are necessary as the viscosity term dropped out because $\nabla \cdot \mathbf{q} = 0$. The result, therefore, applies to either a viscous or nonviscous fluid so long as the conditions of incompressibility, steadiness, and irrotationality hold. With viscous fluids, of course, velocity gradients always result in viscous shear and only in a few special cases is this compatible with irrotational motion. The irrotational vortex described in Section 6–8 is such a case. Otherwise, when for initially irrotational motions (such as those generated from a condition of rest), viscosity causes a spread of vorticity, the flow becomes rotational and H in Eq. (6–62) varies throughout the fluid.

An irrotational motion can never become rotational as long as only gravitational and pressure forces act upon the fluid particles. As a rule of thumb it may be stated that nearly irrotational flows may be generated in real fluids if the motion is primarily a result of pressure and gravity forces. Consider as examples the generation of a free-surface wave motion produced by the oscillation of a hinged plate, as shown in Fig. 6–8. If the wave motion is of the "deep-water" type, in which the fluid particle motion at the bottom is zero, the resulting motion is almost exactly irrotational. This can be verified by comparing experimental measurements and theory based on the assumption of irrotational motion. Another example is the flow of a liquid over a vertical plate known as a weir, as

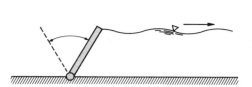

FIG. 6–8. Wave generation by pressure forces.

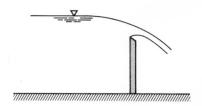

FIG. 6–9. Flow over a weir under gravity forces.

shown in Fig. 6–9. In this case, the fluid motion is generated by the force of gravity and is nearly irrotational. On the other hand, a fluid motion generated through the action of tangential shear stresses will tend to be rotational since shear can only be transmitted in a viscous fluid. For example, consider an open cylindrical container half filled with a liquid. If the container is rotated by placing it on a turntable, the resulting motion of the fluid within the cylinder is entirely a result of viscous shear between the rotating cylinder and the fluid. Another type of fluid motion in circular paths can be generated in a cylindrical container by draining liquid from a hole in the bottom. Here the generating forces are pressure and gravity. These two types of vortex motion illustrate most of the foregoing remarks on rotational and irrotational motions and will be analyzed in detail in Section 6–8.

Modern developments in fluid mechanics have been largely due to an understanding of the role of viscosity in fluid motions. It is obvious that the assumption of irrotationality leads to an important simplification of the equations of motion. The assumption is therefore of considerable advantage in the analysis of many fluid problems in which the rotational characteristics are of secondary importance. In other cases, the rotational flow can be regarded as being confined to thin "boundary layers," with the flow outside of this layer being considered irrotational. This type of motion will be treated in detail in a later chapter.

6–7   EQUATIONS FOR FRICTIONLESS FLOW

6–7.1   Equations for flow along a streamline

In the previous section, the Bernoulli equation was obtained as a special case of the incompressible Navier-Stokes equations for irrotational motion in a real fluid. Several examples of types of real fluid motion which approximately satisfy this requirement were discussed. It is also of interest to show that by neglecting frictional (viscous) effects, the Navier-Stokes equations when integrated along a streamline reduce to the Bernoulli form without assuming irrotational motion. This form also has the advantage of being applicable to compressible fluid flows, whereas the previous development is limited to incompressible flows.

We start with the compressible flow Navier-Stokes equation in vector form (Eq. 6–24a). Setting $\mu = 0$ and dividing through by the density gives

$$\mathbf{g} - \frac{\nabla p}{\rho} = \frac{\partial \mathbf{q}}{\partial t} + (\mathbf{q} \cdot \nabla)\mathbf{q}. \tag{6-63}$$

This is known as *Euler's equation of motion*. Letting $\mathbf{g} = -g\nabla h$, we get from Eq. (6–26a):

$$-g\nabla h - \frac{\nabla p}{\rho} = \frac{\partial \mathbf{q}}{\partial t} + (\mathbf{q} \cdot \nabla)\mathbf{q}. \tag{6-64}$$

Because considerable economy of effort can be achieved, it is convenient to perform the integration along a streamline in vector form. Hence, multiplying each vector in Eq. (6–64) by the element of streamline length $d\mathbf{r}$ and integrating along the streamlines, we have

$$-g\int \nabla h \cdot d\mathbf{r} - \int \frac{1}{\rho} \nabla p \cdot d\mathbf{r} = \int \left(\frac{\partial \mathbf{q}}{\partial t}\right) \cdot d\mathbf{r} + \int [(\mathbf{q} \cdot \nabla)\mathbf{q}] \cdot d\mathbf{r} + C(t). \tag{6-65}$$

The integration of the first two terms may be performed by making use of the vector relation

$$\nabla \eta \cdot d\mathbf{r} = d\eta,$$

where η is any single-valued scalar function of space. Hence

$$-gh - \int \frac{dp}{\rho} = \int \left(\frac{\partial \mathbf{q}}{\partial t}\right) \cdot d\mathbf{r} + \int [(\mathbf{q} \cdot \nabla)\mathbf{q}] \cdot d\mathbf{r} + C(t). \tag{6-66}$$

The last term may be integrated by noting that*

$$[(\mathbf{q} \cdot \nabla)\mathbf{q}] \cdot d\mathbf{r} = d\mathbf{r} \cdot [(\mathbf{q} \cdot \nabla)\mathbf{q}] = \mathbf{q} \cdot [(d\mathbf{r} \cdot \nabla)\mathbf{q}],$$

and also that

$$d\mathbf{r} \cdot \nabla = \frac{\partial (\ )}{\partial x} \, dx + \frac{\partial (\ )}{\partial y} \, dy + \frac{\partial (\ )}{\partial z} \, dz = d(\ ),$$

and hence

$$\int [(\mathbf{q} \cdot \nabla)\mathbf{q}] \cdot d\mathbf{r} = \int \mathbf{q} \cdot d\mathbf{q} = \int d\left(\frac{q^2}{2}\right) = \frac{q^2}{2}.$$

Equation (6–66) is then written as the sum of the scalar terms,

$$\int \frac{dp}{\rho} + gh + \frac{q^2}{2} + \int \left(\frac{\partial \mathbf{q}}{\partial t}\right) \cdot d\mathbf{r} = -C(t). \tag{6-67}$$

---

* This operation may be verified by expanding in Cartesian components and making use of Eq. (2–10) for a streamline.

For steady motion, $\partial \mathbf{q}/\partial t = 0$, and C becomes a constant. Then we have the useful compressible fluid relation between the flow quantities along a streamline,

$$\int \frac{dp}{\rho} + gh + \frac{q^2}{2} = \text{const along a streamline.} \qquad (6\text{--}68)$$

Equation (6–68) can be integrated for a barotropic fluid, that is one in which the density can be expressed as a function of pressure. The isothermal and adiabatic relations for gases discussed in Chapter 1 are the most common examples. Applications of Eq. (6–68) to nonuniform compressible flows are given in Chapter 14.

For incompressible fluids, ρ is constant, and upon dividing by g, we obtain

$$\frac{p}{\gamma} + h + \frac{q^2}{2g} = \text{const along a streamline.} \qquad (6\text{--}69)$$

Equation (6–69) is the Bernoulli equation for steady frictionless incompressible fluid flow. The constant will change from one streamline to another in a rotational flow; it will be invariant throughout the fluid for irrotational flow.

6–7.2 Summary of Bernoulli equation forms

In summary, three forms of the Bernoulli equation have been developed for steady incompressible flows in a gravitational field. For irrotational flow, Eq. (6–62) gives

$$H = \frac{p}{\gamma} + h + \frac{q^2}{2g} = \text{const throughout flow.} \qquad (6\text{--}70)$$

For frictionless flow, Eq. (6–69) gives

$$H = \frac{p}{\gamma} + h + \frac{q^2}{2g} = \text{const along streamline,} \qquad (6\text{--}71)$$

and Eq. (4–25) yields

$$H = \frac{p}{\gamma} + h + K_e \frac{V^2}{2g} = \text{const along finite conduit.} \qquad (6\text{--}72)$$

Equations (6–62) and (6–69) have been developed from the momentum equations. Equation (4–25) was obtained from the conservation-of-energy principle for a finite control volume. In addition, it was also shown in Chapter 4 that for a steady flow with friction (in the absence of shaft work), the total head, H, decreases along a conduit in the direction of flow. This change in the total head between any two sections is known as the "head loss." The frictional effects must be considered in the case of fluid flow in long pipe lines. Therefore, in a later chapter we shall develop a general method of calculating head loss for uniform flow in conduits. Note that we have already done this for laminar flow in a tube

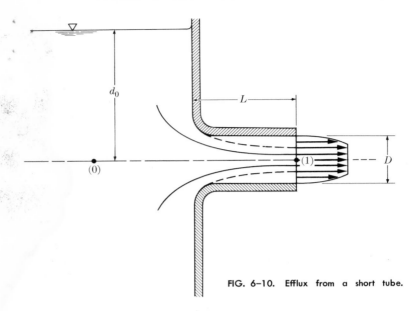

FIG. 6–10. Efflux from a short tube.

of constant area. It is easy to verify from Eq. (6–42) that the total head decreases along the tube at a constant rate.

For nonuniform flows (in general, due to a variable area of flow), the analytical determination of frictional effects is difficult. Fortunately, in many cases the distances involved are small, and frictional effects are also correspondingly small. In these cases, the frictionless-flow Bernoulli equation provides a reasonable method of analysis. Several examples of this kind are discussed below.

6–7.3 Frictionless-flow examples

Efflux. Consider the flow of a liquid discharging through a short tube opening into the side of a large reservoir of constant depth, d_0, as shown in Fig. 6–10. The dashed lines in the tube indicate the lateral extent of significant viscous action. The fluid at the wall must have zero velocity, while in the central core the velocity is constant. If the tube were longer, the zone of viscous action would extend to the centerline, and the assumption of negligible frictional effects would no longer be justified. Within the reservoir and central core of the tube, the flow is acted upon primarily by pressure and gravity forces. Velocity gradients and viscous action are negligible. Hence, since the flow started from rest, this portion is in irrotational motion. We may therefore use either Eq. (6–70) or (6–71). Writing the Bernoulli equation along the centerline streamline between (0) and (1) and neglecting velocities in the reservoir, we have

$$\frac{p_0}{\gamma} + h_0 + 0 = \frac{p_1}{\gamma} + h_1 + \frac{q_1^2}{2g}. \qquad (6\text{–}73)$$

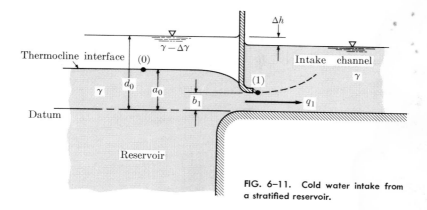

FIG. 6–11. Cold water intake from a stratified reservoir.

If the ambient atmospheric pressure is arbitrarily taken as $p_a = 0$, then the pressure at the pipe exit is $p_1 = p_a = 0$. The pressure at point (0) is hydrostatic (since the reservoir velocities are neglected). Hence $p_0 = \gamma d_0$, and since $h_0 = h_1$, Eq. (6–73) becomes

$$\frac{q_1^2}{2g} = d_0. \tag{6–74}$$

The problem seems simple, but nevertheless we cannot at this time determine the total rate of flow from the tube unless we are willing to neglect the thickness of the zone of viscous influence. In that case,

$$Q = \frac{\pi D^2}{4} q_1. \tag{6–75}$$

The error in Eq. (6–75) will be a function of the length of the tube L.

Stratified flow. During the summer months, large reservoirs and lakes become thermally stratified, that is, at a certain depth (known as the *thermocline*) the temperature changes rapidly with depth. This results in a lighter liquid on top of a heavier one since (above 4°C) the specific weight is inversely related to temperature. Figure 6–11 shows a schematic representation of a stratified reservoir with an intake channel and submerged wall at one side.

The colder water is drawn into the intake channel with a velocity q_1, which will be considered uniform over the height b_1. The surface elevations in the reservoir and intake channel and the thermocline interface elevation are all considered to be constant. We want to find the relation between the surface-elevation change across the wall (Δh) and the velocity of cold water flow under the wall. The Bernoulli equation is written between points (0) and (1) on the interfacial streamline, while neglecting velocities within the reservoir,

$$\frac{p_0}{\gamma} + a_0 + 0 = \frac{p_1}{\gamma} + b_1 + \frac{q_1^2}{2g}. \tag{6–76}$$

The fluid above points (0) and (1) is at rest, and thus the pressures are obtained from the hydrostatic relationship

$$p_0 = (\gamma - \Delta\gamma)(d_0 - a_0), \qquad p_1 = \gamma(d_0 - \Delta h - b_1).$$

Substituting the pressures into Eq. (6–76) and simplifying, we have

$$\frac{q_1^2}{2g} = \Delta h - \frac{\Delta\gamma}{\gamma}(d_0 - a_0). \tag{6–77}$$

For the isothermal case, $a_0 = d_0$, and

$$\frac{q_1^2}{2g} = \Delta h. \tag{6–78}$$

Intake channels of the type shown in Fig. 6–11 have been used to provide cool condenser water for thermal power plants.

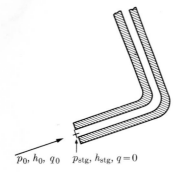

$p_0, h_0, q_0 \quad p_{stg}, h_{stg}, q = 0$

FIG. 6–12. Notation for stagnation tube.

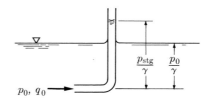

p_0, q_0

$\dfrac{p_{stg}}{\gamma} \quad \dfrac{p_0}{\gamma}$

FIG. 6–13. River velocity measurements with an impact tube.

Velocity measurements with the Pitot tube. For liquids of small viscosity, Eq. (6–71) can be used to compute the velocity from the *stagnation* or *impact* pressure measured on the blunt nose of a probe in a steady flow. Consider the flow around a probe placed in a stream, as shown in Fig. 6–12. At the stagnation point at the center of the blunt face, the velocity is zero, and we have for the stagnation streamline

$$\frac{p_0}{\gamma} + h_0 + \frac{q_0^2}{2g} = \frac{p_{stg}}{\gamma} + h_{stg} = \text{const},$$

or

$$\frac{q_0^2}{2g} = \left(\frac{p_{stg}}{\gamma} + h_{stg}\right) - \left(\frac{p_0}{\gamma} + h_0\right), \tag{6–79}$$

where the terms are defined in Fig. 6–12. When the body force is unimportant the pressure difference will be due solely to the change in velocity, and we may write simply

$$\frac{q_0^2}{2g} = \frac{p_{stg} - p_0}{\gamma}.$$ (6–80)

In 1732 Henri Pitot [3] used Eq. (6–80) to compute the velocity in the Seine from stagnation pressure measurements, using the arrangement shown in Fig. 6–13.

In the more general case of a compressible fluid, the elevation is unimportant, and the velocity can be computed from the measured difference in enthalpy between the undisturbed flow and the stagnation point on an impact tube. Assuming frictionless adiabatic compression and applying Eq. (4–21) along a streamline, we have

$$\frac{q_0^2}{2} = \left(\frac{p}{\rho} + u\right)_{stg} - \left(\frac{p}{\rho} + u\right)_0.$$ (6–81)

6–8 VORTEX MOTION

A fluid motion in which the streamlines are concentric circles is called a *vortex*. Consider the Navier-Stokes equations in cylindrical coordinates for the steady flow of an incompressible fluid about the z-axis. The velocity components v_r and v_z are zero, there is no pressure gradient in the θ-direction, and v_θ is not a function of z. Let z coincide with the vertical direction h. Under these conditions, the continuity equation and equations of motion reduce to the following:

Continuity, Eq. (6–30),

$$\frac{1}{r}\frac{\partial}{\partial \theta} v_\theta = 0;$$ (6–82)

Navier-Stokes, Eq. (6–29),

r-component: $$\frac{v_\theta^2}{r} = \frac{1}{\rho}\frac{\partial p}{\partial r}$$

θ-component: $$0 = \frac{\mu}{\rho}\frac{\partial}{\partial r}\left[\frac{1}{r}\frac{\partial}{\partial r}(rv_\theta)\right]$$ (6–83)

z-component: $$0 = -g - \frac{1}{\rho}\frac{\partial p}{\partial z} = -g - \frac{1}{\rho}\frac{\partial p}{\partial h}.$$

The θ-equation may be integrated to give

$$v_\theta = \frac{c_1 r}{2} + \frac{c_2}{r}.$$ (6–84)

The boundary conditions must be specified in order to evaluate the constants. Note that the pressure varies hydrostatically in the vertical direction.

6–8.1 ˙ Forced vortex

A cylindrical container of radius R is rotated at a constant angular velocity Ω about a vertical axis. The container is partly filled with a liquid of specific weight γ. The boundary conditions are

$$r = 0: \qquad v_\theta = 0; \qquad \text{hence} \quad c_2 = 0;$$
$$r = R: \qquad v_\theta = R\Omega; \qquad \text{hence} \quad c_1 = 2\Omega, \tag{6–85}$$

and

$$v_\theta = \Omega r \qquad \text{(solid-body rotation)}. \tag{6–86}$$

The r-component of Eq. (6–83) then becomes

$$\frac{\partial p}{\partial r} = \rho \Omega^2 r.$$

The z-equation gives the hydrostatic condition

$$\frac{\partial p}{\partial z} = \frac{\partial p}{\partial h} = -\rho g = -\gamma.$$

Substituting into the total differential, we have

$$dp = \frac{\partial p}{\partial r}\, dr + \frac{\partial p}{\partial h}\, dh = \rho \Omega^2 r\, dr - \gamma\, dh,$$

and integrating, we obtain

$$p = \rho\, \frac{\Omega^2 r^2}{2} - \gamma h + c_3. \tag{6–87}$$

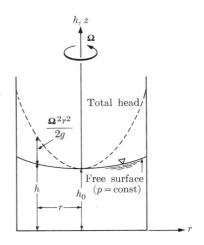

FIG. 6–14. Forced vortex.

Using the boundary conditions

$$r = 0: \qquad h = h_0, \qquad p = p_0,$$

we get

$$p - p_0 = -\gamma(h - h_0) + \rho\, \frac{\Omega^2 r^2}{2}. \tag{6–88}$$

The locus of the free surface is obtained by setting $p = p_0$, and thus $h = h_0 + \Omega^2 r^2 / 2g$. Therefore the free surface is a paraboloid of revolution. The rotation component, ω_z, given by Eq. (6–18) can be evaluated from $v_\theta = r\Omega$ which is the velocity distribution for the forced vortex. Hence

$$\omega_z = \frac{\Omega + \Omega}{2} = \Omega. \tag{6–89}$$

The vorticity is

$$\zeta = 2\omega_z = 2\Omega. \tag{6–90}$$

Thus the motion of a forced vortex is rotational as was expected, since it was generated by the transmission of tangential shear stresses. It can easily be verified that the total head $(p/\gamma + h + v_\theta^2/2g)$ is not constant but increases with the radius, as indicated in Fig. 6–14.

6–8.2 Irrotational or free vortex

It has been shown that in an irrotational flow the total head is a constant throughout the fluid. Hence

$$\frac{p}{\gamma} + h + \frac{v_\theta^2}{2g} = \text{const} \tag{6–91}$$

throughout the flow field. Differentiating with respect to r and rearranging gives

$$\frac{\partial p}{\partial r} = -\rho v_\theta \frac{\partial v_\theta}{\partial r}.$$

Also, from the r-component of the Navier-Stokes equations (6–83),

$$\frac{\partial p}{\partial r} = \rho \frac{v_\theta^2}{r}.$$

Equating the two values of $\partial p/\partial r$ and integrating, we get

$$v_\theta r = \text{const} = c_4. \tag{6–92}$$

This is the condition of constant angular momentum defined by Eq. (4–41). The radial pressure gradient is therefore

$$\frac{\partial p}{\partial r} = \rho \frac{c_4^2}{r^3}.$$

Again, using the hydrostatic condition, we obtain

$$dp = \frac{\partial p}{\partial r}\,dr + \frac{\partial p}{\partial h}\,dh = \rho \frac{c_4^2}{r^3}\,dr - \gamma\,dh,$$

and

$$p = -\rho \frac{c_4^2}{2r^2} - \gamma h + c_5. \tag{6–93}$$

By using the boundary conditions shown in Fig. 6–15,

$$r = \infty: \quad h = h_0, \quad p = p_0,$$

we have

$$p - p_0 = \gamma(h_0 - h) - \rho \frac{c_4^2}{2r^2}. \tag{6–94}$$

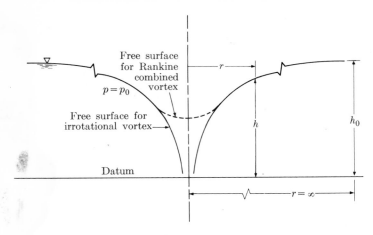

FIG. 6–15.   Free or irrotational vortex.

The locus of the free surface is given when $p = p_0$ as

$$h = h_0 - \frac{c_4^2}{2gr^2}. \tag{6–95}$$

Therefore, the free surface is a hyperboloid of revolution as shown in Fig. 6–15.

The fact that the circulation for a contour enclosing the origin is not zero can be shown by using any of the circular streamlines given by a constant value of v_θ. With the definition of circulation, from Eqs. (6–19) and (6–92), we have

$$\Gamma = \oint \mathbf{q} \cdot d\mathbf{s} = \int_0^{2\pi} v_\theta r \, d\theta = 2\pi c_4. \tag{6–96}$$

Hence c_4 in the above equations is

$$c_4 = \frac{\Gamma}{2\pi}.$$

The stream function for a free vortex also follows from Eq. (6–92) since

$$\frac{\partial \psi}{\partial r} = v_\theta = \frac{c_4}{r} = \frac{\Gamma}{2\pi r}.$$

Therefore

$$\psi = \frac{\Gamma}{2\pi} \int \frac{dr}{r} = \frac{\Gamma}{2\pi} \ln r, \tag{6–97}$$

where, in accordance with the previous sign convention, the direction of rotation will be counterclockwise.

The fact that the motion is irrotational may be verified by computing the vorticity component ω_z, with $v_\theta = \text{const}/r$, which is the velocity distribution

for the free vortex. Hence

$$\omega_z = \frac{\text{const}}{r^2} - \frac{\text{const}}{r^2} = 0, \tag{6–98}$$

provided that r is not equal to zero. Either the fluid does not occupy the space at $r = 0$ or the fluid at $r = 0$ is undergoing a rotational motion (as a forced vortex), as shown by the dashed line in Fig. 6–15. In this case, the profile will become that of a Rankine combined vortex. This is what happens if the drain in the tank bottom is suddenly closed and the fluid motion is ultimately dissipated through viscous action.

PROBLEMS

6–1. Verify that Eq. (6–2a) can be obtained from Eq. (6–1a).

6–2. An incompressible flow has the following velocity components: $v_r = A/r$, $v_\theta = A/r$, $v_z = 0$. Prove that the flow satisfies the continuity equation and find the equation for the streamlines $\psi = f(r, \theta)$. Sketch a few streamlines.

6–3. The x-component of velocity in a two-dimensional incompressible flow is given by $u = Ax^3 + By^2$. (a) Find the equation for the y-component of velocity v, assuming that at $y = 0$, $v = 0$ for all values of x. (b) Is the flow irrotational?

6–4. Consider an incompressible two-dimensional flow of a viscous fluid in the xy-plane in which the body force is due to gravity. (a) Prove that the divergence of the vorticity vector is zero. (This expresses the conservation of vorticity, $\nabla \cdot \zeta = 0$.) (b) Show that the Navier-Stokes equation for this flow can be written in terms of the vorticity as $d\zeta/dt = \nu\nabla^2\zeta$. (This is a "diffusion" equation and indicates that vorticity is diffused into a fluid at a rate which depends on the magnitude of the kinematic viscosity.) Note that $d\zeta/dt$ is the substantial derivative defined in Section 2–1.

6–5. Consider a steady, incompressible laminar flow between parallel plates as shown in Fig. 6–4 for the following conditions: $a = 0.1$ ft, $U = 1.0$ ft/sec, $\mu = 0.01$ lb.sec/ft^2, $\partial p/\partial x = 4$ lb/ft^3 (pressure increases in $+x$-direction). (a) Plot the velocity distribution, u, in the z-direction. (b) In which direction is the net fluid motion? (c) Plot the distribution of shear stress τ_{zx} in the z-direction.

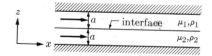

FIGURE 6–16

6–6. Two immiscible and incompressible fluids of equal depth a are flowing in steady motion between fixed horizontal parallel plates as shown in Fig. 6–16. The viscosity and density are μ_1 and ρ_1 for the upper fluid and μ_2 and ρ_2 for the lower fluid. The pressure gradient is $\partial p/\partial x$ and is independent of x and z. Find the velocity distribution between the plates, $u = f(z)$. (Note that at the interface the velocity magnitude has a single value and the shear stress τ_{zx} has a single value.)

6–7. An incompressible liquid of density ρ and viscosity μ flows in a thin film down a glass plate inclined at an angle α to the horizontal. The thickness, a, of the liquid film normal to the plate is constant, the velocity is everywhere parallel to the plate, and the flow is steady. Neglect viscous shear between the air and the moving liquid at the free surface. Determine the variation in longitudinal velocity in the direction normal to the plate, the shear stress at the plate, and the average velocity of flow.

6–8. Consider a semiinfinite body of fluid (in the $+y$-direction) bounded on one side by a horizontal solid plate (the xz-plane). Initially the fluid is at rest, but at time $t = 0$ the solid surface is moved in the $+x$-direction with a constant velocity U. The viscosity and density of the fluid are constant, and it may be assumed that there is no pressure gradient and no gravitational force in the x-direction. Show that the continuity condition and the Navier-Stokes equation result in the differential equation $\partial u/\partial t = \nu(\partial^2 u/\partial y^2)$. [This differential equation is similar to Eq. (16–64), with $U = 0$, and has a solution analogous to Eq. (16–66).]

6–9. For a steady, incompressible laminar flow of $Q = 0.02$ cfs through a circular tube of radius $r_0 = 1$ in., determine (a) the maximum velocity in the tube; (b) the radius at which $|v_z| = |V|$ for $V = Q/A$; (c) the radial thickness of the outermost annulus through which 25% of the flow passes.

6–10. A straight horizontal length of a smooth copper pipe 1-in. in diameter is used to measure the viscosity of liquids in the laboratory. Pressure gauges measure the pressure at two stations 20 ft apart along the pipe. The quantity of liquid flow is measured in a weighing tank over a measured time interval.

During one test with oil of specific gravity of 0.92, the following data were obtained: pressure $p_1 = 10$ psi, $p_2 = 7$ psi; weight measured in 5-min interval is 2000 lb; oil temperature at station where p_2 is measured is 70°F. (a) Find the viscosity indicated from this data. (b) Estimate the effect of friction in the pipe on the viscosity computed from the data, given that the pipe is insulated to eliminate heat transfer between the two pressure-measuring stations.

6–11. Consider steady laminar flow in the horizontal axial direction through the annular space between two concentric circular tubes. The radii of the inner and outer tube are r_1 and r_2, respectively. Derive the expression for the velocity distribution in the radial direction as a function of viscosity, pressure gradient $\partial p/\partial x$, and tube dimensions.

6–12. For the case of the previous problem find: (a) the equation for the discharge rate Q; (b) the location and magnitude of the maximum velocity in the annulus.

6–13. A shaft of 3-in. diameter rotates in a concentric cylindrical sleeve of 3.02-in. diameter. The sleeve length is 12 in. Compute the power needed to rotate the shaft at 1500 rpm given that the lubricant is (a) water at 68°F; (b) lubricating oil of 0.85 specific gravity at 68°F; (c) glycerin at 68°F.

6–14. Given the equation for a family of streamlines $\psi = 2xy$, find the equation for the velocity potential function $\phi = f(x, y)$. Sketch a few lines of constant ϕ and ψ.

6–15. The velocity potential for a steady incompressible flow is given by $\phi = (-a/2)(x^2 + 2y - z^2)$, where a is an arbitrary constant greater than zero. (a) Find the equation for the velocity vector $\mathbf{q} = \mathbf{i}u + \mathbf{j}v + \mathbf{k}w$. (b) Find the equation of the streamlines in the xz ($y = 0$) plane. (c) Prove that the continuity equation is satisfied.

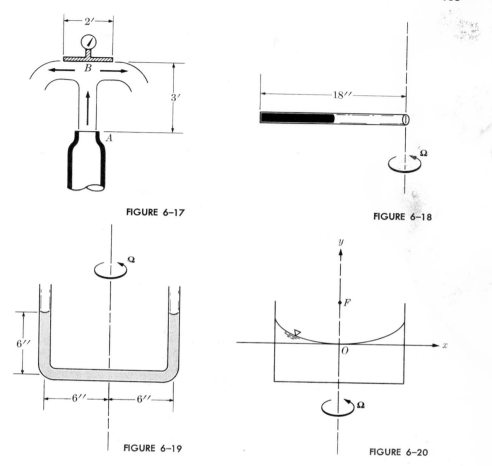

FIGURE 6–17

FIGURE 6–18

FIGURE 6–19

FIGURE 6–20

6–16. A jet of water discharges vertically upward into the atmosphere at section A and strikes a horizontal circular disk 2 ft in diameter, as shown in Fig. 6–17. A pressure gauge is connected to a small hole in the center of the plate at (B). The cross-sectional area of the nozzle at (A) is 0.1 ft^2, and the pressure gauge at (B) registers a pressure of 10 psig. (a) Determine the centerline velocity of the jet at section (A). (b) Estimate the vertical thickness of the jet at the perimeter of the disk. Neglect energy dissipation.

6–17. A slender tube, closed at one end, is 18 in. long (Fig. 6–18). The tube contains a volume of mercury equal to one-half the volume of the tube. If the tube is rotated about its open end in a horizontal plane at a constant speed of 180 rpm, what is the pressure intensity at the closed end?

6–18. A slender U-tube (Fig. 6–19) is initially filled with water (68°F) to a depth of 6 in. in the vertical legs. The tube is rotated about its central axis at a constant angular velocity. At what angular velocity Ω will a vapor-filled cavity form at the center of the U-tube?

6-19. The surface of a cylindrical open tank containing mercury is to be used as a parabolic reflector as Fig. 6–20 shows. A point source of light is to be placed at the focus (F) of the parabola so as to reflect parallel vertical beams of light. The focal distance $OF = p$, where $x^2 = 4py$ is the equation of the parabola. When the focal distance is 6 in., find the angular speed Ω of the tank in radians per second.

6-20. Figure 6–21 shows a cylindrical centrifuge having a diameter of 3 ft and a height of 1 ft which is to be rotated about its axis so that the mid-height thickness of the liquid layer along the vertical wall will be 0.10 ft. Determine the speed of rotation in rpm which will produce the required thickness with a maximum variation of $\pm10\%$ of the mid-height value at the top and bottom of the cylinder.

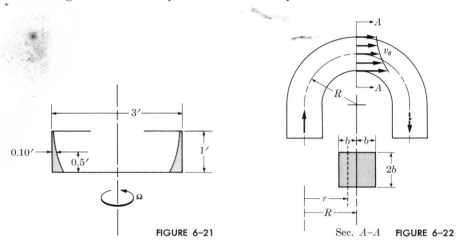

FIGURE 6–21   Sec. A–A FIGURE 6–22

6-21. The velocity variation across the radius of a rectangular bend (Fig. 6–22) may be approximated by a free vortex distribution $v_\theta r = \text{const}$. Derive an expression for the pressure difference between the inside and outside of the bend as a function of the discharge Q, the fluid density ρ, and the geometric parameters R and b, assuming frictionless flow.

REFERENCES

1. MILNE-THOMPSON, L. M., *Theoretical Hydrodynamics*, The Macmillan Co., New York, 1960, 4th ed., Section 2.60.

2. ROUSE, H., *Advanced Mechanics of Fluids*, John Wiley and Sons, Inc., New York, 1959. Chapter II contains axisymmetric and three-dimensional cases.

3. PITOT, H., "Description d'une machine pour mesurer la vitesse des eaux et le sillage des vaisseaux," *Histoire de l'Academie des Sciences*, 1732.

CHAPTER 7

Dynamic Similitude

7-1 INTRODUCTION

The study of dynamically similar fluid motions forms the basis for the theory of models, the design of experiments, and the correlation of experimental data. The emphasis on the word experimental serves to point out that we must rely on a large body of experimental-information to solve many problems in fluid mechanics. Only a limited number of exact analytical solutions are obtainable from the equations of motion, and many important problems cannot be treated analytically by means of the equations of motion for laminar flows or through nonviscous or irrotational assumptions.

As an example we may consider the problem of finding the force exerted by moving fluid upon a stationary sphere. As we shall see in Chapters 8 and 9, if the fluid velocities are very small (creeping motions), the nonlinear acceleration terms in the equations of motion may be neglected and an analytical solution can be found. As the fluid velocities and accelerations become larger, the analytical approach becomes difficult because of our inability to solve the nonlinear equations. We must also be concerned with the possibility that the flow will become unstable and turbulent. In this event, the useful concept of a steady motion requires a new definition which introduces additional terms into the equations of motion. The irrotational assumption is of no help in the sphere problem since the pressure distributions computed for the irrotational flow are symmetric and yield zero net force. If the answer is to be found by experiments, the problem of investigating all possible sphere diameters in all possible fluids at all possible velocities is indeed formidable, if not impossible. As we shall see later in the chapter dealing with forces on immersed bodies, the fact is that many experiments can be represented by a single curve relating a dimensionless net force to a dimensionless parameter expressing the characteristics of the flow and the properties of the fluid. This is a striking demonstration of the power and usefulness of the concepts of dynamic similarity in the correlation of experimental information.

Interest in dynamically similar fluid motions also arises from the desire to investigate flows involving complex boundary conditions by making experiments on a geometrically similar system differing from the original or prototype system only in size. Most often it is expedient to make these investigations on a reduced scale system known as a model. Thus models of aircraft or missile components are tested in wind tunnels, and models of rivers or estuaries are constructed for the purpose of determining the effect of proposed changes on the behavior of the prototype. In these examples, the equations of motion and the boundary conditions are generally too complex to permit purely analytical predictions of prototype behavior.

7–2 GEOMETRIC SIMILARITY

In a strict sense, geometric similarity implies that the ratio of all corresponding lengths in two systems must be the same. Thus, if certain lengths are selected in the x-, y-, and z-directions (and the two systems are designated by the subscripts M and P), the condition for geometric similarity becomes

$$\frac{x_M}{x_P} = \frac{y_M}{y_P} = \frac{z_M}{z_P} = L_r, \tag{7–1}$$

where L_r is the length ratio which describes the relative size of the two systems. Throughout this chapter the subscript r will be used to designate the ratio of corresponding quantities in the two systems. A consequence of exact geometric similarity is that the ratio of areas and volumes in the two systems may be expressed in terms of the square and cube of the length ratio, and hence

$$A_r = \frac{A_M}{A_P} = \frac{x_M y_M}{x_P y_P} = L_r^2, \tag{7–2}$$

$$\mathcal{V}_r = \frac{\mathcal{V}_M}{\mathcal{V}_P} = \frac{x_M y_M z_M}{x_P y_P z_P} = L_r^3. \tag{7–3}$$

In some cases, departures from exact geometric similarity become necessary, particularly in the case of river and oceanographic models, in which the depth is small in relation to the width and length. Such models are frequently distorted in the vertical direction, and it is necessary to specify the length correspondence by means of an additional equation similar to Eq. (7–1). For example,

$$\frac{x_M}{x_P} = \frac{y_M}{y_P} = L_r, \tag{7–4}$$

and

$$\frac{z_M}{z_P} = Z_r. \tag{7–5}$$

7–3 DYNAMIC SIMILARITY

The basic concept of dynamic similarity may be stated as the requirement that two systems with geometrically similar boundaries have geometrically similar flow patterns at corresponding instants of time. Thus all individual forces acting on corresponding fluid elements of mass must have the same ratios in the two systems. Individual forces acting on the fluid element may be due either to a body force such as weight, in a gravitational field, or to surface forces

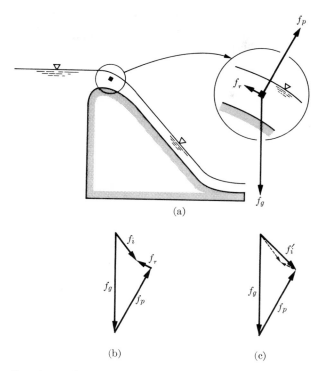

FIG. 7-1. Force/mass polygons for dynamically similar systems: (a) analysis of forces acting on fluid particle; (b) unit force polygon for dynamic similitude; (c) unit force polygon neglecting shear.

resulting from pressure gradients, viscous shear, or surface tension. The resultant or inertial force must also have the same force ratio in the two systems as any of the individual force components. The requirement of a single-scale ratio for forces necessitates that the force polygons for corresponding elements of mass be geometrically similar. This concept may be illustrated by the following example:

Example 7-1: Force polygon for dynamically similar systems

Consider the forces acting on a particle of water near the crest of the spillway of a dam, as shown in Fig. 7-1(a). We may identify the surface forces acting on the particle in terms of the forces due to the normal pressure gradient and the tangential shear. Using the force per unit mass of fluid, we find that these are represented by f_p and f_τ, respectively. The body force per unit mass due to gravitational attraction is f_g. The resultant or inertial force per unit mass, f_i, is found by drawing the polygon shown in Fig. 7-1(b). The direction of the particle acceleration is then identical to the direction of f_i. Figure 7-2 shows a model spillway of the type considered in this example.

Dynamically similar motions in geometrically similar systems require that the magnitude and direction of the resultant acceleration per unit mass be the same in

FIG. 7–2. Model spillway and energy dissipator.

the two systems. Thus the force-per-unit-mass polygons must be identical. We will see later that it is often impossible to obtain exact dynamic similitude, and that it is then necessary to simplify the force polygon by ignoring the least important component. The polygon shown in Fig. 7–1(c) is an example of the effect of eliminating the shear force. This results in only a slight change in the direction and magnitude of the resultant vector f_i'.

In the above discussion of dynamic similarity, the existence of length, time, mass, and force ratios relating to the two flow systems has been implied. It is important to observe that within the realm of Newtonian mechanics, we cannot independently choose more than three of these four ratios. Proof of this statement is obtained by writing the requirement that both systems satisfy Newton's second law,

$$\sum(dF_M) = (dM_M)\mathbf{a}_M, \qquad \sum dF_P = (dM_P)\mathbf{a}_P.$$

Dividing one equation by the other establishes the relation which must exist between the force, mass, and acceleration ratios, that is,

$$F_r = M_r\mathbf{a}_r. \tag{7–6}$$

The acceleration ratio may also be written in terms of the kinematic length and time ratios

$$\mathbf{a}_r = \frac{V_r}{T_r} = \frac{L_r}{T_r^2}. \tag{7–7}$$

Thus we have

$$F_r = M_r \frac{L_r}{T_r^2}.$$ (7–8)

Equation (7–8) indicates the basic interrelation of the length, time, mass, and force ratios. A more convenient form is obtained by noting that the mass ratio may be expressed in terms of the fluid density and volume ratios. Hence, using Eq. (7–3), we have

$$M_r = \rho_r \mathcal{U}_r = \rho_r L_r^3,$$ (7–9)

and

$$F_r = \rho_r \frac{L_r^4}{T_r^2}.$$ (7–10)

The length and time ratio may also be combined into a fluid velocity ratio since $V_r = L_r/T_r$; hence

$$F_r = \rho_r L_r^2 V_r^2.$$ (7–11)

Equation (7–11) must be valid for all dynamically similar systems. Although this is a necessary condition, it is not at the same time a sufficient condition for dynamically similar behavior. For example, Eq. (7–11) provides no information on the effect of fluid viscosity in the two systems. In order to determine both the necessary and sufficient conditions for dynamic similitude, it is appropriate to consider the complete statement of Newton's second law as embodied in the dynamical equations of motion derived in Chapter 6. These differ from the starting point of the above analysis (Eq. 7–6) in that the individual force components (surface and body forces) appear as separate terms in the equation of motion. The conditions under which dynamic similarity for the two systems will be achieved are obtained by writing the dynamical equations of motion in dimensionless form and by making the numerical coefficients identical in the two systems. We will therefore restate the equations of motion developed in Chapter 6 in dimensionless form for both incompressible and compressible fluids.

7–4   SIMILARITY CONDITIONS FOR INCOMPRESSIBLE FLUIDS. FROUDE AND REYNOLDS NUMBERS

Let us consider the motion of an incompressible, constant-viscosity fluid in a gravity field. This will include liquids and gases flowing under conditions in which compressibility is unimportant. The conditions for dynamic similarity of two flow systems can be obtained by writing the Navier-Stokes equation in dimensionless form. Let us begin with the x-component of the incompressible Navier-Stokes equations (Eqs. 6–28),

$$\frac{\partial u}{\partial t} + u\frac{\partial u}{\partial x} + v\frac{\partial u}{\partial y} + w\frac{\partial u}{\partial z} = -g\frac{\partial h}{\partial x} - \frac{1}{\rho}\frac{\partial p}{\partial x} + \frac{\mu}{\rho}\left(\frac{\partial^2 u}{\partial x^2} + \frac{\partial^2 u}{\partial y^2} + \frac{\partial^2 u}{\partial z^2}\right).$$ (7–12)

This equation may be written in terms of a set of dimensionless quantities defined in the following manner:

$$x^0 = \frac{x}{L}, \qquad u^0 = \frac{u}{V_0}, \qquad t^0 = \frac{t}{t_0} = \frac{t}{L/V_0},$$

$$y^0 = \frac{y}{L}, \qquad v^0 = \frac{v}{V_0}, \qquad p^0 = \frac{p}{\rho V_0^2},$$

$$z^0 = \frac{z}{L}, \qquad w^0 = \frac{w}{V_0}, \qquad \rho = \text{const}$$

$$\mu = \text{const}$$

$$h^0 = \frac{h}{L}, \qquad\qquad\qquad g = \text{const}.$$

(7–13)

In the above equations, L and V_0 are constant reference values of length and velocity to be chosen as characteristics of the system. For example, if we were studying the net force on a sphere due to a flow, L would be chosen as the sphere diameter and V_0 as the approach velocity of the fluid far upstream of the sphere.

If Eqs. (7–13) are substituted in Eq. (7–12), the following equation is obtained:

$$\left(\frac{V_0^2}{L}\right)\frac{\partial u^0}{\partial t^0} + \left(\frac{V_0^2}{L}\right) u^0 \frac{\partial u^0}{\partial x^0} + \left(\frac{V_0^2}{L}\right) v^0 \frac{\partial u^0}{\partial y^0} + \left(\frac{V_0^2}{L}\right) w^0 \frac{\partial u^0}{\partial z^0}$$

$$= -(g)\frac{\partial h^0}{\partial x^0} - \left(\frac{V_0^2}{L}\right)\frac{\partial p^0}{\partial x^0} + \frac{\mu V_0}{\rho L^2}\left(\frac{\partial^2 u^0}{\partial (x^0)^2} + \frac{\partial^2 u^0}{\partial (y^0)^2} + \frac{\partial^2 u^0}{\partial (z^0)^2}\right). \quad (7\text{–}14)$$

The above substitutions have not altered the dimensions of each term which are those of an acceleration or force per unit mass. Equation (7–14) may be made dimensionless by dividing each term by V_0^2/L. The latter has the dimensions of inertia force per unit mass since it is the dimensional coefficient of the inertia terms on the left-hand side of the equation. Hence

$$\frac{\partial u^0}{\partial t^0} + u^0 \frac{\partial u^0}{\partial x^0} + v^0 \frac{\partial u^0}{\partial y^0} + w^0 \frac{\partial u^0}{\partial z^0}$$

$$= -\left(\frac{gL}{V_0^2}\right)\frac{\partial h^0}{\partial x^0} - \frac{\partial p^0}{\partial x^0} + \left(\frac{\mu}{\rho V_0 L}\right)\left(\frac{\partial^2 u^0}{\partial (x^0)^2} + \frac{\partial^2 u^0}{\partial (y^0)^2} + \frac{\partial^2 u^0}{\partial (z^0)^2}\right). \quad (7\text{–}15)$$

Since all of the quantities with superscript zero are dimensionless, each of the two coefficient groups of reference quantities must be a dimensionless combination. The physical significance of these two groups is of particular interest in understanding dynamic similitude.

The first group was formed by dividing the gravity force per unit mass (acceleration due to gravity) by the inertia force per unit mass:

$$\frac{\text{gravity force/mass}}{\text{inertia force/mass}} = \frac{g}{V_0^2/L} = \frac{gL}{V_0^2}.$$

The square root of the inverse of this dimensionless group is called the *Froude number.* Therefore

$$\mathbf{F} = \frac{V_0}{\sqrt{gL}} = \left(\frac{\text{inertia force}}{\text{gravity force}}\right)^{1/2}. \tag{7–16}$$

The Froude number is an important parameter whenever gravity is a factor which influences fluid motion. Similarly we find for the second group,

$$\frac{\text{viscous force/mass}}{\text{inertia force/mass}} = \frac{\mu V_0/\rho L^2}{V_0^2/L} = \frac{\mu}{\rho V_0 L}.$$

The inverse of this group is known as the *Reynolds number.* Therefore

$$\mathbf{R} = \frac{\rho V_0 L}{\mu} = \frac{\text{inertia force}}{\text{viscous force}}. \tag{7–17}$$

The *Reynolds number* is important whenever viscous forces influence fluid motions.

The dimensionless equation of motion may then be written,

$$\frac{\partial u^0}{\partial t^0} + u^0 \frac{\partial u^0}{\partial x^0} + v^0 \frac{\partial u^0}{\partial y^0} + w^0 \frac{\partial u^0}{\partial z^0}$$

$$= -\frac{1}{\mathbf{F}^2} \frac{\partial h^0}{\partial x^0} - \frac{\partial p^0}{\partial x^0} + \frac{1}{\mathbf{R}} \left[\frac{\partial^2 u^0}{\partial (x^0)^2} + \frac{\partial^2 u^0}{\partial (y^0)^2} + \frac{\partial^2 u^0}{\partial (z^0)^2} \right]. \tag{7–18}$$

The same dimensionless groups would be obtained from the *y*- and *z*-components of the equation of motion.

The continuity equation for incompressible fluids may also be written in dimensionless form by substituting Eqs. (7–13) in Eq. (6–5). The result is

$$\frac{\partial u^0}{\partial x^0} + \frac{\partial v^0}{\partial y^0} + \frac{\partial w^0}{\partial z^0} = 0, \tag{7–19}$$

which introduces no additional similitude conditions.

Equation (7–18) has the same solution for two geometrically similar systems if the dimensionless coefficients are numerically the same for the two systems. Therefore one may conclude that the general requirement for dynamically similar fluid motions of an incompressible viscous fluid in a gravity field is equality of Froude and equality of Reynolds numbers in both systems. Since dynamic similarity implies kinematic similarity, the time scale in an unsteady motion, and hence all the accelerations, must be established in accordance with the Froude and Reynolds numbers requirements and Newton's second law. For special cases, one of the dimensionless coefficients may be unimportant or may have no bearing as discussed in the following paragraphs.

7–4.1 Enclosed systems

In equations of motion the pressure *change* is due to a combination of the dynamic effects produced by acceleration, viscosity, and gravity. In some cases, the gravity effect is simply a hydrostatic pressure distribution superposed on a variable pressure due to other effects. This will be true for constant-density fluids in what we will call an *enclosed system*. An enclosed system may be defined as one in which the fluid is contained entirely within fixed boundaries or one in which the extent of the fluid field is so large as to be considered infinite. The former might be the flow of a liquid in a closed conduit such as a circulating water tunnel. The latter might be the low-speed motion of an object immersed in a gas (so that compressibility is not important). For such enclosed flows the pressure would be distributed hydrostatically if the motion were suddenly frozen into rest. In these cases, we can express the pressure as the sum

$$p = (p_d + p_s), \qquad (7\text{–}20)$$

where p_s is determined by the hydrostatic relation $(p_s + \gamma h) = $ const, and p_d is the part responding to the "dynamic" effects. The value of the constant depends only on the datum selected. The introduction of Eq. (7–20) into Eq. (7–12) cancels the gravity term and converts the pressure term into $(1/\rho)(\partial p_d/\partial x)$. Then

$$\frac{\partial u}{\partial t} + u\,\frac{\partial u}{\partial x} + v\,\frac{\partial u}{\partial y} + w\,\frac{\partial u}{\partial z} = -\frac{1}{\rho}\frac{\partial p_d}{\partial x} + \frac{\mu}{\rho}\left(\frac{\partial^2 u}{\partial x^2} + \frac{\partial^2 u}{\partial y^2} + \frac{\partial^2 u}{\partial z^2}\right), \quad (7\text{–}21)$$

and the dimensionless equation of motion becomes

$$\frac{\partial u^0}{\partial t^0} + u^0\,\frac{\partial u^0}{\partial x^0} + v^0\,\frac{\partial u^0}{\partial y^0} + w^0\,\frac{\partial u^0}{\partial z^0} = -\frac{\partial p_d^0}{\partial x^0} + \frac{1}{\mathbf{R}}\left(\frac{\partial^2 u^0}{\partial (x^0)^2} + \frac{\partial^2 u^0}{\partial (y^0)^2} + \frac{\partial^2 u^0}{\partial (z^0)^2}\right).$$

$$(7\text{–}22)$$

Equation (7–21) shows that the effect of viscosity is the same as would be found in a zero gravity field. The subscript d is often omitted from the pressure term, since it is understood that the equation refers only to the "dynamic" component of the pressure gradient.

Equation (7–22) shows that for an incompressible fluid in enclosed systems only Reynolds numbers must be equal in order for the two geometrically similar systems to be dynamically similar. The time scale for unsteady motion must also satisfy the requirements of a constant Reynolds number and Newton's second law.

For two systems having dynamic similarity determined by equality of Reynolds numbers, we may determine the ratios or "scales" of velocity, time, force, and all other quantities derivable from them as follows.

For two systems denoted by subscript M and P, we may write the equality of Reynolds numbers as

$$\frac{\rho_M V_{0M} L_M}{\mu_M} = \frac{\rho_P V_{0P} L_P}{\mu_P}.$$

The velocity ratio V_r becomes

$$V_r = \frac{V_{0M}}{V_{0P}} = \frac{\mu_M \rho_P L_P}{\mu_P \rho_M L_M} = \frac{\mu_r}{\rho_r L_r}. \qquad (7\text{–}23)$$

The *time ratio* T_r is obtained by noting that

$$T_r = \frac{L_r}{V_r} = \frac{\rho_r L_r^2}{\mu_r}. \qquad (7\text{–}24)$$

The force ratio follows by substitution of Eq. (7–23) in the dimensionless Newtonian equation (7–11),

$$F_r = \frac{\mu_r^2}{\rho_r}. \qquad (7\text{–}25)$$

For Reynolds-number similitude, there are two independent choices, or degrees of freedom, in achieving dynamic similarity. These are: (1) the choice of L_r (i.e., the physical size relation between the two systems), and (2) the choice of fluid to be used in one system, where we assume that the fluid in the other system is specified. The latter choice determines the fluid property ratios μ_r and ρ_r. Thus, with Eqs. (7–24) and (7–25), the values for L_r, T_r, and F_r are determined.

Any measurable quantity in one system which can be expressed in terms of force, length, and time can be determined in the other system. For example, the fluid pressure ratio p_r becomes

$$p_r = \frac{F_r}{A_r} = \frac{F_r}{L_r^2} = \frac{\mu_r^2}{\rho_r L_r^2}. \qquad (7\text{–}26)$$

Note that the restriction to constant-density fluids in what we have called enclosed systems rules out flows involving multiphase fluids (including cavitation due to hydrodynamically induced vaporization) and systems of two or more dissimilar and immiscible fluids. When the interface is between a liquid and a gas, it is known as a free surface and gravity plays a special role.

Example 7–2: Reynolds model

The towing force is to be determined for an underwater listening device in the shape of an ellipsoid of revolution. The device is 5 ft long, and it is to be towed deeply submerged in water (60°F) at a speed of 10 ft/sec. A geometrically similar model 1 ft long is tested in a low-speed variable-pressure wind tunnel having a maximum velocity of 100 ft/sec (about 10% of the speed of sound) at 60°F.

Assuming that the tunnel is operated at 100 fps, we find that the required density of the wind-tunnel air is determined by the condition that the Reynolds numbers of the two systems be equal. Thus, from Eq. (7–23), we have

$$\rho_r = \frac{\mu_r}{V_r L_r},$$

and

$$\mu_r = \frac{\mu_{\text{air}}}{\mu_{\text{water}}} = \frac{3.77 \times 10^{-7}}{2.36 \times 10^{-5}} = 1.6 \times 10^{-2},$$

$$V_r = \frac{V_{\text{air}}}{V_{\text{water}}} = \frac{100}{10} = 10, \qquad L_r = \frac{L_{\text{air}}}{L_{\text{water}}} = \frac{1}{5}.$$

Hence

$$\rho_r = \frac{1.6}{(100)(10)(1/5)} = 0.008,$$

Then since

$$\rho_{\text{water}} = 1.94 \text{ slugs/ft}^3,$$

we must have

$$\rho_{\text{air}} = (0.008)(1.94) = 0.015 \text{ slug/ft}^3.$$

Since the density of standard atmospheric air at 60°F is 0.0024 slug/ft^3, the tunnel is to be operated at a pressure of about 6 atm in order to achieve the desired air density.

For the above conditions, the total force component on the object in the wind tunnel measured in the direction of the free-stream flow is 0.65 lb. The force ratio from Eq. (7–25) is

$$F_r = \frac{\mu_r^2}{\rho_r} = \frac{(1.6)^2}{(100)^2(0.008)} = 0.032.$$

Therefore, the towing force for the prototype in water is

$$F_P = \frac{0.65}{0.032} = 20.2 \text{ lb.}$$

In general, it is permissible to neglect compressibility effects for air speeds up to the order of 10% of the speed of sound.

Example 7–3: Correlation of experimental data

The wind-tunnel testing program of the previous example is to be extended to collect experimental data on the total force on the ellipsoid at various velocities and air densities. Determine the dimensionless form in which the experimental data should be presented.

The total force should be made dimensionless by using the reference quantities of Eq. (7–13),

$$F^0 = \frac{F}{\rho V_0^2 L^2}.$$

A characteristic area, such as the maximum projected area in the direction of motion, may be used for L^2, and V_0 would then be the free-stream velocity of the tunnel. The dimensionless force F^0 should therefore be a function only of the Reynolds number of the flow,

$$\mathbf{R} = \frac{V_0 L}{\nu},$$

where L may be taken as the diameter of the maximum cross section. The functional relation

$$F^0 = f(\mathbf{R}),$$

determined by plotting the experimental data, is valid for all geometrically similar objects in incompressible fluids in enclosed systems.

7–4.2   Free-surface systems

In a system with a free surface, the pressure (measured with respect to the ambient atmospheric pressure) at any point in a liquid cannot be changed arbitrarily without also affecting the geometry of the free surface. Thus, the conditions for the use of the dynamic pressure of the previous section, which eliminated the gravitational term, are not satisfied for free-surface flows.* It is therefore necessary to return to the more general condition for dynamic similarity expressed by Eq. (7–18). Thus, equality of Reynolds numbers and equality of Froude numbers are both necessary for *exact* dynamic similitude in free-surface flows.

Specifying the equality of Froude numbers, we have

$$\frac{V_M}{\sqrt{g_M L_M}} = \frac{V_P}{\sqrt{g_P L_P}}, \tag{7-27}$$

or

$$V_r = \sqrt{g_r L_r}. \tag{7-28}$$

From the equality of Reynolds numbers, Eq. (7–23), we obtain

$$V_r = \frac{\mu_r}{\rho_r L_r}.$$

Since the velocity ratios must be the same, and since we recognize that for terrestrial events $g_r = 1$, then

$$L_r = \left(\frac{\mu_r}{\rho_r}\right)^{2/3} = \nu_r^{2/3}. \tag{7-29}$$

---

* Strictly speaking, this also applies to flows where cavitation exists, although the cavitation process involves in addition the dynamics of vaporization which depends on the special thermodynamic properties of the liquid and on the effects of dissolved material and other contaminants.

FIG. 7–3. General view of movable bed model for studying proposed harbor at Memphis, Tennessee. Current direction is demonstrated by use of confetti. (Courtesy U. S. Army Corps of Engineers, Waterways Experimental Station, Vicksburg, Miss.)

For dynamic similitude of both viscous and gravity forces, there is only one degree of freedom; that is, the choice of fluid to be used determines the length ratio through Eq. (7–29). Conversely, if the length ratio is chosen independently, the fluid properties in the two systems are no longer independent. Since the range of kinematic viscosities among the common liquids is very limited, the requirement of Eq. (7–29) usually cannot be satisfied unless the scale ratio is close to unity.

In certain free-surface flows, viscous effects are fortunately very small in comparison with gravity effects. Examples are free-surface wave motion and flow over weirs, as cited in Chapter 6 (Figs. 6–8 and 6–9). The dam spillway discussed in Example 7–1 is also an illustration of a case in which differences in the friction effect introduce only small changes in the dynamic flow pattern. In this type of free-surface flow, it is common practice to require only Froude number similarity when two systems are compared experimentally. Corrections for viscous effects can be made, if necessary, by using models of different scale ratios and extrapolating the results to full prototype scale. The difficulty arises in ensuring that viscous effects do not become too important in the smaller system. This places a lower limit on the physical size of the model; for example, the flow in the model must not become laminar, if the flow in the prototype is turbulent. Another, perhaps more important, class of incompressible free-surface

FIG. 7–4. General view of the Savannah River estuary model. (Courtesy U. S. Army Corps of Engineers, Waterways Experimental Station, Vicksburg, Miss.)

problems includes cases in which fluid friction is important but molecular viscosity effects are negligible. Examples of such cases are highly turbulent, high Reynolds-number flows. The question of similitude of frictional effects then becomes one of geometric similitude of boundary roughness rather than equality of Reynolds numbers. The majority of hydraulic-model studies of open channels and rivers and of tidal estuaries fall into this category. Since both model and prototype use the same fluid (water), Eq. (7–29) cannot be satisfied, and such models are designed on the basis of Froude number similarity alone. Figure 7–3 shows a model of a portion of the Mississippi River. The large-scale boundary roughness used to simulate the resistance of the overbank areas which are submerged during floods is shown clearly. Figure 7–4 gives an impression of the large area required for estuary models used in studying shoaling problems. Since exact dynamic similitude is rarely achieved in models of free-surface systems, a considerable amount of judgment is necessary both in designing models and in interpreting the results.

If only Froude similitude is required, the velocity ratio is given by Eq. (7–28). The corresponding time ratio follows immediately,

$$V_r = L_r/T_r = \sqrt{g_r L_r}$$

or

$$T_r = \sqrt{(L_r/g_r)}, \tag{7–30}$$

and the force ratio is obtained from Eq. (7–11) as

$$F_r = \rho_r g_r L_r^3 = \gamma_r L_r^3. \tag{7–31}$$

Another form of Eq. (7–31), obtained by use of Eq. (7–3), is

$$F_r = \gamma_r \mho_r = W_r,$$
(7–31a)

where W_r is the weight ratio. As would be expected, if gravitational forces are dominant, the force ratio must equal the weight ratio. As in the Reynolds models, two degrees of freedom (size and fluid) are again obtained.

Example 7–4: Froude model

The filling and emptying gates of a canal lock extend the full height of the lock chamber. When a vessel is lowered in the prototype lock, the gates at the outlet end are programmed to open at a rate of 10 in./min. The waves and currents produced by the outflow cause the vessel to pull at its moorings. In a $\frac{1}{25}$-scale model of the system using water, the maximum tension in the moorings is 1.6 lb when the gates are opened at the proper rate. Determine the proper gate-opening rate in the model and the maximum mooring-line tension in the prototype.

Since this is a free-surface system with negligible viscous effects, gravitational-force similarity will be obtained by equality of Froude numbers. From Eq. (7–28), we have

$$V_r = \sqrt{g_r L_r}.$$

Since $g_r = 1$, the rate of opening (velocity) ratio is

$$V_r = \sqrt{L_r} = 1/5.$$

Thus the proper rate for the model is 2 in./min. Since $\gamma_r = 1$, the force ratio (Eq. 7–31) is

$$F_r = (1)(1/25)^3 = 1/15{,}625,$$

and hence the maximum tension in the prototype mooring lines is

$$F_P = (1.6)15{,}625 = 25{,}000 \text{ lb.}$$

In systems in which density stratification is important, interfaces may occur between two fluids which differ in density and other physical properties (Fig. 6–11, for example). Dynamic similarity of internal motions in stratified flows are governed by a more general form of the gravitational-similitude parameter known as the densimetric Froude number [1]. This is given by

$$\mathbf{F} = \frac{V_0}{\sqrt{g(\Delta\gamma/\gamma)L}},$$
(7–32)

where $\Delta\gamma$ is the difference in specific weight between adjacent fluid layers, and L is a characteristic vertical dimension such as a_0 in Fig. 6–11. In Fig. 6–11 only the lower layer is moving. If two or more layers are in motion, a densimetric Froude number is needed for each layer with L the depth of each layer.

Another interesting class of free-surface problems is that involving the motion of objects on or near the air-water interface. If the ellipsoid of Example 7–2 is brought near enough to the surface to cause the generation of surface waves, then we should expect the gravitational forces to become significant. Hence the towing force would be a function of both the Reynolds number and the Froude number. The characteristic length in the latter would be the depth of submergence. These problems, along with the classical ship-model problem, are discussed in Chapter 15.

We conclude this section on free-surface systems by mentioning a possible boundary condition for the dynamic equation of motion which involves an independent force not previously considered. Under some conditions, for example in the generation of very small surface waves, it may be necessary to recognize the existence of surface tension as a boundary condition at the free surface or interface between immiscible fluids. Without going into a formulation involving normalization of the differential equations of motion, we see that the proper similitude parameter may be written in the form of Eqs. (7–16) and (7–17). We form the ratio of inertia force per unit mass (V_0^2/L) to the surface tension force per unit mass. Since σ (see Chapter 1) is the surface tension force per unit length, then

$$\frac{\sigma}{L^2} = \frac{\text{surface tension force}}{\text{volume}},$$

and hence

$$\frac{\sigma}{\rho L^2} = \frac{\text{surface tension force}}{\text{mass}}.$$

The dimensionless ratio is known as the Weber number,

$$\mathbf{W} = \frac{V_0^2/L}{\sigma/\rho L^2} = \frac{\rho V_0^2 L}{\sigma} = \frac{\text{inertia force}}{\text{capillary force}}. \tag{7–33}$$

Equality of Weber numbers would then ensure similitude with respect to surface tension effects. This may be important in the study of the formation of droplets from nozzles.

7–5   SIMILARITY CONDITIONS FOR COMPRESSIBLE FLUIDS. MACH NUMBER

For compressible fluids, the compressibility effects are dependent on the magnitude of the local velocity variations. In particular, the distribution of pressure and density depends on the local velocity magnitude relative to the local velocity of sound in the fluid. To treat compressibility, we return to the full Navier-Stokes equations for compressible fluids with constant viscosity. We also introduce a dimensionless density $\rho^0 = \rho/\rho_0$, a dimensionless pressure $p^0 = p/p_0$, and corresponding reference quantities. Introducing these quantities together with the other pertinent dimensionless quantities from Eqs. (7–13), into Eqs. (6–24) gives for the x-direction:

$$\rho^0 \frac{du^0}{dt^0} = -\left(\frac{1}{\mathbf{F}^2}\right)\rho^0\frac{\partial h^0}{\partial x^0} - \left(\frac{p_0}{\rho_0 V_0^2}\right)\frac{\partial p^0}{\partial x^0} + \left(\frac{1}{\mathbf{R}}\right)\left(\frac{\partial^2 u^0}{\partial (x^0)^2} + \frac{\partial^2 u^0}{\partial (y^0)^2} + \frac{\partial^2 u^0}{\partial (z^0)^2}\right)$$

$$+ \left(\frac{1}{3\mathbf{R}}\right)\left(\frac{\partial^2 u^0}{\partial (x^0)^2} + \frac{\partial^2 v^0}{\partial x^0 \partial y^0} + \frac{\partial^2 w^0}{\partial x^0 \partial z^0}\right). \tag{7–34}$$

Thus a new dimensionless group $p_0/\rho_0 V_0^2$ appears in addition to the familiar Froude and Reynolds numbers. When one deals with the flow of gases, gravitational effects are usually not relevant and Froude similarity is generally ignored.

The local speed of sound is given by

$$c = \sqrt{kp/\rho}, \tag{7–35}$$

where the ratio of specific heats is

$$k = c_p/c_v.$$

If we define the reference speed of sound as

$$c_0 = \sqrt{kp_0/\rho_0}, \tag{7–36}$$

then the dimensionless pressure group can be written as

$$\frac{p_0}{\rho_0 V_0^2} = \frac{c_0^2}{kV_0^2} = \frac{1}{k}\frac{1}{\mathbf{M}_0^2}, \tag{7–37}$$

where

$$\mathbf{M}_0 = \frac{V_0}{c_0} = \text{Mach number.} \tag{7–38}$$

Dynamic similitude in a compressible gas flow requires equality of both Reynolds and Mach numbers and equal values of the ratio of specific heats, k.

Specifying equality of Mach numbers, we have, for the velocity ratio,

$$V_r = c_r, \tag{7–39}$$

and for the Reynolds number equality, Eq. (7–23),

$$V_r = \frac{\mu_r}{\rho_r L_r}.$$

Hence

$$L_r = \frac{\mu_r}{\rho_r c_r}. \tag{7–40}$$

This corresponds to a single degree of freedom as in the case of Reynolds-Froude similitude. Fortunately, due to the expedient of variable-pressure wind tunnels, considerable latitude in the fluid property ratios exists. For common gases, such as air, the absolute viscosity and the speed of sound* depend on temperature and are essentially independent of pressure. Therefore, for two air systems at the same temperature,

$$k_r = 1, \quad \text{and} \quad L_r = \frac{1}{\rho_r}.$$

Viscous effects may become negligible at very high speeds, and Mach-number similitude may be sufficient. In the supersonic range, heat-transfer effects become important.

7–6   SUMMARY

It should now be evident that the conditions for dynamic similitude are obtained from the dynamic equations of motion which are appropriate to the flow system under investigation. The greater the number of individual forces required to describe the flow, the fewer the independent choices in the design of a dynamically similar system. In selecting the dimensionless groups which are to be used in establishing dynamic similarity, one should place the emphasis on the relative importance of the individual forces with respect to the general nature of the flow. Table 7–1 aids in the identification of these forces and their corresponding ratios with the inertial force. It is interesting to note that the time and length ratios are related by exponents which differ by values of $\frac{1}{2}$, starting with the Froude relation, $T_r \sim L_r^{1/2}$ and ending with the Reynolds relation $T_r \sim L_r^2$.

TABLE 7–1

IDENTIFICATION OF FORCES AND DIMENSIONLESS GROUPS
IN DYNAMIC SIMILITUDE

Force	Dimensionless group	Time ratio T_r	Force ratio F_r
Viscous shear	Reynolds number $\mathbf{R} = V_0 L/\nu$	$T_r = \rho_r L_r^2/\mu_r$	$F_r = \mu_r^2/\rho_r$
Gravity	Froude number $\mathbf{F} = V_0/\sqrt{gL}$	$T_r = \sqrt{L_r/g_r}$	$F_r = \gamma_r L_r^3$
Surface tension	Weber number $\mathbf{W} = \rho V_0^2 L/\sigma$	$T_r = L_r^{3/2}\sqrt{\rho_r/\sigma_r}$	$F_r = \sigma_r L_r$
Elastic	Mach number $\mathbf{M} = V_0/c_0$	$T_r = L_r/c_r$	$F_r = \rho_r c_r^2 L_r^2$

* For the perfect gas, $p/\rho = RT$, $c = \sqrt{kRT}$, with R in units of (ft-lb/slug-°R).

Many introductory texts in fluid mechanics use the technique of dimensional analysis in treating the subject of dynamic similarity and the formation of dimensionless groups. This approach is primarily concerned with the formal procedures for establishing the minimum number of dimensionless groups of variables from an assumed assemblage of pertinent quantities [2]. The method used in this chapter has been called inspectional analysis [3]. It appears to give a more rational basis for dynamic similitude by focusing attention on the physical forces as separate terms in the appropriate equation of motion and associated boundary conditions. It is not necessary that we be able to solve the equations since, in inspectional analysis, we use them only to determine the necessary and sufficient conditions for similitude.

The term "dynamic similitude" implies a similarity in the dynamic behavior of fluids. The term "thermal similitude" may be used to describe similarity in heat transfer. The conditions for thermal similitude are obtained by normalizing the energy equation rather than the equation of motion. The dimensionless groups such as the Prandtl and Nusselt numbers appropriate to heat transfer problems are found in this manner [4].

PROBLEMS

7–1. The design of a large Venturi meter for air flow is tested with water by means of a model one-fifth the size of the prototype. At a water flow rate of 3 cfs the differential pressure for the model meter is 14 psi. (a) Determine the corresponding prototype air flow rate. (b) Determine the corresponding prototype pressure differential. (c) Is there any advantage in using water instead of air in the model test?

7–2. Because of the relative ease in making internal velocity traverses, the flow pattern in a proposed water turbine is studied at a 1:10 scale reduction using air instead of water. If the operating temperature in both model and prototype is assumed to be 60°F and the air in the turbine is at a pressure of 20 psia, what discharge ratio must be used to obtain dynamically similar conditions? What will be the conversion ratio for (a) velocities, and (b) changes in pressure intensity?

7–3. Liquid sodium in a closed circuit is to be used as a heat-transfer medium in a nuclear-powered submarine. We wish to determine the pressure drop caused by flow through a specially designed check valve in the liquid sodium loop. The temperature of the liquid sodium is 700°C, the specific gravity is 0.78, and μ, the absolute viscosity, is 0.18 centipoise (μ in lb·sec/ft^2 = μ in centipoises/47,870). The pipeline has a diameter of 8 in., and the design flow rate is 1000 gal/min.

A one-half size geometrically similar model of the check valve has been constructed. An experimental program is to be designed to predict the pressure drop in the prototype system by measurements on the dynamically similar flow in the smaller system. (a) State the requirements to be satisfied in order to achieve dynamic similarity. (b) Discuss the relative advantages and disadvantages of using water (60°F) or air (60°F) in the model test. (c) For the fluid chosen, determine the volume rate of flow required for the model test.

7–4. A popular demonstration experiment (Fig. 7–5) consists of supporting a spherical ball in a vertical jet of air discharging into the atmosphere. Assume the following data for the air experiment: d = diameter of jet at nozzle = 6 in., D = diameter of ball = 8 in., W = weight of ball = 0.3 lb, V = velocity of jet at nozzle = 30 ft/sec, temperature of air = 60°F, atmospheric pressure = 14.7 psia, a = distance of ball above nozzle = 20 in. (a) Describe a dynamically similar experiment using a water jet of diameter d = 0.5 in. also at 60°F. What would be the required weight of the ball and its specific gravity (assuming uniform density)? (b) Should the water jet be discharged into a large tank of water or into the atmosphere?

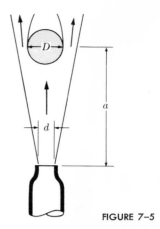

FIGURE 7–5

7–5. A sphere of steel (specific gravity = 7) has a 0.01-in. diameter. When dropped into a tank of water at 68°F it reaches a steady terminal velocity of 0.265 ft/sec. What diameter must a steel sphere have to attain a velocity of 0.265 ft/sec when dropped into a tank of glycerin at 68°F if the flows around the sphere are to be dynamically similar in the two cases?

7–6. A ship model 4 ft long is tested in a towing tank at a speed that will produce waves which are dynamically similar to those observed around the prototype. The test speed of the model is 3 ft/sec. To what speed of the prototype ship does this correspond, given that the prototype is 150 ft long?

7–7. A submarine which is deeply submerged moves at a speed of 30 knots (30 nautical miles per hour). A 1:25 model is towed in sea water. What must be the towing speed to produce dynamically similar flows for the model and prototype? What will be the ratio of the total resistance force of the model to that of the prototype? Would you consider this a practical test?

7–8. The model submarine of Problem 7–7 is tested in a wind tunnel where the tunnel air stream is at p = 100 psia, 125°F and a velocity of 200 knots. Will the test give dynamic similarity with the prototype condition?

7–9. An open cylindrical tank of water, 3 ft in diameter, is rotated about its central, vertical axis at a constant rate of 60 rpm. After a period of time this rotation produces a constant difference in water-surface elevation between the tank wall and the axis of rotation. (a) Write the equation for the steady-state locus of the free surface h in dimensionless form, using the water depth at the axis, h_0, and the tangential velocity of the tank, V_t, as reference values. Show that this equation can be written in the form

$$h/h_0 = f(\mathbf{F}, r/r_0),$$

where r_0 is the radius of the tank and $\mathbf{F} = V_t/\sqrt{gh_0}$. (b) Using the conditions for dynamic similarity specified in (a), determine the angular velocity which would produce a geometrically similar surface-elevation difference in a tank of mercury 6 in. in diameter. (c) The dimensionless equation developed in part (a) shows that the steady-state free-surface configuration is independent of the viscosity of the fluid in the tank. Would you expect that the time required to reach the steady state would also be independent of the viscosity? Explain.

7–10. Certain harbors exhibit standing waves which are excited by wave energy entering the harbor from the ocean. Under certain conditions these waves may cause damage by impact of moored ships against docks. A model of a harbor is built in order to study this problem. The length ratio is $L_r = 1/100$ and a standing-wave period of 3 sec is measured in the model. Determine the corresponding period of oscillation in the prototype harbor.

7–11. A model test is to be made of a submarine (cruising under an icepack) whose maximum diameter is 25 ft and whose top speed in the ocean is 20 fps. The water temperature is 4°C. Two tunnels are available for the test: (1) A water tunnel with a maximum velocity of 100 fps in which the model may have a maximum diameter of 3 in. (2) A wind tunnel with a maximum velocity of 150 mph in which models with a maximum diameter of 18 in. may be used. The wind tunnel can be pressurized up to three atmospheres. The temperature of both the air and water is 25°C. Which of the two tunnels would you use for the testing? Give reasons for your choice.

7–12. The energy equation for a steady-state frictionless process in an ideal gas is

$$\rho c_p\left(u\,\frac{\partial T}{\partial x} + v\,\frac{\partial T}{\partial y} + w\,\frac{\partial T}{\partial z}\right) = k\left(\frac{\partial^2 T}{\partial x^2} + \frac{\partial^2 T}{\partial y^2} + \frac{\partial^2 T}{\partial z^2}\right) + u\,\frac{\partial p}{\partial x} + v\,\frac{\partial p}{\partial y} + w\,\frac{\partial p}{\partial z},$$

where

$$
\begin{aligned}
c_p &= \text{specific heat at constant pressure,}\\
T &= \text{absolute temperature,}\\
k &= \text{coefficient of thermal conductivity,}\\
p &= \text{pressure,}\\
\rho &= \text{density,}\\
u, v, w &= x\text{-, } y\text{-, } z\text{-components of velocity.}
\end{aligned}
$$

Rewrite the equation in dimensionless form, using a characteristic length L, a characteristic velocity U, and a characteristic temperature T_0. Indicate the dimensionless groups appropriate to the class of similitude problems described by this equation.

7–13. We wish to study the general circulation pattern produced by a propeller 1 ft in diameter rotating at a speed of 180 rpm immersed in a large tank of glycerin (68°F). In particular it is planned to study the current pattern for two conditions: (1) a closed tank completely filled with glycerin; and (2) an open tank with the rotating propeller shaft extending vertically through the free surface of glycerin.

   Discuss the feasibility of making model studies for both of the conditions described above, using water and a propeller 3 in. in diameter. If feasible, state the rotational speed and water temperature to be used in both model experiments.

7–14. A large tank contains a layer of salt water of specific weight γ_2 and depth z_0 underneath a layer of fresh water of specific weight γ_1, as shown in Fig. 7–6. An outlet pipe of diameter D is located in the bottom of the tank. Experiments have been performed to determine the velocity V_c in the outlet pipe at which fresh water from the upper layer begins to be drawn into the outlet. The experimental results are given by the equation

$$V_c = 2.05 g^{1/2}\left(\frac{\Delta\gamma}{\gamma}\right)^{1/2}\frac{z_0^{5/2}}{D^2},$$

where $\Delta\gamma = \gamma_2 - \gamma_1$. (a) Show that this equation may be written in dimensionless form in terms of the densimetric Froude number. Specify the characteristic length to be used in the densimetric Froude number. (b) When the tank contains sea water (68°F) underneath fresh water (68°F) and the depth of the sea water is 2 ft, determine the maximum rate of withdrawal of sea water in cfs through a 6-in. outlet pipe.

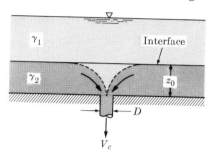

FIGURE 7-6

7-15. An outlet pipe leads from a large oil-storage tank (crude oil, sp. gr. = 0.855 at 68°F) to a pump and thence to an oil pipeline. When the depth of oil in the tank approaches the level of the outlet pipe, difficulty in pump operation is experienced due to the periodic formation of a vortex above the outlet. This leads to air being drawn into the pump and pipeline. Because of the unsteady nature of the vortex formation at the free surface, it is believed that both viscous and gravitational effects are important. A one-fourth scale model is to be constructed. Determine the required fluid properties for the model fluid. Can you suggest a convenient liquid which meets the requirement?

7-16. Consider a liquid flowing under conditions in which viscous, gravity, and surface-tension effects are of approximately equal importance. What interrelationship between the fluid properties is required to obtain a dynamically similar model?

7-17. The conditions for dynamic similarity may be derived from a specific equation of motion by writing it in dimensionless form. Consider the equation for the velocity of propagation, c, of small-amplitude surface waves in a liquid of uniform depth,

$$c^2 = \left(\frac{\sigma}{\rho}\frac{2\pi}{\lambda} + \frac{g\lambda}{2\pi}\right)\tanh\frac{2\pi h}{\lambda},$$

where

σ = surface tension,

λ = wave length,

h = depth of undisturbed liquid,

ρ = density.

(a) Using L as a characteristic length and V_0 as a characteristic velocity, write the above equation in dimensionless form and indicate the conditions for similitude in terms of dimensionless groups. (b) If water plus detergent (surface tension equals one-third that of pure water) is used in a model and pure water in the prototype, what is the L_r for dynamically similar wave motion?

7-18. An experimental investigation of the prototype flow system described in Problem 6-16 is to be made to determine the free-surface configuration of the water jet after impingement on the disk. A geometrically similar system using a disk 1 ft in diameter is set up in the laboratory. (a) It is planned to measure the vertical thickness of the water jet at the perimeter of the disk. When the flow rate in the smaller system is properly adjusted, what should be the reading of the pressure gauge at point B? What is the ratio of water velocities in the two systems at section A? Neglect surface tension effects. (b) We wish to determine the radius at which the thin jet of water breaks up into droplets. If we assume that the prototype operates with water at 68°F, is it possible to operate the smaller system so that a dynamically similar behavior of the jet may be observed?

7-19. Compute the speed of sound in water at a pressure of 15 psi and a temperature of 68°F (see Table 1-5). Assume that compressibility effects are negligible at Mach numbers less than 0.1. Determine the speed that would have to be achieved by a missile traveling under water for compressibility effects to be important in the total resistance.

7-20. A high-speed airplane wing is tested in a wind tunnel. The prototype flies at 1000 mph in standard air at an altitude of 25,000 ft. At what wind-tunnel velocity should a 1:50 model be tested to properly model the shock waves, given that the tunnel air temperature is 185°F. Discuss the feasibility of testing this model in a water tunnel.

7-21. Consider the feasibility of making a wind-tunnel test to determine the aerodynamic characteristics of a missile in which dynamic similarity is required with respect to both viscous and compressibility effects. The prototype is to travel within the atmosphere of the earth at an altitude of 25,000 ft (see Table 1-7). From the dimensions of the wind-tunnel test section the desirable scale ratio for the model is $L_r = 1/5$. If such a test is feasible, determine the proper pressure for the wind-tunnel test section, assuming an operating air temperature of 60°F.

REFERENCES

1. HARLEMAN, D. R. F., "Stratified Flow," *Handbook of Fluid Dynamics* (V. L. Streeter, ed.), McGraw-Hill Book Co., New York, 1961, Chapter 26.

2. HOLT, M., "Dimensional Analysis," *Handbook of Fluid Dynamics* (V. L. Streeter, ed.), McGraw-Hill Book Co., New York, 1961, Chapter 15; D. C. IPSEN, *Units, Dimensions and Dimensionless Numbers*, McGraw-Hill Book Co., New York, 1961; H. L. LANGHAAR, *Dimensional Analysis and Theory of Models*, John Wiley and Sons, Inc., New York, 1951; G. MURPHY, *Similitude in Engineering*, Ronald Press Co., 1950. All contain treatments of dimensional analysis.

3. BIRKHOFF, G., *Hydrodynamics, a Study in Logic, Fact and Similitude*, Dover Publications, Inc., New York, 1955 (Princeton University Press, 1950), p. 90.

4. LANDAU, L., and E. LIFSHITZ, *Fluid Mechanics*, Addison-Wesley Publishing Co., Inc., Reading, Mass., 1959; R. BIRD, W. STEWART, and E. LIGHTFOOT, *Transport Phenomenon*, John Wiley and Sons, Inc., New York, 1960.

CHAPTER 8

Some Fundamental Concepts and Specialized Equations in Fluid Dynamics

8–1 FLOW CLASSIFICATIONS

We consider a viscous fluid with its field of motion corresponding to specified rigid boundaries and subjected to a particular body force. Depending on the relative magnitudes of the viscous and inertia forces, we find that the character of the flow and the detailed velocity and pressure distributions show large differences. This fact leads to two important fundamental concepts in classifying types of flow and in treating these types analytically. One is the distinction between *laminar flows* and *turbulent flows* as two possible modes of motion. The other distinguishes between *creeping flows* and *boundary layer flows* as two extremes of viscosity effects. Let us examine these concepts.

8–1.1 Laminar versus turbulent flow

The existence of two modes of motion was observed very early by experimental investigators. For example, in 1839 Hagen [1] called attention to the fact that the mode of flow in a cylindrical tube changes when the velocity exceeds a certain limit. He observed that below this velocity the outflowing jet was as smooth as a solid bar of glass; above this velocity, the jet surface oscillated and the flow appeared to come in spurts. These behaviors are shown in Fig. 8–1 [2]. These were evidences of laminar and turbulent flows which are now recognized as modes encountered throughout fluid flow applications.

In 1883 Osborne Reynolds [3] demonstrated the two modes very clearly and presented the parameter (which now bears his name) as a criterion for use in determining which mode should exist. He injected a fine stream of dye into water flowing from a large tank into a glass tube. With low flow rates through the tube, the dye stream persisted as a straight streak showing that the water moved in parallel streamlines or laminas. The velocities of adjacent laminas were not the same, but there was no macroscopic mixing across laminas. This is the simplest case of laminar flow. As the flow was increased above a certain critical rate, the dye streak in Reynolds' demonstration broke into irregular vortices and then mixed laterally throughout the cross section of the tube. This mixing across laminas is evidence of turbulent flow. Examples of these two cases are shown in Fig. 8–2.

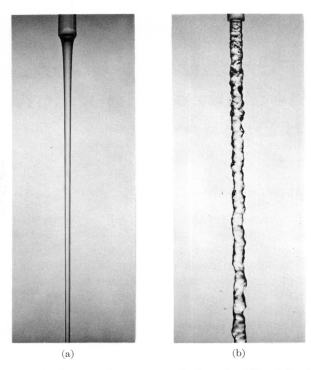

(a)   (b)

FIG. 8–1. Efflux from a circular tube showing (a) laminar and (b) turbulent flow [2].

In general, laminar flow occurs when adjacent layers of fluid move relative to each other, forming smooth streamlines, not necessarily straight, without macroscopic mixing. It is the mode that occurs when viscous shear, which is caused by molecular momentum exchange between fluid layers, is the predominant influence in establishing the flow field.

Turbulent flow is characterized by fluid particles having irregular, near random, fluctuating motions and erratic paths. Macroscopic mixing results both lateral to, and in the direction of, the main flow. It is the mode that occurs when viscous shear forces are secondary to inertia forces in establishing the flow field.

We can identify the criterion for the existence of laminar motion by noting the consequences of small accelerations and high fluid viscosity. First, the shear stresses tend to become very large, even for small velocity gradients. Second, the inertia force per unit mass of the fluid tends to become of secondary importance relative to the viscous shear force per unit mass. We can characterize this condition by noting from the ratio

$$\frac{\text{inertia force/mass}}{\text{frictional force/mass}} \propto \text{Reynolds number}$$

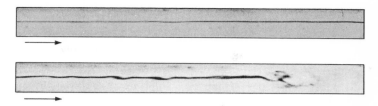

FIG. 8–2. Laminar flow and development of turbulence in a glass tube.

that we are talking about low Reynolds numbers. Laminar motion will exist when the Reynolds number is below a certain critical value. Above this critical value, the laminar motion becomes unstable and turbulence may develop. The numerical value of the critical Reynolds number depends on the flow geometry and on the characteristic length and velocity used to define it.

We should note at this point that both laminar and turbulent motions are the result of the viscosity property, and that neither would occur in the absence of viscosity. Consequently, while turbulence results in an effective shear due to momentum exchange between laterally fluctuating particles, as we will discuss in a later chapter, this turbulent shear is evidence of viscosity effects.

8–1.2 Creeping motions

The extreme of laminar motion occurs when inertia forces can be neglected completely and the Reynolds number tends toward zero. Then the relation between pressure gradients, body forces, and velocity field is governed solely by the transmission of shear from the boundaries into and through the fluid medium. The fall of light-weight objects through a mass of molasses and the filtration of a liquid through a densely packed bed of fine solid particles bear witness to this statement.

It is the nature of these motions that the relatively high viscous shear forces exert a major influence completely across the fluid space. Thus, for the falling sphere shown in Fig. 8–3, the fluid undergoes a measurable "deformation" due to the viscous action, which extends a considerable distance in all directions from the sphere. This effect is shown by the streamlines and velocity distributions

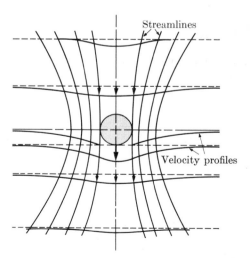

FIG. 8–3. Deformation flow around a falling sphere. (Streamlines and velocity profiles are shown for observer at rest.)

in Fig. 8–3.   Creeping motions of this type are often called *deformation flows.*

The conditions imposed by neglecting inertia mean that the motion is so very slow that (1) convective accelerations have no significant inertial effect, and (2) unsteadiness may be neglected.

We should point out that in Chapter 6 some examples were given for using the Navier-Stokes equations of motion to solve cases of parallel flow.   The steady uniform flows described have the characteristics of creeping motion in that acceleration and inertia are absent and that the viscous effect is pronounced completely across the fluid space.   They differ from creeping motion, however, in that the flow need not be limited to very slow motion.   The only restriction is that the flow remain laminar.

8–1.3   The boundary-layer concept

With real fluids there is no "slip" at the rigid boundaries.*   The fluid velocity relative to the boundary is zero.   As a result, the velocity gradient and shear stress have maximum values at the boundary and decrease into the fluid.   This was true, as illustrated in Fig. 8–3, for deformation flow caused by the sphere falling slowly in a very viscous fluid.   It is also true, but even more pronounced, for the other extreme of very low viscosity and high velocities or accelerations of the fluid motion.   Then the steepness of the velocity gradient close to the wall becomes very large, and the only significant viscous shear occurs within a relatively thin layer next to the boundary.   Outside this layer, the velocity gradients rapidly taper off and the viscous shear becomes small.   This narrow zone is called a *boundary layer.*   Inside the layer, the viscous effects override inertia effects in determining the fluid deformation.   Outside this layer, the flow will suffer only a minor influence of the viscous forces and will be determined primarily by the relation among inertia, pressure gradient, and body forces interacting with the geometry of the solid boundaries.   Consequently, the streamlines of the main flow beyond the boundary layer conform essentially to a potential flow.

This boundary-layer concept is illustrated in Fig. 8–4, which shows flow past a thin flat plate and flow past a circular cylinder.   If the fluid were completely nonviscous and slip occurred at the boundaries, the streamlines and velocity distributions would be the same as those shown by the dashed lines. With a small viscosity and the no-slip condition, the velocity distributions would be altered as shown by the solid lines.   The fluid in front of the plate and the cylinder has a uniform velocity.   The velocity of the fluid passing over the rigid surfaces of these objects is zero at the boundary surface and varies rapidly to essentially the nonviscous flow velocity.   The distance over which the main viscosity influence is felt is the boundary-layer thickness, δ.   Outside the bound-

---

* Strictly speaking, this statement applies only to fluids that can be treated as a continuum as defined in Chapter 1.

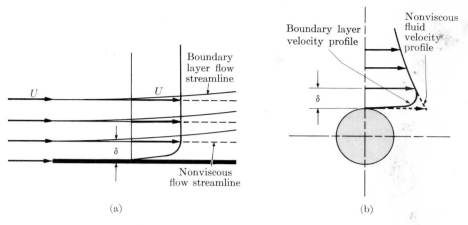

FIG. 8–4. Boundary layer versus slip flow: (a) flat plate; (b) cylinder.

ary layer of the flat plate the only modification of the initial uniform flow is a slight displacement of streamlines. Due to flow retardation within the boundary layer, displacement is necessary to satisfy continuity. Beyond the boundary layer of the cylinder, this displacement effect will occur as well as a continued small viscous shear effect associated with the small remaining velocity gradient.

The flow in a boundary layer may be two-dimensional as suggested by Fig. 8–4(a) and (b). More generally, however, the boundary-layer velocity vectors at different distances from the boundary may have components in all three coordinate directions, and the boundary-layer flow will be three-dimensional.

Boundary layers form along conduit walls as well as over immersed bodies. For example, the flow into a slot develops boundary layers on the slot walls, as shown in Fig. 8–5. After a certain "entrance length," the two layers converge to establish uniform flow. In the entrance length, the layers are like the non-

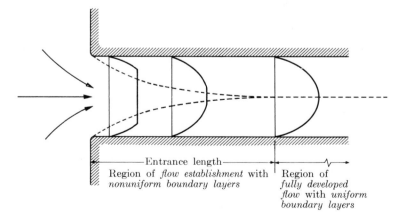

FIG. 8–5. Boundary layers in ducts.

uniform layers on immersed bodies. The uniform flow is said to be composed of uniform boundary layers. Thus, the example in Section 6–3 for uniform laminar flow between parallel walls gives the solution for a uniform laminar boundary layer. Similar boundary-layer growth to uniform flow occurs in pipes and other ducts.

The boundary-layer concept was developed by the German engineer-mathematician L. Prandtl, in a series of publications starting in 1904 [4]. It is one of the most significant discoveries in the development of fluid mechanics and permits an understanding of many seeming paradoxes in real fluid behavior. It reveals ways to analyze problems much too complicated to be solved by direct integration of the full set of equations of motion and continuity. Creeping motion and boundary-layer flow are extreme manifestations of viscosity effects. Broadly speaking, the former occurs for very viscous fluids and the latter for slightly viscous fluids. On the other hand, while creeping flow is uniquely laminar, boundary layers may be laminar or turbulent as we will discuss in later chapters.

8–2 EQUATIONS FOR CREEPING MOTION AND TWO-DIMENSIONAL BOUNDARY LAYERS

The concepts of creeping motion and boundary-layer flow described in the previous paragraphs make possible the modification of the equations of motion to approximate the special physical conditions that nature produces. Simplifications result which facilitate reaching analytical solutions to important problems in both categories of motions. For the incompressible isothermal fluid we begin with the Navier-Stokes equations of Chapter 6, Eqs. (6–28), and introduce changes as described in the following sections. For creeping motion we will consider the general three-dimensional equations. For the boundary layer we will restrict ourselves to two-dimensional flow on surfaces of small curvature.

8–2.1 Creeping motion

When the inertia terms are omitted, the incompressible Navier-Stokes equations for flow in a gravity body-force field become

$$\frac{\partial(p + \gamma h)}{\partial x} = \mu \left(\frac{\partial^2 u}{\partial x^2} + \frac{\partial^2 u}{\partial y^2} + \frac{\partial^2 u}{\partial z^2} \right),$$

$$\frac{\partial(p + \gamma h)}{\partial y} = \mu \left(\frac{\partial^2 v}{\partial x^2} + \frac{\partial^2 v}{\partial y^2} + \frac{\partial^2 v}{\partial z^2} \right), \tag{8–1}$$

$$\frac{\partial(p + \gamma h)}{\partial z} = \mu \left(\frac{\partial^2 w}{\partial x^2} + \frac{\partial^2 w}{\partial y^2} + \frac{\partial^2 w}{\partial z^2} \right).$$

Upon adding Eqs. (8–1), we have

$$\nabla(p + \gamma h) = \mu\nabla^2\mathbf{q}. \tag{8–1a}$$

In these equations, the pressure change is due to the combination of viscous effects and gravity. When we are dealing with incompressible fluids in an enclosed system, we can, as described in Chapter 7, let

$$p = (p_d + p_s) = (p_d + \text{const} - \gamma h).$$

Then Eq. (8–1a) becomes

$$\nabla p_d = \mu\nabla^2\mathbf{q}. \tag{8–2}$$

Thus, for such cases, the effect of viscosity is the same as would be found in a zero-gravity field.

The above equation must be solved together with the constant-density continuity equation $\nabla \cdot \mathbf{q} = 0$. The entire system of equations above must satisfy the same boundary conditions as must be satisfied for the full Navier-Stokes equations, namely the vanishing of the normal and tangential components of the relative velocity on the surface of rigid boundaries.

Reexamination of the examples of laminar flow between parallel plates and through circular tubes in Section 6–5 will show that they are based on the solution of Eq. (8–1a) with the kinematic and physical boundary conditions for prescribed plane rigid walls.

8–2.2   Equations for two-dimensional boundary layers

Prandtl showed how to simplify the Navier-Stokes equations using order-of-magnitude arguments stemming from the physical concept of the boundary layer. We will present his results for two-dimensional flow along plane or slightly curved boundaries.

We begin with the Navier-Stokes equations in dimensionless form as represented by Eq. (7–15) for enclosed systems. For the two dimensions x and y, (omitting gravity) these are:

$$\frac{\partial u^0}{\partial t^0} + u^0\frac{\partial u^0}{\partial x^0} + v^0\frac{\partial u^0}{\partial y^0} = -\frac{\partial p^0}{\partial x^0} + \frac{1}{\mathbf{R}}\left(\frac{\partial^2 u^0}{\partial(x^0)^2} + \frac{\partial^2 u^0}{\partial(y^0)^2}\right),$$

$$1 \qquad 1 \quad 1 \qquad \delta^0\frac{1}{\delta^0} \qquad\qquad \delta^{0^2} \quad 1 \qquad \frac{1}{\delta^{0^2}}$$

$$\frac{\partial v^0}{\partial t^0} + u^0\frac{\partial v^0}{\partial x^0} + v^0\frac{\partial v^0}{\partial y^0} = -\frac{\partial p^0}{\partial y^0} + \frac{1}{\mathbf{R}}\left(\frac{\partial^2 v^0}{\partial(x^0)^2} + \frac{\partial^2 v^0}{\partial(y^0)^2}\right). \tag{8–3}$$

$$\delta^0 \qquad 1 \quad \delta^0 \qquad \delta^0 \quad 1 \qquad\qquad \delta^{0^2} \quad \delta^0 \qquad \frac{1}{\delta^0}$$

For continuity

$$\frac{\partial u^0}{\partial x^0} + \frac{\partial v^0}{\partial y^0} = 0.$$

$$1 \qquad\qquad 1$$

First, we note that the boundary layer is thin and of small curvature, and that within this layer

$$u \gg v, \qquad x \gg y,$$

$$\frac{\partial u}{\partial y} \gg \frac{\partial u}{\partial x}, \qquad \frac{\partial p}{\partial y} \text{ is small.} \tag{8–4}$$

The dimensionless boundary-layer thickness is $\delta^0 = \delta/L$ and is assumed to be small relative to unity, $\delta^0 \ll 1$. We thus establish a scale for decreasing orders of magnitude of

$$\frac{1}{\delta^{0^2}}, \frac{1}{\delta^0}, 1, \delta^0, \delta^{0^2}. \tag{8–5}$$

With the notation $\sim 0(\ )$ expressing "is the order of," we use the relations (8–4) to set down the relative magnitudes of distances and velocities as follows:

$$x^0 \sim 0(1), \qquad y^0 \sim 0(\delta^0)$$
$$u^0 \sim 0(1), \qquad v^0 \sim 0(\delta^0) \tag{8–6}$$

Now for $\partial u^0/\partial x^0 \sim 0(1)$, continuity gives $\partial v^0/\partial y^0 \sim 0(1)$. It follows that $\partial^2 u^0/\partial(x^0)^2$ remains $\sim 0(1)$ while $\partial^2 v^0/\partial(y^0)^2$ becomes $\sim 0(1/\delta^{0^2})$. From this we see that differentiating with respect to y^0 increases the order of magnitude by one, while there is no change when differentiating with respect to x^0. In order to rule out very sudden accelerations, we assume $\partial u^0/\partial t^0$ is of the same order as $u^0(\partial u^0/\partial x^0)$. We can now establish for Eqs. (8–3) the orders of magnitude of the quantities as shown below each individual term. In doing this, we note from the first equation that the Reynolds number must be of the order $1/\delta^{0^2}$ to satisfy the boundary-layer concept postulate that the viscous term is to be comparable to the inertia term inside. We also note that $\partial p^0/\partial x^0$ may or may not be negligible. In the second equation, $\partial p^0/\partial y^0$ is not zero but may not exceed the largest order, δ^0, of the other terms.

Eliminating all terms of order less than unity and reverting to dimensional terms, we obtain Prandtl's two-dimensional boundary-layer equations

$$\frac{\partial u}{\partial t} + u\frac{\partial u}{\partial x} + v\frac{\partial u}{\partial y} = -\frac{1}{\rho}\frac{\partial p}{\partial x} + \frac{\mu}{\rho}\frac{\partial^2 u}{\partial y^2}, \qquad \frac{\partial u}{\partial x} + \frac{\partial v}{\partial y} = 0. \tag{8–7}$$

These must satisfy the boundary conditions

$$(1) \quad y = 0, \qquad (2) \quad y = \infty,$$
$$u = 0, \qquad\qquad u = U(x).$$
$$v = 0, \qquad\qquad\qquad\qquad (8\text{-}8)$$

For three-dimensional boundary layers, it is not possible to make a generalized simplification similar to the preceding one. In certain specific cases, terms in the Navier-Stokes equations will vanish of their own accord either due to symmetry or for other reasons. However, as a rule the treatment of three-dimensional layers requires the full set of equations.

8-2.3 Boundary-layer thickness definitions

The thickness δ cannot be established with precision since the point separating the boundary layer from the zone of negligible viscous influence is not a sharp one. It is common, therefore, to define δ as the distance to the point where the velocity is within some arbitrary percentage (usually 1%) of the free-stream velocity.

As already noted, the velocity retardation in the boundary layer causes a "defect" of the mass-flow rate adjacent to the boundary as compared to the mass that would pass through the same zone in the absence of a boundary layer. This is illustrated in Fig. 8-6. By continuity, this defect is equivalent to displacing the streamlines outside the boundary layer by a definite amount known as the *displacement thickness* δ^* and defined by

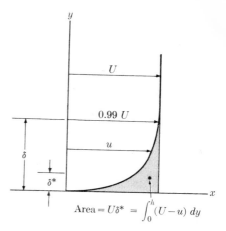

$$\rho U \, \delta^* = \rho \int_0^h (U - u) \, dy, \quad (8\text{-}9a)$$

or

$$\delta^* = \int_0^h \left(1 - \frac{u}{U}\right) dy, \quad (8\text{-}9b)$$

FIG. 8-6. Definition diagram for displacement thickness δ^*.

where $h \geq \delta$. Equation (8-9a) shows that δ^* is the thickness of an imaginary layer of fluid of velocity U and of mass flux rate equal to the amount of the defect. The displacement thickness can be determined with much better precision than the overall thickness δ.

The flow retardation within δ also causes a reduction in the rate of momentum flux. It is useful to define a *momentum thickness* θ as the thickness of an imaginary layer of fluid of velocity U for which the momentum flux rate equals the reduction caused by the velocity profile. It is defined by

$$\rho \theta U^2 = \rho \int_0^h (Uu - u^2)\, dy,$$
(8–10a)

or by

$$\theta = \int_0^h \frac{u}{U}\left(1 - \frac{u}{U}\right) dy.$$
(8–10b)

8–2.4 Integral momentum equation for two-dimensional boundary layers

Prandtl's equations can be integrated to obtain a useful relation between boundary shear stress and velocity distribution characteristics for the constant-density, constant-temperature case.

Let us consider *steady motion with* $\partial p / \partial x = 0$. Integrating Eq. (8–7), we obtain

$$\int_{y=0}^{y=h \geq \delta} \left(u \frac{\partial u}{\partial x} + v \frac{\partial u}{\partial y}\right) dy = \frac{\mu}{\rho} \int_{y=0}^{y=h} \frac{\partial^2 u}{\partial y^2}\, dy,$$
(8–11)

in which

$$\mu \int_0^h \frac{\partial^2 u}{\partial y^2}\, dy = \int_0^h \frac{\partial \tau}{\partial y}\, dy = [0 - \tau_0] = -\tau_0,$$
(8–12)

and

$$\int_0^h v \frac{\partial u}{\partial y}\, dy = \int_0^h \frac{\partial uv}{\partial y}\, dy - \int_0^h u \frac{\partial v}{\partial y}\, dy,$$
(8–13)

with

$$\int_0^h \frac{\partial uv}{\partial y}\, dy = [uv]_0^h = Uv_h;$$
(8–14)

and from continuity

$$\frac{\partial v}{\partial y} = -\frac{\partial u}{\partial x}$$
(8–15)

and

$$v_h = -\int_0^h \frac{\partial u}{\partial x}\, dy.$$

Then we get the integral momentum equation for wall shear stress,

$$\int_0^h u \frac{\partial u}{\partial x}\, dy - U \int_0^h \frac{\partial u}{\partial x}\, dy + \int_0^h u \frac{\partial u}{\partial x}\, dy = -\frac{\tau_0}{\rho}.$$
(8–16)

Assuming that $\partial p/\partial y = 0$, we can write

$$\rho \frac{\partial U}{\partial t} + \rho U \frac{\partial U}{\partial x} = -\frac{\partial p}{\partial x} \tag{8-17}$$

for the flow outside the boundary layer where the viscous influence is negligible. For steady motion with $\partial p/\partial x = 0$, $U = $ const, and Eq. (8-16) can be written

$$\frac{\tau_0}{\rho} = \frac{\partial}{\partial x} \int_0^h u(U - u) \, dy, \tag{8-18}$$

or

$$\frac{\tau_0}{\rho} = U^2 \frac{\partial \theta}{\partial x}, \tag{8-18a}$$

where θ is the momentum thickness. Substituting in the equation for the local surface resistance coefficient, we have

$$c_f = \frac{\tau_0}{\rho U^2/2}, \tag{8-19}$$

$$c_f = 2 \frac{\partial \theta}{\partial x} = \frac{\tau_0}{\rho U^2/2}. \tag{8-20}$$

For both *unsteady motion and $\partial p/\partial x \neq 0$*, we again assume that $\partial p/\partial y = 0$ and that the flow outside the boundary layer obeys Eq. (8-17). Using Eq. (8-17) in Eq. (8-7), integrating, and employing the definitions for displacement and momentum thickness, δ^* and θ, we get

$$\frac{\tau_0}{\rho} = \frac{\partial}{\partial x} (U^2\theta) + U \frac{\partial U}{\partial x} \delta^* + \frac{\partial}{\partial t} (U \delta^*). \tag{8-21}$$

The integral momentum equations were originally derived by von Kármán by applying the momentum principle of Chapter 4 directly to flat-plate boundary-layer flow. Equation (8-18) and its generalized counterpart, Eq. (8-21), are frequently called Kármán's integral momentum equations.

We should note here that in applying order-of-magnitude arguments to obtain Prandtl's equations, Eqs. (8-7), and hence the integral momentum equation, Eq. (8-21), the contribution of any turbulent fluctuations were neglected. Nevertheless, as we shall see in later chapters, the momentum form is used for turbulent as well as laminar flow. This is permissible so long as the net flux of momentum of turbulence is small compared to the flux of momentum of the mean flow velocities.

8-3 THE NOTION OF RESISTANCE, DRAG, AND LIFT

The notion of resistance to motion or drag of a fluid on an immersed body is an intuitive concept easy to grasp. However, the quantitative determination requires that it be carefully defined.

A fluid moving relative to a rigid boundary exerts a dynamic force* on the boundary that is caused by two factors. First, shear stresses due to viscosity and velocity gradients at the boundary surface cause forces tangential to the surface. Second, pressure intensities which vary along the surface due to dynamic effects result in forces normal to the boundary.

For an immersed body, the vector sum of the normal and tangential surface forces integrated over the complete surface gives a resultant force vector, as illustrated in Fig. 8–7. The component of this resultant force in the direction of the relative velocity V_0 past the body is the *drag*. The component normal to the relative velocity is a *lift* or lateral force.

Both drag and lift include frictional and pressure components. For the total drag we write

$$D = D_f + D_p, \tag{8–22}$$

with the components

$$\text{frictional drag,} \quad D_f = \int_S \tau_0 \sin \Phi \, dS, \tag{8–23}$$

$$\text{pressure drag,} \quad D_p = - \int_S p \cos \Phi \, dS, \tag{8–24}$$

where

$S = $ total surface area,

$\Phi = $ angle between the normal to the surface element and the flow direction.

The frictional drag is also known as *surface resistance* or *skin-friction drag*. The pressure drag depends largely on the shape or form of the body and is known as *form drag*. Bodies like airfoils, hydrofoils, and slim ships have large and sometimes completely dominant surface resistance. Bluff objects like spheres, bridge piers, and automobiles have large form drag relative to surface resistance.

Useful drag coefficients are defined by the relations

$$D_f = C_f \rho \, \frac{V_0^2}{2} \, A_f, \tag{8–25}$$

$$D_p = C_{D_p} \rho \, \frac{V_0^2}{2} \, A_p, \tag{8–26}$$

where A_f, A_p are suitably chosen reference areas. For surface resistance, A_f is usually the actual area over which shear stresses act to produce D_f, or a logical

---

* The hydrostatic buoyancy of objects immersed in a fluid is considered separately from the dynamic lift and drag forces due to the motion of the fluid relative to the object.

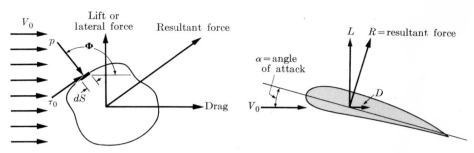

FIG. 8–7. Definition diagram for flow-induced forces.   **FIG. 8–8.** Lift and drag on a hydrofoil section.

representation of this area such as the planform area of a wing or hydrofoil. For form drag, A_p is usually the frontal area normal to the velocity V_0.

One can compute the coefficient C_f from Eqs. (8–23) and (8–25), using experimental or theoretical values of the local surface resistance coefficient, Eq. (8–19). Similarly, one can obtain the coefficient C_{D_p} from Eqs. (8–24) and (8–26), using experimentally determined distributions of pressure p and in some cases from theoretical calculations of p.

It is customary to define a total drag coefficient C_D by the relation

$$D = C_D \rho \frac{V_0^2}{2} A, \qquad (8\text{–}27)$$

with

$$C_D = C_{D_f} + C_{D_p}, \qquad (8\text{–}28)$$

where A is the frontal area normal to V_0. Thus, when $A = A_p$, the coefficient

$$C_f = C_{D_f}\left(\frac{A_p}{A_f}\right)$$

and C_{D_p} is the same in Eqs. (8–26) and (8–28).

For the lift force, it is not customary to separate the frictional and pressure components. For bodies like the hydrofoil in Fig. 8–8, which are designed particularly to produce useful lift, the lift force is primarily a pressure-component effect. The total lift and lift coefficient are defined by

$$L = C_L \rho \frac{V_0^2}{2} A. \qquad (8\text{–}29)$$

Here, the characteristic area A is usually either the largest projected area of the body (e.g., the planform area of a wing) or the projected area normal to V_0. As Fig. 8–8 illustrates, a good lifting device exhibits a high lift-drag ratio. Forces on immersed bodies are discussed in detail in Chapter 15.

REFERENCES

1. HAGEN, G., "On the Motion of Water in Narrow Cylindrical Tubes" (German), *Poggendorff's Ann. Phys. Chem.*, **46,** 423 (1839).

2. SEARS, F. W., and M. W. ZEMANSKY, *University Physics*, Addison-Wesley Publishing Co., Inc., Reading, Mass., 3rd ed., 1963, p. 328.

3. REYNOLDS, O., "An Experimental Investigation of the Circumstances which Determine Whether the Motion of Water Shall Be Direct or Sinuous, and of the Law of Resistance in Parallel Channels," *Trans. Cambridge Phil. Soc.*, **8** (1883).

4. PRANDTL, L., "Über Flüssigkeitsbewegung bei sehr Kleiner Reibung," *Proc. Third International Math. Congress*, Heidelberg, 1904.

CHAPTER 9

Creeping Motions

9–1 STOKES' MOTION

In two papers published in 1845 and 1851, Stokes gave the first known solution for a case of creeping motion. In the latter [1] he used the approximate relation given by Eq. (8–2) to solve the very slow motion case of flow past a fixed sphere and its counterpart, the case of a solid sphere falling through a very viscous infinite fluid. Constant-density continuity and the usual boundary condition of a vanishing relative velocity at the surface of the sphere were satisfied in addition to the modified Navier-Stokes equations. The mathematical details of the theory involve concepts beyond the scope of the present text [2]; however, the results of the computations are given in the following.

Consider a sphere of radius a, centered at the origin of a coordinate system, as shown in Fig. 9–1, and subject to a flow of velocity V_0 in the positive x-direction. Using Eq. (8–2), we find that the velocity and dynamic pressure fields which satisfy continuity and the boundary conditions are

$$u = V_0 \left[\frac{3}{4} \frac{ax^2}{r^3} \left(\frac{a^2}{r^2} - 1 \right) - \frac{1}{4} \frac{a}{r} \left(3 + \frac{a^2}{r^2} \right) + 1 \right], \tag{9-1}$$

$$v = V_0 \frac{3}{4} \frac{axy}{r^3} \left(\frac{a^2}{r^2} - 1 \right), \tag{9-2}$$

$$w = V_0 \frac{3}{4} \frac{axz}{r^3} \left(\frac{a^2}{r^2} - 1 \right), \tag{9-3}$$

$$p_d = - \frac{3}{2} \mu \frac{ax}{r^3} V_0. \tag{9-4}$$

These can be checked readily by substitution of the above, together with the boundary conditions, into Eqs. (8–1) and (8–2). When the sphere is in a gravity field, pressure values from Eq. (9–4) are added algebraically to the hydrostatic pressure p_s that would exist at the point if the fluid were at rest. The three velocity relations apply to the relative motion, i.e., the conditions as viewed by an observer having no motion relative to the sphere. The pressure relation applies to either the stationary or moving sphere so long as the motion is steady, a condition implicit in the assumption applying to Eq. (8–2) that all inertia terms are negligible.

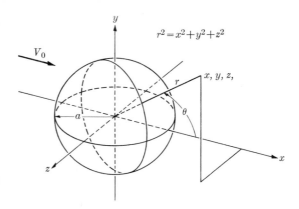

FIG. 9–1. Sphere centered at origin of a coordinate system.

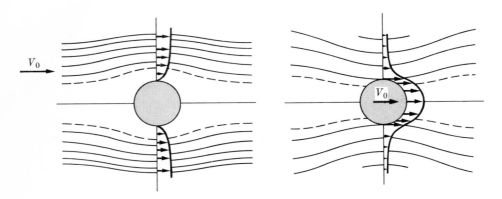

FIG. 9–2. Streamlines for Stokes' creeping flow past a fixed sphere.

FIG. 9–3. Streamlines for Stokes' creeping flow caused by a moving sphere.

The motion in any $r\theta$-plane (see Fig. 9–1) is described by

$$v_r = V_0 \cos \theta \left(1 - \frac{3}{2}\frac{a}{r} + \frac{1}{2}\frac{a^3}{r^3}\right), \tag{9–5}$$

$$v_\theta = -V_0 \sin \theta \left(1 - \frac{3}{4}\frac{a}{r} - \frac{1}{4}\frac{a^3}{r^3}\right). \tag{9–6}$$

The equation for the streamlines is given by constant values of

$$\psi = -\frac{1}{2}V_0 r^2 \sin^2 \theta \left(1 - \frac{3}{2}\frac{a}{r} + \frac{1}{2}\frac{a^3}{r^3}\right). \tag{9–7}$$

In spherical polar coordinates,

$$v_\theta = [1/(r \sin \theta)]\,d\psi/dr, \qquad v_r = -[1/(r^2 \sin \theta)]\,d\psi/d\theta. \tag{9–8}$$

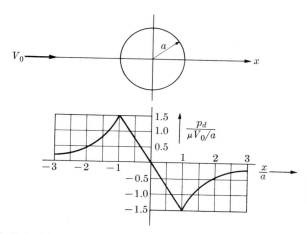

FIG. 9–4.   Pressure distribution for Stokes' flow around a sphere of radius a.

Streamlines and velocity distributions computed from the preceding equations are shown in Fig. 9–2. This figure emphasizes the departure of this creeping or deformation flow from the nonviscous flow and boundary-layer flow illustrated schematically in Fig. 8–4(b). The streamlines in Fig. 9–2 are the same ahead of and behind the sphere, a peculiarity which will not hold for viscous motion if inertia is important. Streamlines for the sphere in motion relative to the observer are given in Fig. 9–3.

From Eq. (9–4), we get the pressure at the surface of the sphere. Dropping the subscript d, we have

$$p|_{r=a} = -\frac{3}{2}\,\mu\,\frac{x}{a^2}\,V_0 = -\frac{3}{2}\,\frac{\mu V_0}{a}\cos\theta, \qquad (9\text{--}9)$$

and the maximum and minimum values are

$$p_{\max}|_{r=a} = \frac{3}{2}\,\frac{\mu V_0}{a}, \qquad (9\text{--}10)$$

which occurs at the upstream stagnation point,

$$p_{\min}|_{r=a} = -\frac{3}{2}\,\frac{\mu V_0}{a} \qquad (9\text{--}11)$$

which occurs at the downstream stagnation point. The pressure distribution along the x-axis and over the sphere surface is plotted dimensionlessly in Fig. 9–4.

Because of symmetry, the resultant shear stress at any point along the sphere surface is in the $r\theta$-plane containing that point. The polar coordinate form of the shear stress in terms of rate of strain can be obtained by transforming Eqs. (5–30). For the $r\theta$-plane, we find the stress in the fluid for this symmetry case to be

$$\tau_{r\theta} = \mu\left(\frac{1}{r}\,\frac{\partial v_r}{\partial\theta} + \frac{\partial v_\theta}{\partial r}\right). \qquad (9\text{--}12)$$

Using Eqs. (9–5) and (9–6), we get the magnitude of the stress at $r = a$,

$$\tau_{r\theta}\big|_{r=a} = \frac{3}{2} \frac{\mu V_0}{a} \sin \theta. \tag{9–13}$$

The drag on the sphere can be computed by means of Eq. (8–22);

$$D = + \int_0^\pi \tau_{r\theta} \sin \theta \, dS - \int_0^\pi p \cos \theta \, dS, \tag{9–14}$$

where $dS = 2\pi a^2 \sin \theta \, d\theta$. The computation gives

$$D = 4\pi a\mu V_0 + 2\pi a\mu V_0 = 6\pi a\mu V_0. \tag{9–15}$$

We see that the frictional drag equals twice the pressure drag. It is interesting also that the drag is proportional to the first power of the velocity, a fact peculiar to this very slow motion. Writing the drag in terms of the coefficient defined by Eq. (8–27), we have

$$D = C_D\rho \frac{V_0^2}{2} A = \frac{24}{\mathbf{R}} \rho \frac{V_0^2}{2} \pi a^2, \tag{9–16}$$

or

$$C_D = \frac{24}{\mathbf{R}}, \tag{9–17}$$

where

$$\mathbf{R} = \frac{V_0 2a\rho}{\mu}. \tag{9–18}$$

A comparison of Stokes' theoretical drag coefficient with experiment, given in Fig. 9–5 shows Eq. (9–17) to hold if $\mathbf{R} < 1$. Above this value, inertia can no longer be neglected and Stokes' solution does not apply.

If the liquid is at rest and the sphere is moving, the solution gives the same pressure and shear distributions over the sphere surface, and hence the same drag. The equation for the streamlines in this case is given by constant values of

$$\psi = -\frac{1}{2} V_0 r^2 \sin^2 \theta \left(-\frac{3}{2} \frac{a}{r} + \frac{1}{2} \frac{a^3}{r^3} \right). \tag{9–19}$$

Streamlines and velocity distributions for this case are shown in Fig. 8–3 and are reproduced as Fig. 9–3 for direct comparison with the stationary case given in Fig. 9–2.

For a solid sphere falling under gravity we can calculate the steady state or terminal velocity by noting that

$$\text{weight of sphere} = \text{drag} + \text{buoyant force}, \tag{9–20}$$

or

$$\tfrac{4}{3}\pi a^3 \gamma_s = 6\pi a\mu V_0 + \tfrac{4}{3}\pi a^3 \gamma, \tag{9–21}$$

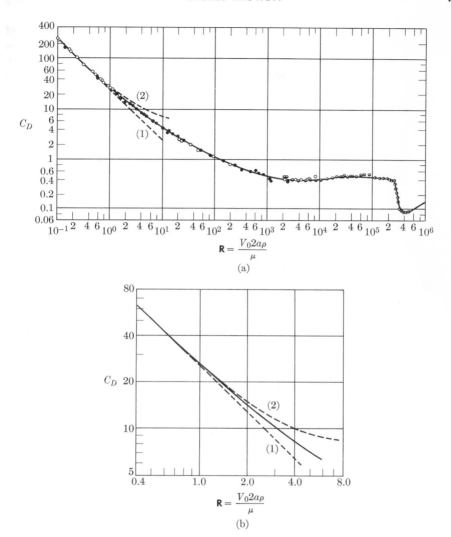

FIG. 9–5. Drag coefficient for spheres versus Reynolds number. Solid line: curve for experimental data. Dashed lines: (1) Stokes' equation, (9–17); (2) Oseen's equation, (9–23).

where

$$\gamma_s = \text{specific weight of solid,}$$
$$\gamma = \text{specific weight of fluid.}$$

Then the terminal fall velocity is

$$V_{0_{\text{term}}} = \frac{2}{9} \frac{a^2}{\mu} (\gamma_s - \gamma). \qquad (9\text{–}22)$$

An extension of Stokes' solution given by Oseen takes the inertia terms of the Navier-Stokes equations partially into account. The resulting streamlines for the moving sphere are no longer the same in front of and behind the solid. The velocities behind are higher than those ahead of the sphere and some fluid follows the sphere, a fact observed experimentally at large Reynolds numbers. The refined drag coefficient is

$$C_D = \frac{24}{\mathbf{R}}\left(1 + \frac{3}{16}\,\mathbf{R}\right),$$ (9–23)

which, as seen in Fig. 9–5, holds up to $\mathbf{R} \approx 2$.

9–2 VISCOUS EFFECTS ON IMPACT TUBES

Henri Pitot's stagnation-pressure method for measuring velocity assumes a negligible effect of viscosity. As we saw in Section 6–7.3, Bernoulli's equation relates the approach velocity to the stagnation pressure on an impact probe. For an incompressible fluid with approach velocity V_0, we have

$$\frac{V_0^2}{2g} + \frac{p_0}{\gamma} + h_0 = \frac{p_{\text{stg}}}{\gamma} + h_{\text{stg}}.$$ (9–24)

For a gravity field, this can be written

$$\frac{V_0^2}{2g} = \frac{(p_{\text{stg}} - p_0)_d}{\gamma},$$ (9–25)

where the hydrostatic pressure component cancels the elevation term. Dropping the subscript d we define a pressure coefficient as

$$C_p = \frac{p_{\text{stg}} - p_0}{\rho V_0^2/2},$$ (9–26)

and we see from Eq. (9–25) that $C_p = 1.0$ if viscous effects are absent.

The stagnation pressure is the maximum pressure that will be registered on an obstacle in the flow. As we have seen in the previous section, when the Reynolds number is very low, corresponding to relatively large viscous effects, the maximum pressure on a sphere is not independent of viscosity (Eq. 9–10). The stagnation pressure on impact tubes also depends on viscosity if the Reynolds number is very small.

The actual readings from impact tubes will depend both on the viscous effect and on the size of the pressure-sensing hole in the probe tip. Unless the hole diameter is a small fraction of the tip diameter, the average pressure over the hole cannot be expected to equal the point-pressure intensity. The average,

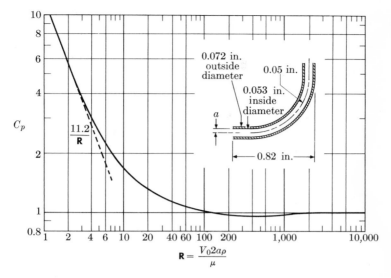

FIG. 9–6.   Pressure coefficient versus Reynolds number for impact tube [3].

in turn, will depend on the probe tip shape, i.e., flat faced, hemispherical, etc. In general, then, the pressure coefficient will be a function of the Reynolds number and tube geometry, or

$$C_p = C_p(\mathbf{R}, \text{ geometry}). \tag{9–27}$$

The viscous effects in the readings from impact tubes of the simple cylindrical type shown in Fig. 6–12 have been investigated experimentally. The results of one set of experiments [3] are given in Fig. 9–6. There the pressure coefficient C_p is plotted against the Reynolds number based on the diameter $2a$ of the cylindrical portion of the probe:

$$\mathbf{R} = \frac{V_0 2a\rho}{\mu}, \tag{9–28}$$

It can be seen that for $\mathbf{R} > 2000$, C_p is essentially unity. Any viscous effects are confined to a thin boundary layer which has a negligible effect on the stagnation pressure. For $100 < \mathbf{R} < 2000$, viscosity causes C_p to fall below unity. For $\mathbf{R} < 100$, C_p rises rapidly to the asymptotic value

$$C_p \approx \frac{11.2}{\mathbf{R}}, \tag{9–29}$$

when $\mathbf{R} < 1$. It will be noted that the upstream stagnation-point pressure on a sphere in Stokes' motion (Eq. 9–10) will give $C_p = 6/\mathbf{R}$.

9–3 FLOW THROUGH POROUS MEDIA

9–3.1 Creeping-motion equations for porous media

Seepage at low Reynolds numbers through various types of porous media constitutes an important class of hydrodynamic problems. The more common examples include flow of water, oil, and other fluids through filter beds, surface soil, and porous rock. In certain manufacturing processes, such as the dewatering of paper sheets during formation on a Fourdrinier machine, the principles of porous media flow are important.

In Chapter 8 we showed that for creeping motion of liquids where inertia is negligible, the Navier-Stokes equations could be reduced to the form of Eq. (8–1a), namely,

$$\nabla(p + \gamma h) = \mu \nabla^2 \mathbf{q}.$$

Taking the divergence of both sides and noting that

$$\nabla^2 \mathbf{q} = \nabla(\nabla \cdot \mathbf{q}) - \nabla \times (\nabla \times \mathbf{q}),$$

we have

$$\nabla^2(p + \gamma h) = \mu \nabla \cdot \nabla^2 \mathbf{q} = \mu\{\nabla^2(\nabla \cdot \mathbf{q}) - \nabla \cdot [\nabla \times (\nabla \times \mathbf{q})]\}. \quad (9\text{–}30)$$

Since $\nabla \cdot [\nabla \times (\nabla \times \mathbf{q})] = 0$* and continuity makes $\nabla \cdot \mathbf{q} = 0$, we have

$$\nabla^2(p + \gamma h) = 0. \quad (9\text{–}31)$$

Equation (9–31) is Laplace's equation, and its solution for particular boundary conditions yields the spatial distribution of $(p + \gamma h)$. In Section 6–6 we noted that Laplace's equation was obtained for incompressible fluids in irrotational motion, and that the function satisfying Laplace's equation was the velocity potential. More generally, we shall see that for certain flows of viscous liquids, the quantity $(p + \gamma h)$ will serve as a velocity potential. Equation (9–31) merely requires for each direction of differentiation that, when we take the gradient $\nabla[\nabla(p + \gamma h)]$, either $(p + \gamma h)$ or $\nabla(p + \gamma h)$ be constant. We have seen cases where these conditions hold in the laminar-flow examples of Section 6–5. For these cases, inertia effects were zero and solutions of Eqs. (6–31a) and (6–38) showed the velocities and flow rate proportional to the gradient of $(p + \gamma h)$. For example, from Eq. (6–41), for flow through a tube in the z-direction, the mean velocity is

$$V_z = -\frac{\text{const}}{\mu} \frac{\partial(p + \gamma h)}{\partial z}, \quad (9\text{–}32)$$

where the constant of proportionality depends on the flow-passage geometry. Equations (6–31a) and (6–38) are Eq. (8–1a) specialized for particular cases.

---

* The divergence of the curl of a continuous vector function of position is identically zero.

For porous-media flow where we are interested in the average discharge velocities rather than the local velocities in the pores, Laplace's equation is also applicable. To show this, we begin by noting that this is a case of laminar flow through small irregular pore passages instead of through uniform ducts. Then, by analogy to the flow through tubes, for a liquid (or of a gas under small pressure differentials so its density does not vary), we write

$$V_z = -\frac{k_0}{\mu}\frac{\partial(p + \gamma h)}{\partial z}.$$
(9–33)

In Eq. (9–33), V_z is the *superficial* or *discharge velocity* defined as the local flow rate averaged over a finite area of the porous medium. Thus, for a given cross-sectional area ΔA of porous material through which ΔQ is flowing, the superficial or discharge velocity is

$$V_z = \frac{\Delta Q}{\Delta A}.$$
(9–34)

For ΔA finite but of the order of several pore channels, V_z approaches a "local-average" value. When a local-average velocity is used in the equations, the physical system is replaced by a mathematical continuum. Sometimes use is made of a *seepage velocity* through the pores defined as

$$\text{seepage velocity} = \frac{\Delta Q}{m\,\Delta A}.$$
(9–35)

where m, the porosity, is the ratio of the volume of the voids to the total volume of the porous medium and the voids. The factor k_0 is called the *intrinsic* or *physical permeability*. It has the dimensions of length squared, and assuming saturated conditions, we note that it depends on the flow geometry, and hence on the type of porous medium and on the density, shape, and arrangement of the pores. It is a constant if the medium is incompressible and isotropic. Typical values of k_0 are given in Table 9–1. A *coefficient of permeability* k is defined by the relation

$$k = \frac{k_0\gamma}{\mu},$$
(9–36)

so that

$$V_z = -k\frac{\partial(p/\gamma + h)}{\partial z}.$$
(9–37)

The coefficient k has the units of a velocity and, again assuming saturated conditions, we find that it depends on the flow geometry (type of medium and pore characteristics) and also on the fluid specific weight and viscosity. It is constant for a given fluid at a fixed temperature if the porous medium is incompressible and isotropic. Typical values for a few media with water and other fluids are given in Table 9–1. Equation (9–37) is known as Darcy's law after the French engineer who published an equivalent relation based on experiments in connection with water supplies for the fountains of the city of Dijon [4].

TABLE 9–1

TYPICAL VALUES OF PERMEABILITY FOR
INCOMPRESSIBLE POROUS MEDIA

Medium	Physical permeability k_0, ft^2	Coefficient of permeability		
		Fluid	Fluid temperature	k, ft/sec
Gravel	10^{-6} to 10^{-8}	Water	60°F	3 to 3×10^{-2}
Sand	10^{-8} to 10^{-11}	Water	60°F	3×10^{-2} to 3×10^{-5}
Silt	10^{-11} to 10^{-15}	Water	60°F	3×10^{-5} to 3×10^{-9}
Randomly packed uniform spheres (d = diameter)	$(6.15 \times 10^{-4})\, d^2$			
Randomly packed uniform spheres ($d = \frac{1}{8}$ in.)	6.68×10^{-8}	Water	60°F	0.196
		Atmospheric air	60°F	0.0134
		Glycerin	68°F	0.00029
		Linseed oil	68°F	0.00427

If we define an *isotropic* medium as one having the same permeability in all flow directions, we can write

$$V_x = -\frac{k_0}{\mu}\frac{\partial(p + \gamma h)}{\partial x}, \qquad V_y = -\frac{k_0}{\mu}\frac{\partial(p + \gamma h)}{\partial y},$$

$$V_z = -\frac{k_0}{\mu}\frac{\partial(p + \gamma h)}{\partial z}, \qquad (9\text{--}38)$$

or

$$\mathbf{V} = -\frac{k_0}{\mu}\nabla(p + \gamma h), \qquad (9\text{--}38a)$$

where $\mathbf{V}$ is the vector superficial velocity. Taking the divergence of Eq. (9–38a) and using

$$\nabla \cdot \mathbf{V} = \frac{\partial V_x}{\partial x} + \frac{\partial V_y}{\partial y} + \frac{\partial V_z}{\partial z} = 0, \qquad (9\text{--}39)$$

results in

$$- \frac{k_0}{\mu} \nabla^2 (p + \gamma h) = 0. \tag{9-40}$$

Thus the seepage problem for isotropic materials can be reduced to the solution of Laplace's equation for particular boundary conditions. If the distribution of $(p + \gamma h)$ is known, the superficial velocities can be obtained from the gradient using Darcy's law in the form of Eq. (9–38a). Any of the methods for solving Laplace's equation may be chosen.

In many cases, the porous medium is *anisotropic*, and the permeability depends on the flow direction. In such cases, Darcy's law still holds, giving the expressions

$$V_x = - \frac{k_{0_x}}{\mu} \frac{\partial(p + \gamma h)}{\partial x}, \qquad V_y = - \frac{k_{0_y}}{\mu} \frac{\partial(p + \gamma h)}{\partial y},$$

$$V_z = - \frac{k_{0_z}}{\mu} \frac{\partial(p + \gamma h)}{\partial z}. \tag{9-41}$$

Adding and taking the divergence, we have

$$\nabla \cdot \mathbf{V} = 0 = - \frac{k_{0_x}}{\mu} \frac{\partial^2(p + \gamma h)}{\partial x^2} - \frac{k_{0_y}}{\mu} \frac{\partial^2(p + \gamma h)}{\partial y^2} - \frac{k_{0_z}}{\mu} \frac{\partial^2(p + \gamma h)}{\partial z^2}. \tag{9-42}$$

Hence, in general, Laplace's equation does not result from continuity as for the isotropic materials. However, if k_{0_x}, k_{0_y}, k_{0_z} are each constant, and we use

$$X^2 = x^2 k_{0_y}/k_{0_x}, \qquad Z^2 = z^2 k_{0_y}/k_{0_z}, \tag{9-43}$$

Eq. (9–42) can be written as

$$\frac{\partial^2(p + \gamma h)}{\partial X^2} + \frac{\partial^2(p + \gamma h)}{\partial y^2} + \frac{\partial^2(p + \gamma h)}{\partial Z^2} = 0. \tag{9-44}$$

Thus, by the geometric transformations of Eq. (9–43), we obtain Laplace's equation and can consider the true physical case to be represented by a fictitious isotropic case in the transformed coordinates. The use of this device in connection with graphical solutions for two-dimensional anisotropic materials will be described in a later paragraph.

The Darcy law linear relationship between velocity and gradient of $(p + \gamma h)$ used in the seepage-flow equations of the previous paragraphs holds only so long as the flow remains laminar and inertia effects remain unimportant. For sand beds a Reynolds number may be defined by

$$\mathbf{R} = \frac{V d_{50}}{\nu}, \tag{9-45}$$

where we have

$$V = \text{superficial seepage velocity},$$
$$\nu = \text{fluid kinematic viscosity},$$
$$d_{50} = \text{grain size for which 50\% of the sample by weight is smaller.}$$

Experiments with sand beds give **R** ≈ 10 as an upper limit for the Darcy law. This Reynolds number may be exceeded in rock aquifers and in the vicinity of well casings. The evidence with sand beds is that beyond **R** = 10, departure from Darcy's law occurs while the flow is still laminar and is due to fluid accelerations causing inertia effects. The transition from laminar into the fluctuating flow condition which is called turbulence is gradual, somewhere between **R** = 60 and **R** = 600. The flow resistance seems to become independent of Reynolds number at about **R** = 1000.

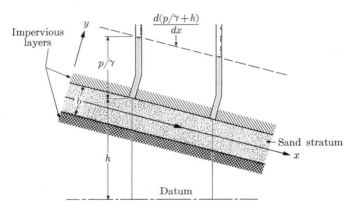

FIG. 9–7. Sand stratum of constant thickness.

9–3.2 Examples for isotropic materials

One-dimensional steady flow between impermeable layers. Many seepage problems can be reduced to one-dimensional or two-dimensional cases. As a one-dimensional example, consider the steady flow of ground water through a sand stratum of uniform thickness between impermeable layers of hardpan. Assume that the resultant gradient and all flow is in the direction of incline of the stratum (Fig. 9–7). Equation (9–31) becomes simply

$$\frac{d^2(p + \gamma h)}{dx^2} = 0, \tag{9–46}$$

which has as its general solution

$$(p + \gamma h) = Ax + B. \tag{9–47}$$

Equation (9–38) gives

$$\frac{d(p + \gamma h)}{dx} = - \frac{V_x \mu}{k_0},$$

or since γ is a constant,

$$\frac{d(p/\gamma + h)}{dx} = - \frac{q_0 \mu}{b k_0 \gamma}, \tag{9–48}$$

where q_0 is the discharge per unit width measured normal to the flow direction and b is the stratum thickness. Then

$$\left(\frac{p}{\gamma} + h \right) = - \frac{q_0 \mu}{b k_0 \gamma} x + \text{const}, \tag{9–49}$$

and $(p/\gamma + h)$ decreases linearly with increasing x as shown in Fig. 9–7.

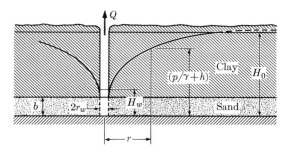

FIG. 9–8.   Radial flow due to pumping from a stratum of constant thickness.

Two-dimensional radial flow into a well. A simple two-dimensional example is the radial flow from an extensive sand stratum of uniform thickness into a well as shown in Fig. 9–8. Assume that there are impervious layers above and below the sand. Before pumping begins, $(p/\gamma + h)$ is constant throughout the stratum, and the water level in the well is at the elevation H_0. After a period of pumping, the well level will drop to H_w and $(p/\gamma + h)$ will fall toward the well. When we introduce plane polar coordinates for the resulting horizontal radial flow, the Darcy equation reduces to

$$V_{\text{rad}} = - \frac{k_0}{\mu} \frac{d(p + \gamma h)}{dr} = - \frac{k_0 \gamma}{\mu} \frac{d(p/\gamma + h)}{dr}. \tag{9–50}$$

At any distance r from the well, the discharge flowing toward the well is

$$Q = 2\pi r b \frac{k_0 \gamma}{\mu} \frac{d(p/\gamma + h)}{dr}. \tag{9–51}$$

If we assume that Q is constant, integrating will give the distribution of $(p/\gamma + h)$ as follows:

$$\int_{r_w}^r d\left(\frac{p}{\gamma} + h\right) = \frac{Q\mu}{2\pi b k_0 \gamma} \int_{r_w}^r \frac{dr}{r}, \tag{9–52}$$

where r_w is the radius of the cylindrical well, or

$$\left(\frac{p}{\gamma} + h\right)_r - \left(\frac{p}{\gamma} + h\right)_{r_w} = \frac{Q\mu}{2\pi b k_0 \gamma} \ln \frac{r}{r_w}. \tag{9–53}$$

This result is plotted in Fig. 9–8. There it is seen that the potential $(p/\gamma + h)$ rises logarithmically, with the greatest gradient close to the well. Also, the solution gives an ever increasing value of $(p/\gamma + h)$ with r, whereas the physical limit is H_0. Actually, the steady-flow (constant Q) assumption is only an approximation to what can be realized for an infinite aquifer under a finite head. The solution given by Eq. (9–53) is useful only near the well. [In the above it should be noted, of course, that $(p/\gamma + h)$ equals the piezometric head relative to the bottom of the pervious stratum. Thus, at $r = r_w$, $(p/\gamma + h) = H_w$.]

9–3.3   Graphical method for two-dimensional flows

Isotropic materials. The two-dimensional hydrodynamic theory provides a basis for a useful graphical solution for incompressible fluid flows described by a velocity potential such as we employ here. As discussed in Section 6–6, it is the nature of such flows that the velocities computed from the gradient of the potential function will result in streamlines that are perpendicular to lines of constant potential. Beginning with known boundary streamlines and with known boundary potential lines, a *flow net* consisting of a grid of orthogonally intersecting velocity and potential lines can be systematically constructed. A basic hydrodynamic theorem of potential flow is that every set of boundary conditions has its own *unique* flow net. Hence the graphical solution obtained is *the* solution. The procedure for drawing a flow net is described below.

For any two-dimensional case which does not involve separation of the flow from the boundaries, the boundaries represent the limiting streamlines. Then all equipotential lines must meet both boundaries and all streamlines orthogonally. Sketch a system of streamlines, using the boundaries as guides. Space the streamlines so the incremental flow rate between any pair is the same. This can be done by starting with equal spacing in a region where the velocity is known to be constant. Then sketch in orthogonal equipotential lines, beginning in the regions of greatest flow curvature (such as at A in the example of seepage under a barrier in Fig. 9–9) and working outward throughout the system. Space the equipotential lines so as to make approximate squares throughout. In the theoretical limit of closely spaced streamlines and potential lines, all the quadrangles would become squares with the spacings Δn and Δs equal. To obtain reasonable accuracy in regions of large curvature, it may be neces-

sary to subdivide the flow space into
smaller units by drawing additional
streamlines and potential lines.   As
the potential lines are drawn, it will be
necessary to make adjustments of the
sketched streamlines to bring both
into a systematic orthogonal system.
Check the result by drawing diagonal
lines through all squares in both directions. These diagonal lines must also
form an orthogonal system. It must be
remembered that the result is for the
flow whose velocity field would be
described by the gradient of a poten-

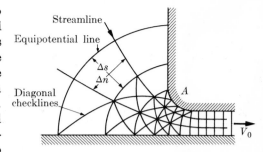

FIG. 9–9. Flow-net solution for seepage under a barrier by graphical construction of orthogonal sets of streamlines and equipotential lines.

tial. If the true flow is not described by such a potential, this graphical procedure
will not give the true flow net.

In the flow net, the velocity between streamlines is inversely proportional to
their spacing Δn. The gradient of the potential is inversely proportional to the
spacing Δs of the equipotential lines. Thus at any point the velocity v relative
to v_0 at another point is

$$v = v_0 \frac{\Delta n_0}{\Delta n} = v_0 \frac{\Delta s_0}{\Delta s} = \frac{\text{incremental discharge}}{\Delta s}. \qquad (9\text{--}54)$$

Since in the limit the spacings of streamlines and potential lines are the same
(i.e., for squares in the limit), the velocity is proportional to the potential
gradient as required of flows satisfying Laplace's equation. Applied to seepage
through isotropic materials, the flow net permits numerical evaluation by means
of Darcy's law of flow rates, required head drop or necessary permeability
depending on which data are known and which desired.

Anisotropic materials. Where the porous medium is *anisotropic* so that k_0 has
different values in different directions, we must make use of the transformed
Laplace equation, Eq. (9–44). For two-dimensional flow to which flow nets
apply, we would use only the first two terms of Eq. (9–44). To obtain a flow
net, it is necessary to first distort the given boundary geometry by the transformation of Eq. (9–43). The flow net is then constructed for this *transformed
section* and the solution satisfies Eq. (9–44). Finally, the completed flow net
and distorted boundary geometry are restored to original scale to give the desired
streamlines.   The resulting diagram will not have orthogonal potential and
streamlines, since the final solution is not a solution to Laplace's equation.

As an example, a porous bed is layered with materials of three different
permeabilities. The flow in directions normal to and parallel to the layers will
be different for identical drops in potential.   In such cases, the average permeability coefficients can be computed in analogy to series and parallel electric

circuits. Let k_{0_1}, k_{0_2}, and k_{0_3} be the coefficients of three layers having thicknesses b_1, b_2, and b_3. Then, parallel to the layers, we will have

$$k_{0_I} = \frac{k_{0_1}b_1 + k_{0_2}b_2 + k_{0_3}b_3}{b_1 + b_2 + b_3}, \tag{9-55}$$

and normal to the layers,

$$k_{0_{II}} = \frac{b_1 + b_2 + b_3}{b_1/k_{0_1} + b_2/k_{0_2} + b_3/k_{0_3}}. \tag{9-56}$$

This is a case of *transverse anisotropy*, a condition which can be compensated for by reducing the boundary dimensions in the direction parallel to the stratification in the ratio $(k_{0_{II}}/k_{0_I})^{1/2}$ before constructing the streamline vs. potential line grid. Reconversion of the completed grid will give the desired flow net.

PROBLEMS

9–1. Polystyrene particles (SG = 1.05 referred to water) are observed in a settling tank. Determine the largest diameter that a particle can have whose fall velocity will obey Stokes' solution when the fluid is (a) water at 60°F, (b) air at 60°F and 14.7 psia.

9–2. Compare the terminal settling velocities of the maximum sizes of polystyrene spheres (SG = 1.05) and dust particles (SG = 2.7) that will settle in 60°F water according to Stokes' fall-velocity solution.

9–3. Determine the terminal fall-velocity relation from Oseen's solution for the drag coefficient of a sphere in creeping motion.

9–4. Using the sphere diameter determined in Problem 9–1(b), compare the terminal settling velocities of a polystyrene particle (in air, 60°F, 14.7 psia) as predicted by Stokes' and Oseen's solutions for the drag coefficient.

9–5. Consider a spherical air bubble having a mass of 1.3×10^{-13} slug rising through still water (60°F) according to Stokes' velocity relation. (a) When the air bubble is at the maximum diameter at which it will obey Stokes' equation, determine the pressure in the water. (b) Estimate whether or not the bubble will deform or remain spherical under these conditions, considering the relative magnitude of pressure forces and surface-tension forces.

9–6. A spherical particle 1-in. in diameter whose specific weight is 150 lb/ft³ is dropped into a tank of castor oil. Specific weight of the oil is 58 lb/ft³; dynamic viscosity μ is 3.16×10^{-2} slug/ft-sec. (a) Calculate the terminal velocity of the sphere in the oil. (b) A second sphere is dropped into water at 68°F. What is the ratio of diameters if the flows around the two spheres are dynamically similar when the velocity of the sphere in water equals the terminal velocity of the first sphere in oil?

9–7. Compare the minimum sizes of impact probes that can be used in standard sea-level air and in 68°F water if $C_p = 1.0 \pm 0.05$ and the air and water velocities are the same.

9–8. Determine the minimum diameter of a total pressure (impact) tube that can be used in a 5 ft/sec stream of standard air at sea level if viscous effects are not to affect the pressure measurement.

9–9. Discuss the reasons for the difference in the pressure coefficient C_p as given by Eq. (9–29) for an impact tube and the value of C_p computed from Stokes' equation (9–10) for the stagnation pressure on a sphere.

9–10. Consider two wells each of radius r_w and spaced at a distance $2x$ center to center. Both wells penetrate the same horizontal sand stratum of constant vertical thickness b. One well takes water from the stratum, the second is used to recharge the stratum with surface water run-off. Assume that the pumping discharge and recharge rates are the same and no other wells are in the vicinity. Draw a flow net showing streamlines and equipotential lines. [*Note:* Streamlines can be drawn very accurately. Potential lines then will have a simple form.]

9–11. Derive an expression for the discharge-recharge flow velocity through the stratum in Problem 9–10, when the flow rate passing from one well to the other is q cfs/ft of vertical thickness.

9–12. Determine the shortest time in which recharge water will reach the discharging well of Problem 9–10, when

$$r_w = 6 \text{ in.}, \qquad x = 100 \text{ ft}, \qquad Q = q \cdot b = 0.2 \text{ cfs}, \qquad b = 10 \text{ ft}.$$
$$\text{Porosity of sand} = \text{fraction of voids} = 0.25.$$

9–13. The physical permeability of the sand in the stratum is $k_0 = 10^{-10} \text{ ft}^2$. Compute the difference in water levels in the two wells for the conditions of Problem 9–12.

9–14. The total seepage rate per foot of width in an isotropic, two-dimensional, confined aquifer may be expressed in terms of the number of flow paths and equipotential drops in the flow net. Prove that this expression is of the form

$$q_0 = \frac{n_f}{n_d} k \, \Delta\!\left(\frac{p}{\gamma} + h\right),$$

where the values involved are

$$q_0 = \text{seepage per unit width normal to flow,}$$
$$n_f = \text{number of flow channels,}$$
$$n_d = \text{number of equipotential drops,}$$
$$\Delta\!\left(\frac{p}{\gamma} + h\right) = \text{total head drop through the porous media,}$$
$$k = \text{coefficient of permeability.}$$

9–15. A rectangular concrete wall rests on a stratum of sand of thickness $T = 50$ ft and permeability $k = 5 \times 10^{-8}$ ft/sec. The water depths on either side of the wall are $H_1 = 20$ ft and $H_2 = 5$ ft. The flow net is shown in Fig. 9–10. (a) Using the results of Problem 9–14, determine the quantity of seepage per unit width of the wall. (b) Plot the pressure distribution along the base of the wall.

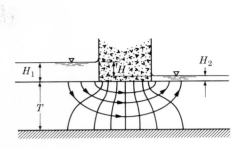

FIGURE 9–10

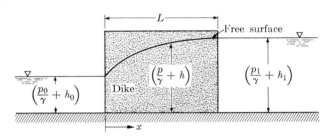

FIGURE 9–11

FIGURE 9–12

9–16. Draw the flow net for the sheet-pile wall shown in Fig. 9–11. Use a scale of 1 in. = 20 ft and four flow channels in the flow net. Using the results of Problem 9–14, determine the seepage rate per unit width of the wall.

9–17. Solutions of problems of flow in unconfined aquifers (see Fig. 9–12) usually involve two assumptions due to Dupuit: (i) The gradient of piezometric head is constant over any vertical section and is equal to $\partial(p/\gamma + h)/\partial x$ at the free surface. (ii) The velocity is horizontal and uniform from top to bottom of any vertical section, and equal to $-k[\partial(p/\gamma + h)/\partial x]$.

For the flow through a dike shown in Fig. 9–12, show that the form of the free surface is given by

$$\frac{2q_0 x}{k} = \left(\frac{p}{\gamma} + h\right)^2 - \left(\frac{p_0}{\gamma} + h_0\right)^2,$$

where q_0 is the discharge per unit length of dike. Determine q_0 in terms of L and the heads at the two boundaries of the dike.

9–18. A ground-water seepage problem involves flow under a sheet-pile wall for the conditions shown in Fig. 9–13. The ground is composed of alternate horizontal layers of silt and sand of thickness 1 ft and 2 ft, respectively. Water temperature is 68°F. (a) Calculate an estimated seepage velocity, assuming straight vertical flow down to sheet-pile depth and back up to the exposed ground surface. (b) A more accurate solution may be obtained graphically by constructing a flow net for a transformed section. Determine the necessary distortion of the transformed section.

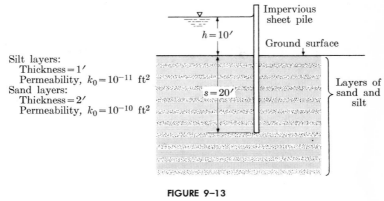

Silt layers:
  Thickness $= 1'$
  Permeability, $k_0 = 10^{-11}$ ft^2
Sand layers:
  Thickness $= 2'$
  Permeability, $k_0 = 10^{-10}$ ft^2

FIGURE 9–13

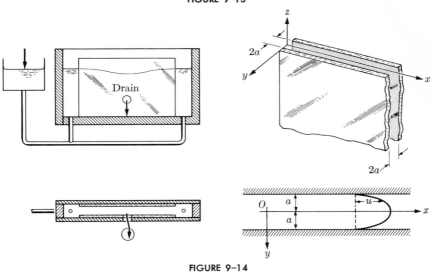

FIGURE 9–14

9–19. A Hele-Shaw apparatus is frequently used as an experimental analogy for the study of seepage in porous media. The apparatus consists of two closely spaced parallel glass plates through which a viscous fluid flows in laminar motion. The x-axis is horizontal and midway between the plates, the z-axis is vertical, and the y-axis is perpendicular to the plates, as shown in Fig. 9–14. The spacing between the plates is $2a$ (in actual cases this is of the order of 1 mm). Since the motion is confined to the xz-plane, the velocity component in the y-direction is zero. Because of the small spacing of the plates, the first and second derivatives of u and w with respect to x and z are negligible in comparison with derivatives of u and w in the y-direction. (a) Assume that the flow is steady and show that the Navier-Stokes equations (6–28) for this type of flow reduce to:

$$\frac{\partial}{\partial x}\left(\frac{p}{\gamma} + h\right) = \frac{\mu}{\gamma}\frac{\partial^2 u}{\partial y^2}, \qquad \frac{\partial}{\partial z}\left(\frac{p}{\gamma} + h\right) = \frac{\mu}{\gamma}\frac{\partial^2 w}{\partial y^2}, \qquad \frac{\partial}{\partial y}\left(\frac{p}{\gamma} + h\right) = 0.$$

(b) By integrating and applying appropriate boundary conditions, obtain the expressions for the local mean velocities $\bar{u}$ and $\bar{w}$ at any point in the xz-plane. (c) Prove that these velocity equations are analogous to the Darcy equations for two-dimensional seepage problems and show how the permeability k is related to the Hele-Shaw flow parameters.

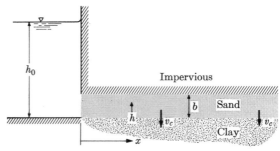

FIGURE 9-15

9-20. Water seeps from a large reservoir into a thin horizontal stratum of sand of thickness b, as shown in Fig. 9-15. Water from the sand stratum leaks into the underlying clay layer at a uniform velocity v_c. Set up a control volume of length dx and height b in the sand stratum and write the continuity condition. Determine the horizontal extent x_L of water penetration into the sand stratum. At $x = x_L$, $u = 0$ (hence $\partial h/\partial x = 0$) and $(p/\gamma + h) = 0$.

REFERENCES

1. STOKES, G. G., "On the Effect of the Internal Friction of Fluids on the Motion of Pendulums," *Trans. Cambridge Phil. Soc.*, 9 (1851).

2. MILNE-THOMPSON, L. M., *Theoretical Hydrodynamics*, The Macmillan Co., New York, 1960, 4th ed. Articles 19.61 through 19.64 are a good reference.

3. HURD, C. W., K. P. CHESKY, and A. H. SHAPIRO, "Influence of Viscous Effects on Impact Tubes," *J. Appl. Mech.*, **20**, 2, 253 (June, 1953).

4. DARCY, H., *Les Fontaines Publiques de la Ville de Dijon*, Paris, 1856.

CHAPTER 10

Laminar Boundary Layers

10–1 INTRODUCTION

Boundary-layer flows may be two-dimensional or three-dimensional as illustrated by the examples in Figs. 10–1. The velocity vectors of two-dimensional layers fall in planes (Fig. 10–1a), which, in turn, are all parallel. The velocity vectors of three-dimensional cases (Figs. 10–1b through 1g) may be coplanar but diverging laterally or, more generally, not coplanar but skewed. Many practical cases are two-dimensional and can be treated by simplified forms of the equations of motion. For example, the Prandtl boundary-layer equations (8–7) and the Kármán integral-momentum equation (8–21) are applicable when the boundary curvature in the flow direction is not great (as might occur near sharp edges). Three-dimensional layers are even more numerous, but similarly simple boundary-layer equations cannot be formulated. As a rule, the full set of Navier-Stokes equations must be used. In this chapter, we will present solutions for the simplest case of two- and three-dimensional layers. We will use the two-dimensional case to illustrate some general properties of boundary layers as well as to give an exact solution of the relations among layer thickness, shear stress, the fluid characteristics, and the flow variables for a case having wide application. We will discuss the flow induced by a rotating disk as an example of a three-dimensional skewed boundary and as a case of practical importance which is encountered frequently.

10–2 TWO-DIMENSIONAL LAMINAR LAYERS ON FLAT SURFACES

10–2.1 General properties

There are many practical applications that can be treated as flow over a flat surface. Moreover, there is a qualitative generality about the boundary-layer properties of this case which makes it worth while to summarize them before proceeding with a detailed analysis.

Let us consider the flow over a smooth flat plate, as illustrated in Fig. 10–2(a). Let x be measured from the leading edge of the plate and let U be the velocity of the mean flow outside the boundary layer.* Experimentally it is found that

---

* For the flat plate, U will equal the approach velocity (usually designated by V_0). For bodies with finite thickness, U will differ from V_0 and be a function of position along the body.

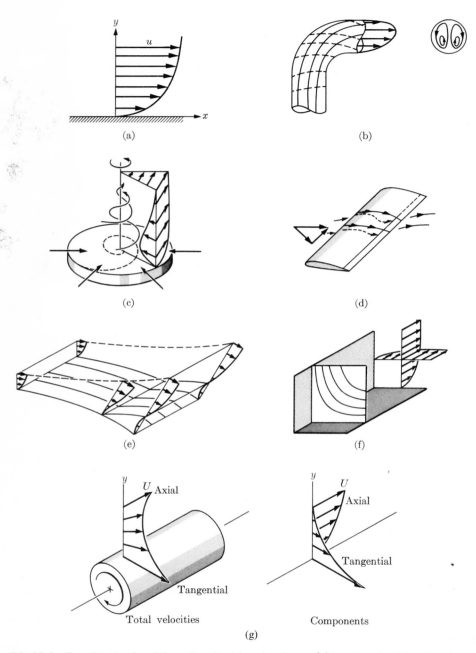

FIG. 10–1. Two-dimensional and three-dimensional boundary layers: (a) two-dimensional boundary layer, velocities relative to walls; (b) secondary flows in curved pipe; (c) rotation above the ground; (d) cross flows on yawed wing; (e) divergent flow; (f) corner flow; (g) axial velocity *U* past rotating cylinders.

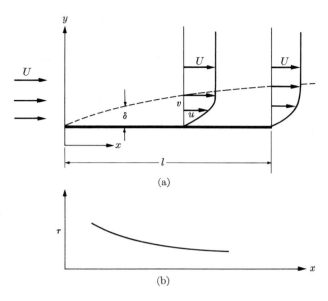

FIG. 10–2. Boundary layer along a flat plate.

the thickness δ depends on the variables U, ρ, μ, and the position x along the plate. The dependence is expressed by:

δ tends to *thicken* as: x increases, μ increases, ρ decreases, U decreases;

δ tends to *thin* as: x decreases, μ decreases, ρ increases, U increases.

The correlation for laminar boundary layers is observed to be

$$\delta \sim \sqrt{\frac{\mu x}{\rho U}} \sim \frac{x}{\mathbf{R}_x^{1/2}} , \tag{10–1}$$

where $\mathbf{R}_x = Ux\rho/\mu =$ a Reynolds number. For a plate of length l, the thickness at the end of the plate is

$$\delta \sim \sqrt{\frac{\mu l}{\rho U}} \sim \frac{l}{\mathbf{R}_l^{1/2}} . \tag{10–2}$$

The proportionality factor for Eqs. (10–1) and (10–2) depends on how U varies along x. The value of this factor for the case of flow with $U =$ const will be derived in the next section.

The Reynolds numbers $\mathbf{R}_x$ and $\mathbf{R}_l$ are especially useful in describing and correlating boundary-layer properties and effects. Another useful form is

$$\mathbf{R}_\delta = \frac{U \, \delta \rho}{\mu} . \tag{10–3}$$

Using Eq. (10–1), we see that

$$\mathbf{R}_\delta \sim \mathbf{R}_x^{1/2}. \tag{10–4}$$

For the two-dimensional boundary layer the shear stress can be expressed as

$$\tau = \mu \left(\frac{\partial v}{\partial x} + \frac{\partial u}{\partial y} \right) \approx \mu \frac{du}{dy}. \tag{10–5}$$

We note that for any fixed set of the variables U, ρ, μ, $\mathbf{R}_x$ varies as x and the thickness δ increases with $\sqrt{x}$. Hence du/dy and the boundary shear stress decrease with increasing x. This trend is shown in Fig. 10–2(b).

If we note that the passage time for a particle to move along the plate is $t \sim x/U$, Eq. (10–1) shows that

$$\delta \sim \sqrt{(\mu/\rho)t}. \tag{10–6}$$

We may apply this to motions starting from rest and conclude that in the starting period the boundary layer increases with the square root of time.

10–2.2   The Blasius solution for laminar flow

For steady laminar flow over a flat plate with $U = \text{const}$, the pressure gradient dp/dx is zero and the relations (8–7) and (8–8) will reduce to the Prandtl boundary-layer equations

$$u \frac{\partial u}{\partial x} + v \frac{\partial u}{\partial y} = \frac{\mu}{\rho} \frac{\partial^2 u}{\partial y^2}, \qquad \frac{\partial u}{\partial x} + \frac{\partial v}{\partial y} = 0, \tag{10–7}$$

where the boundary conditions are

$$
\begin{array}{ll}
\text{at the wall} & \text{beyond the boundary layer} \\
y = 0, & y = \infty, \\
u = 0, & u = U, \\
v = 0. &
\end{array}
\tag{10–8}
$$

A solution to the above, known as the *Blasius solution* after its originator, is obtained by assuming similar profiles along the plate at every x. Blasius assumed that

$$\frac{u}{U} = F\left(\frac{y}{\delta}\right), \tag{10–8a}$$

where $F(y/\delta)$ is the same for all x. Now we make use of the functional relationship of Eq. (10–1) and let

$$\frac{y}{\delta} \sim \frac{y}{x/\sqrt{\mathbf{R}_x}} = \eta. \tag{10–8b}$$

Next we introduce a stream function

$$\psi = -\sqrt{\nu \times U}\, f(\eta). \tag{10–9}$$

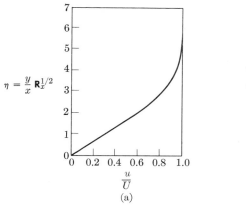

FIG. 10–3.   Velocity distributions on a flat plate. (After Blasius [1].)

Then $u = -\partial\psi/\partial y = Uf'$, $v = \partial\psi/\partial x = \frac{1}{2}\sqrt{\nu U/x}\,(\eta f' - f)$, and the first of Eqs. (10–7) reduces to the ordinary differential equation

$$ff'' + 2f''' = 0, \tag{10–10}$$

where, for $\eta = y/\delta$, we have

$$f = f(\eta), \qquad f' = \frac{\partial f}{\partial \eta} \text{ etc.}, \qquad f''' = \frac{\partial^3 f}{\partial \eta^3}.$$

Blasius [1] obtained the solution of Eq. (10–10) in the form of a power series expanded about $\eta = 0$. The results for longitudinal and lateral velocity distributions are given in Table 10–1 and Fig. 10–3.

TABLE 10–1

SOLUTION FOR STEADY LAMINAR BOUNDARY LAYERS
ON FLAT PLATE WITH ZERO PRESSURE GRADIENT

$\dfrac{y}{x/\sqrt{R_x}}$	$\dfrac{u}{U}$	$\dfrac{v}{U}\sqrt{R_x}$	$\dfrac{\partial u/\partial y}{\sqrt{R_x}\,U/x}$
0	0	0	0.3321
1	0.3298	0.0821	0.3230
2	0.6298	0.3005	0.2668
3	0.8461	0.5708	0.1614
4	0.9555	0.7581	0.0642
5	0.9916	0.8379	0.0159
6	0.9990	0.8572	0.00240
7.2	0.99996	0.8604	0.00013
7.8	1.00000	0.8604	0.00002
8.4	1.00000	0.8604	0.00000

It is seen that the velocity u approaches its asymptotic value U very rapidly. Simultaneously the slope $\partial u/\partial y$ approaches zero. It is interesting to note the finite v-component of the velocity at the outer edge of the boundary layer, a fact we would expect from continuity. Note also that the viscosity μ is assumed to be independent of the integrating variables, which implies isothermal conditions.

From these profile results we can evaluate the boundary-layer thickness as

$$\delta = \frac{5x}{\mathbf{R}_x^{1/2}} \quad \text{at } \frac{u}{U} = 0.992. \tag{10–11}$$

Note that this agrees with the form of Eq. (10–1). If we use Eqs. (8–9) and (8–10), the displacement and momentum thicknesses become

$$\delta^* = \frac{1.73x}{\mathbf{R}_x^{1/2}}, \tag{10–12}$$

$$\theta = \frac{0.664x}{\mathbf{R}_x^{1/2}}. \tag{10–13}$$

The local wall shear is given by the value of the velocity gradient at the wall, which the solution gives as

$$\tau_0(x) = \mu\left(\frac{\partial u}{\partial y}\right)_{y=0} = 0.332\mu\mathbf{R}_x^{1/2}\frac{U}{x} = \frac{0.664}{\mathbf{R}_x^{1/2}}\rho\frac{U^2}{2}.$$

Then the local wall shear-stress coefficient in the equation

$$\tau_0 = c_f\rho\frac{U^2}{2} \tag{10–14}$$

is

$$c_f = \frac{0.664}{\mathbf{R}_x^{1/2}}. \tag{10–15}$$

Using Eq. (10–11) in Eq. (10–15), we have

$$c_f = \frac{3.32}{\mathbf{R}_\delta}. \tag{10–15a}$$

Drag for *one side* of a plate of width b is computed to be

$$D = b\int_0^l \tau_0(x)\,dx = \frac{1.328}{\mathbf{R}_l^{1/2}}\rho\frac{U^2}{2}bl. \tag{10–16}$$

Using the relation

$$D = C_f\rho\frac{U^2}{2}S, \tag{10–17}$$

where $S = $ surface area bl over which τ_0 is integrated in Eq. (10–16), we get

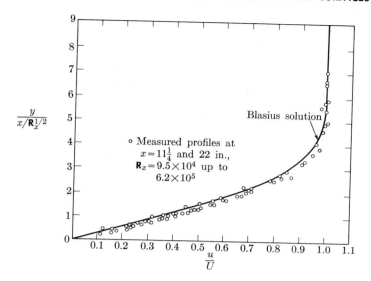

FIG. 10–4. Velocity profiles in a laminar boundary layer on a flat plate [2].

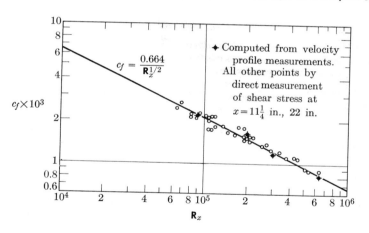

FIG. 10–5. Local shear stress coefficients for laminar boundary layer on flat plate [3].

the coefficient of surface resistance,

$$C_f = \frac{1.328}{\mathbf{R}_l^{1/2}}. \qquad (10\text{–}18)$$

We obtained these theoretical results, assuming the boundary layer to be definitely established at each x. Experiments show that very near the leading edge, there will be a zone in which the transition from zero to finite thickness takes place, and the theoretical results will not hold exactly because δ is not small compared to x. This transition effect on velocity profiles and local shear stress becomes noticeable for $\mathbf{R}_x < 2 \times 10^4$. Measurements of velocity profiles for larger Reynolds numbers are compared with the theory in Fig. 10–4 [2]. The agreement is good. Note also the accuracy with which Eq. (10–11) gives the boundary-layer thickness. Local shear-stress coefficients may be determined directly by measuring the force on a small isolated area of the wall surface, or they may be computed from the measured velocity gradients near the wall using Eq. (10–5). Figure 10–5 [3] compares local shear-stress coefficients from Eq. (10–15) with some experimental values obtained by these two methods at large Reynolds numbers. It can be seen that the theoretical velocity and local shear stress agree very well with experiment making them "exact" in both the mathematical and the physical sense.

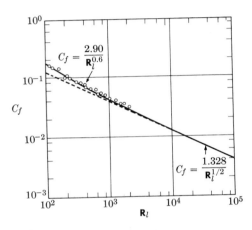

FIG. 10–6. Coefficient of total resistance for laminar boundary layer on a flat plate. Note departures from Blasius' solution for $\mathbf{R}_l < 10^4$ [4].

Since the total drag is the integral of the local shear stress from the very leading edge of the plate, the theoretical coefficient of Eq. (10–18) will always be in some error. This error becomes negligibly small for Reynolds number $\mathbf{R}_l > 10^4$, as shown by comparison of Eq. (10–18) with experiment in Fig. 10–6 [4]. The flat plate drag at very low Reynolds numbers is discussed further in Chapter 15.

10–3 EFFECTS OF BOUNDARY CURVATURE IN TWO-DIMENSIONAL BOUNDARY LAYERS

Boundary curvature in the flow direction can result in nonzero pressure gradients both parallel and normal to the wall. However, unless the curvature is very great and the boundary layer very thick, the normal gradient $\partial p / \partial y$ generally will be of secondary influence. In most cases, therefore, the pressure is treated as a constant across the layer even for curved boundaries. On the

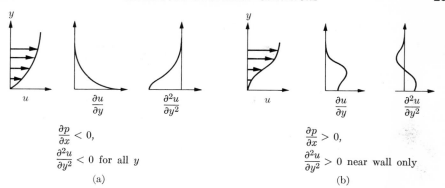

$$\frac{\partial p}{\partial x} < 0,$$

$$\frac{\partial^2 u}{\partial y^2} < 0 \text{ for all } y$$

(a)

$$\frac{\partial p}{\partial x} > 0,$$

$$\frac{\partial^2 u}{\partial y^2} > 0 \text{ near wall only}$$

(b)

FIG. 10–7.   Effect of pressure gradient on boundary-layer velocity profiles.

other hand, even very small gradients in the direction of motion may modify the entire boundary-layer flow. The role of $\partial p/\partial x$ can be seen from Prandtl's boundary-layer equations (Eqs. 8–7) using a qualitative argument described below.

10–3.1   The role of pressure gradients

From Eqs. (8–7), we have at $y = 0$ where $u = v = 0$,

$$\mu \left(\frac{\partial^2 u}{\partial y^2}\right)_{y=0} = \frac{dp}{dx}, \tag{10–19}$$

so that the curvature of the velocity profile in the neighborhood of the wall depends only on the pressure gradient. Let us consider $dp/dx < 0$ (favorable pressure gradient); then we have

$$\left(\frac{\partial^2 u}{\partial y^2}\right) < 0 \qquad \text{at } y = 0,$$

and since the velocity profile is monotonic,

$$\left(\frac{\partial^2 u}{\partial y^2}\right) < 0 \qquad \text{for all } y,$$

as indicated in Fig. 10–7(a). If a boundary flow moves into a region where $dp/dx > 0$ (adverse pressure gradient), we have

$$\left(\frac{\partial^2 u}{\partial y^2}\right) > 0 \qquad \text{at } y = 0.$$

However, near the outer edge of the boundary layer, we also have in this case

$$\left(\frac{\partial^2 u}{\partial y^2}\right) < 0 \qquad \text{at } y \approx \delta.$$

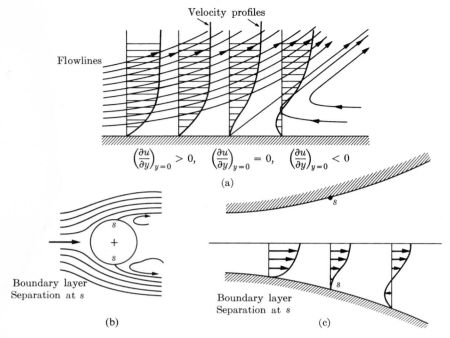

FIG. 10–8. Schematic representation of boundary-layer separation in an adverse pressure gradient: (a) on flat surface; (b) on a cylinder; (c) on diffuser walls.

Therefore, whenever $dp/dx > 0$, an inflection point will develop in the velocity profile as illustrated in Fig. 10–7(b).

With an adverse pressure gradient, $dp/dx > 0$, the fluid near the boundary will be retarded continually and may have its velocity reversed. Reversal will occur when $\partial u/\partial y = 0$ at $y = 0$. Continuity will then require the boundary-layer fluid to be deflected away from the wall. This is called boundary-layer *separation* and is shown schematically in Fig. 10–8(a). This kind of separation occurs whenever the adverse pressure gradient is severe or exists over a long distance. Two common examples of separation are on the rear portions of immersed bodies and on the walls of diffusing ducts as sketched in Fig. 10–8(b) and (c).

10–3.2 Example for nonzero pressure gradient

The pressure gradient dp/dx is superposed on the boundary layer by the main flow outside the boundary layer. With viscous effects negligible in the outside flow, the pressure gradient depends on the main flow velocity as given by Eq. (8–17). Thus, for steady flow,

$$\rho U \frac{\partial U}{\partial x} = -\frac{\partial p}{\partial x}, \qquad (10\text{--}20)$$

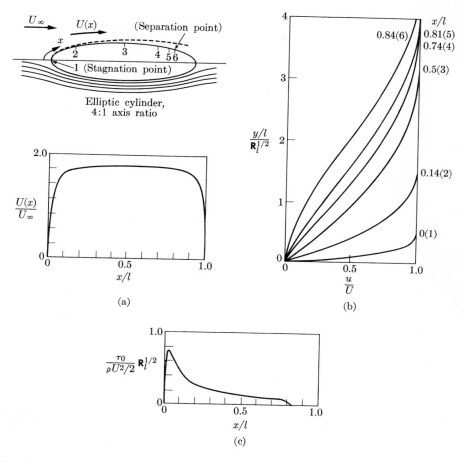

FIG. 10–9. Laminar boundary layer computed for elliptic cylinder of axis ratio 4 : 1. (a) Potential-flow streamwise velocity; (b) boundary-layer profiles up to separation point; (c) boundary shear stress [5].

where $U = U(x)$ at the outer edge of the boundary layer. Therefore, with $U(x)$ given, boundary-layer solutions have been obtained for a variety of cases, either by solving the Prandtl equations (8–7), or by using the Kármán integral-momentum equation (8–21). The common methods which use the latter are approximate. Each employs an assumed velocity-distribution function, $u/U = f(y/\delta(x))$, which satisfies all the necessary physical boundary conditions and allows profiles with inflections.

Computations with nonzero gradients are beyond the scope of this monograph. However, the results of an approximate method for the steady laminar boundary layer around an elliptic cylinder in a stream of velocity U_∞ are given in Fig. 10–9 [5]. Figure 10–9(a) shows the cross section of the 4:1 ellipse and the distribution of the streamwise velocity $U(x)$ at the boundary-layer edge.

In this example, $U(x)$ is assumed to be the frictionless potential-flow value. Figure 10–9(b) shows the computed dimensionless velocity profiles from $x = 0$ at the stagnation point up to the separation point. Note how an inflection develops as distance x/l increases. Separation is predicted at the location where $(du/dy)_{y=0} = 0$. Figure 10–9(c) shows the boundary shear which goes to zero at the separation point.

10–4 THREE-DIMENSIONAL BOUNDARY LAYERS

Several cases of flow around bodies and through ducts where three-dimensionality exists are illustrated in Fig. 10–1. The most pronounced cases involve secondary flows with skewed boundary layers. A secondary flow with skewing results whenever a pressure gradient exists which has a component lateral to the direction of the main, or outer, relative flow. As an example, in the flow around a bend, the main stream tends to be in equilibrium with the centrifugal effect, but the retarded fluid next to the boundary is not. High-velocity fluid moves out with a compensating circulation back along the walls and two secondary spirals result, as shown in Fig. 10–1(b). A similar explanation applies to the cyclonic inflow of a fluid rotating near the ground (Fig. 10–1c). The retarded boundary layer will not be in equilibrium with the radial pressure gradient superposed by the rotating main flow. Radial inflow results at the ground with upflow at the axis to preserve continuity. For flow over an aircraft wing (Fig. 10–1d), the pressure difference between the under and upper surfaces causes a "leakage" at the wing tip and lateral flows in the boundary layers. For the convergent and divergent cases (Fig. 10–1e), three-dimensionality is initially merely a divergence effect. However, if flow separation occurs, the resulting unstable condition gives rise to asymmetry and to the development of secondary motions. For a nonuniform laminar boundary layer in a corner (Fig. 10–1f),

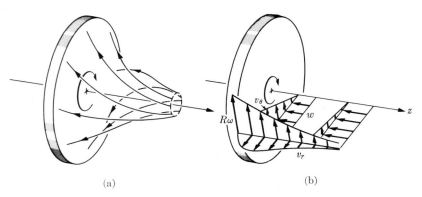

(a)   (b)

FIG. 10–10. Skewed boundary-layer flow near a rotating disk: (a) streamlines; (b) velocity components.

the velocity vectors diverge but do not form secondary motions unless separation occurs. For fully developed laminar flow in a duct, all the velocity vectors near the duct corners remain parallel, even though the velocity gradients and wall shear vary markedly around the duct periphery.

Skewed boundary layers and secondary flows appear on rotating bodies due to the shear at the moving wall and the centrifugal forces imparted to the rotating boundary layer. Flow past the rotating cylinder (Fig. 10–1g) corresponds to a spinning projectile. A different case is presented by the rotating disk (Fig. 10–10) which is selected for analysis in the following paragraphs.

10–4.1   Skewed boundary layers on a rotating disk

The flow in the neighborhood of a disk rotating in a large body of fluid is a case of a three-dimensional skewed boundary with axial symmetry. Fluid very near the disk is given a rotary motion due to friction and then is thrown out radially by centrifugal action. Continuity is satisfied by axial flow inward toward the disk replacing the radially ejected fluid. Figure 10–10 shows a schematic picture of the flow.

For the two-dimensional laminar boundary layer, we noted that $\delta \sim \sqrt{\nu t}$ (Eq. 10–6). For the rotating disk the time characteristic of the motion is $1/\omega$ and $\delta \sim \sqrt{\nu/\omega}$, which can be shown as follows [6].

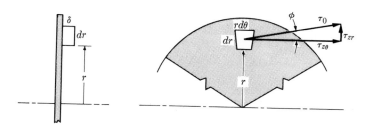

FIG. 10–11.   Definition diagram for fluid element on disk surface.

Consider an element clinging to the disk (Fig. 10–11). The total centrifugal force on the element is approximately $F_c = \rho r \omega^2 \, \delta \, dr \, r \, d\theta$, which is resisted by the radial component of the shear force, or

$$\tau_{zr} \, dr \, r \, d\theta = \tau_0 \sin \phi \, dr \, r \, d\theta = \rho r \omega^2 \, \delta \, dr \, r \, d\theta, \qquad (10\text{–}21)$$

also

$$\tau_{z\theta} = \tau_0 \cos \phi \sim \mu r \omega / \delta, \qquad (10\text{–}22)$$

so that on eliminating τ_0, we have

$$\delta \sim \sqrt{\nu/\omega} \, \sqrt{\tan \phi}. \qquad (10\text{–}23)$$

Experimentally, it is known that ϕ is independent of radius, and we have merely

$$\delta \sim \sqrt{\nu/\omega}. \tag{10-24}$$

Thus the boundary-layer thickness is a constant over the radius.

Using the above relations, we find that the moment or torque due to friction on a disk of radius R can be expressed as

$$dM = \tau_{z\theta} r 2\pi r \, dr \sim \rho r \omega \sqrt{\nu\omega} \, r^2 \, dr, \tag{10-25}$$

or the total moment up to $r = R$ as

$$M \sim \rho R^4 \omega \sqrt{\nu\omega}. \tag{10-26}$$

An exact solution for the fluid motion can be obtained for an infinite disk. We begin with the Navier-Stokes equations in cylindrical coordinates (Eqs. 6–29). With $\partial/\partial\theta = 0$, these become

$$v_r \frac{\partial v_r}{\partial r} - \frac{v_\theta^2}{r} + w \frac{\partial v_r}{\partial z} = -\frac{1}{\rho}\frac{\partial p}{\partial r} + g_r + \frac{\mu}{\rho}\left[\frac{\partial^2 v_r}{\partial r^2} + \frac{\partial}{\partial r}\left(\frac{v_r}{r}\right) + \frac{\partial^2 v_r}{\partial z^2}\right],$$

$$v_r \frac{\partial v_\theta}{\partial r} + \frac{v_r v_\theta}{r} + w \frac{\partial v_\theta}{\partial z} = +\frac{\mu}{\rho}\left[\frac{\partial^2 v_\theta}{\partial r^2} + \frac{\partial}{\partial r}\left(\frac{v_\theta}{r}\right) + \frac{\partial^2 v_\theta}{\partial z^2}\right],$$

$$v_r \frac{\partial w}{\partial r} + w \frac{\partial w}{\partial z} = -\frac{1}{\rho}\frac{\partial p}{\partial z} + g_z + \frac{\mu}{\rho}\left[\frac{\partial^2 w}{\partial r^2} + \frac{1}{r}\frac{\partial w}{\partial r} + \frac{\partial^2 w}{\partial z^2}\right]. \tag{10-27}$$

Continuity from Eq. (6–30) is

$$\frac{\partial v_r}{\partial r} + \frac{v_r}{r} + \frac{\partial w}{\partial z} = 0. \tag{10-28}$$

In the above, v_r, v_θ, and w are the velocity components in the r-, θ-, and z-directions, with z the axis of symmetry. Solutions given by Kármán [7], and Cochran [8] put v_r and v_θ proportional to $r\omega$, and w proportional to $\delta\omega$. The factor of proportionality is a function of z/δ. Noting that $\delta \sim \sqrt{\nu/\omega}$, (Eq. 10-24), we can write

$$\frac{v_r}{r\omega} = F\left(\frac{z}{\sqrt{\nu/\omega}}\right), \qquad \frac{v_\theta}{r\omega} = G\left(\frac{z}{\sqrt{\nu/\omega}}\right),$$

$$\tag{10-29}$$

$$\frac{w}{\sqrt{\nu\omega}} = H\left(\frac{z}{\sqrt{\nu/\omega}}\right), \qquad \frac{p(z)}{\rho\nu\omega} = P\left(\frac{z}{\sqrt{\nu/\omega}}\right).$$

Inserting these into the Navier-Stokes and

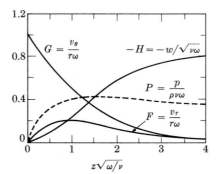

FIG. 10–12. Laminar-flow velocity distributions near a rotating disk.

TABLE 10–2

Kár.mán-Cochran Solution for Laminar-Flow
Velocity Distributions Near a Rotating Disk

$\dfrac{z\sqrt{\omega/\nu}}{(\propto z/\delta)}$	$F = \dfrac{v_r}{r\omega}$	$G = \dfrac{v_\theta}{r\omega}$	$H = \dfrac{w}{\sqrt{\nu\omega}}$	$P = \dfrac{p}{\rho\nu\omega}$	$F' = \dfrac{dF}{d(z/\sqrt{\nu/\omega})}$	$G' = \dfrac{dG}{d(z/\sqrt{\nu/\omega})}$
0	0	1.0	0	0	0.510	−0.616
0.5	0.154	0.708	−0.092	0.312	0.147	−0.533
1.0	0.180	0.468	−0.266	0.395	−0.016	−0.391
1.5	0.156	0.313	−0.435	0.406	−0.070	−0.268
2.0	0.118	0.203	−0.572	0.401	−0.074	−0.177
2.5	0.084	0.131	−0.674	0.395	−0.061	−0.116
3.6	0.036	0.050	−0.802	0.394	−0.030	−0.044
4.0	0.026	0.035	−0.826	0.393	−0.022	−0.031
∞	0	0	−0.886	0.393	0	0

continuity equations with the boundary conditions

$$\text{at } z = 0: \qquad v_r = 0, \qquad v_\theta = r\omega, \qquad w = 0,$$
$$\text{at } z = \infty: \qquad v_r = 0, \qquad v_\theta = 0, \tag{10–30}$$

we obtain a system of ordinary differential equations. Their solution gives the numerical results of Table 10–2. These values are plotted in Fig. 10–12. Note that the dimensionless distributions shown there hold for all radii. If the disk radius R is large so $R \ggg \delta$, we can use the solution for finite disks. This neglects edge effects.

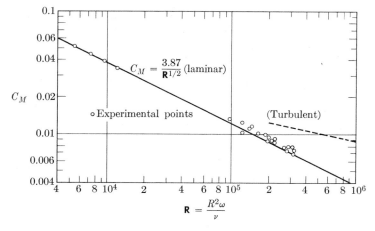

FIG. 10–13. Frictional torque coefficient for a rotating disk.

These results lead to the torque due to shear as follows. Let

$$\tau_{z\theta} = \mu \left(\frac{\partial v_\theta}{\partial z}\right)_{z=0} = \mu \left\{\frac{\partial}{\partial z}\left[r\omega G(z\sqrt{\omega/\nu})\right]\right\}_{z=0},$$
$$= \rho r \nu^{1/2} \omega^{3/2} G'(0). \tag{10-31}$$

Then, if we ignore the edge effects, for two sides of a disk of finite radius R, we have

$$2M = -4\pi \int_0^R r^2 \tau_{z\theta} \, dr$$
$$= -\pi \rho R^4 (\nu\omega^3)^{1/2} G'(0) = 0.616\pi \rho R^4 (\nu\omega^3)^{1/2} \tag{10-32}$$
$$= C_M \tfrac{1}{2}\rho\omega^2 R^5, \tag{10-33}$$

with

$$C_M = \frac{3.87}{\mathbf{R}^{1/2}}, \tag{10-34}$$

$$\mathbf{R} = \frac{R^2\omega}{\nu}. \tag{10-35}$$

The theoretical coefficient (10–34) is compared with experimental values in Fig. 10–13.

PROBLEMS

10–1. A laminar boundary layer develops along a flat surface. When the free-stream velocity is 2 ft/sec, determine for a Reynolds number $\mathbf{R}_x = 300{,}000$ the boundary-layer velocities u and v at distances 0.4δ and 0.8δ from the flat surface.

10–2. A thin plate 1 ft wide and 2 ft long is towed through 68°F water at 2 ft/sec. Compute the drag due to surface resistance on the two surfaces.

10–3. For the plate in Problem 10–2 compute and plot the total drag from the leading edge to a point x on the plate as a function of x.

10–4. A laminar boundary layer on a smooth plate results from benzene at 68° flowing past the plate at a free-stream velocity of 1 ft/sec. For a distance of 1 ft from the leading edge, compute the thicknesses δ, δ^*, and θ.

10–5. The data of Table 10–1 for u/U along a flat plate with $dp/dx = 0$ (Blasius' exact solution) can be approximated by the parabolic equation

$$\frac{u}{U} = 2\frac{y}{\delta} - \left(\frac{y}{\delta}\right)^2.$$

Compute δ^*/δ and θ/δ and compare them with the same ratios for the exact solution as found from Eqs. (10–11) through (10–13).

10–6. A laminar boundary layer results from the steady motion of a stream of ethyl alcohol (68°F) past a slightly curved surface. At a certain position $x = x_1 = 1$ ft, the free-stream velocity is given by $U = 3 + x/3$. In the vicinity of x_1 we find that approximately

$$\delta^* = 1.74\sqrt{\nu x/U}, \qquad \theta = 0.657\sqrt{\nu x/U}.$$

(a) Determine the local pressure gradient dp/dx at x_1. Explain in physical terms why dp/dx is greater or less than the case in which $U = $ const. (b) Use the Kármán integral momentum equation (Eq. 8–21) to estimate the increase in τ_0 over the case for $U = $ const.

10–7. Prove that the integral momentum equation in the form of Eq. (8–18) can be obtained for a two-dimensional boundary layer by applying the momentum principle to the fluid within a control volume which is adjacent to a flat surface as shown in Fig. 10–14. Note that the boundary shear force must equal the net flux of momentum out of the control volume.

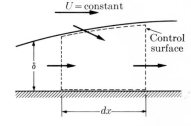

FIGURE 10–14

10–8. Beginning with the Prandtl boundary-layer equations, verify Eq. (8–21).

10–9. Apply the momentum principle to the flow through a control volume enclosing the boundary-layer fluid described in Problem 10–7 and verify Eq. (8–21).

10–10. Verify the value of the constant in Eq. (10–12).

10–11. Verify the value of the constant in Eq. (10–13).

TABLE 10–3

Conditions	Section (0)	Section (1)	Section (2)	Section (3)
Pressure intensity	15 psi		13 psi	7.5 psi
Boundary-layer thickness			0.01 ft	
Maximum velocity (on pipe centerline)	0	10 ft/sec	11 ft/sec	
Velocity distributions		$u_1 = u_{1\max}$ = const	Boundary layer: $u_2 = 2200(y - 50y^2)$ Core flow: $u_2 = u_{2\max}$ = const	
Exact velocity head $K_e \dfrac{V^2}{2g} (V = Q/A)$	0	$\dfrac{V_1^2}{2g}$	$1.12\left(\dfrac{V_2^2}{2g}\right)$	$2\left(\dfrac{V_3^2}{2g}\right)$

Problems 10–12, 10–13, and 10–14 are related to the following information about the flow of an oil from a reservoir into a circular pipe. The solutions of the three problems are independent of one another.

In the zone of *flow establishment* near the pipe entrance, the flow consists of a boundary layer at the pipe wall and a central core across which the velocity is constant. In this example, the boundary layer remains laminar; therefore, the *established flow* is laminar also. The known conditions are described in Fig. 10–15 and listed in Table 10–3. The properties of oil are: $\mu = 0.004$ lb-sec/ft^2, $\gamma = 54.7$ lb/ft^3.

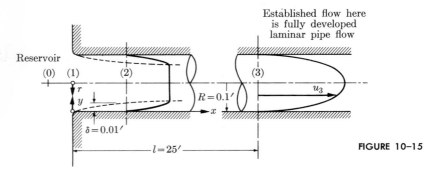

FIGURE 10–15

10–12. (a) Verify with suitable calculations that the established flow at section (3) must be laminar. (b) Compute the wall shear stress τ_0 at sections (2) and (3). Show with qualitative sketches how τ varies across the entire passage at both sections. Give reasons to justify your sketch.

10–13. Determine the exact head loss between (0) and (2) in feet of oil.

10–14. Apply the one-dimensional momentum equation to compute the average value of τ_0 over the length l.

10–15. Measurements are made of the steady-flow velocity distribution in the boundary layer along a hydrofoil surface. In a certain region along the surface the pressure gradient $dp/dx = 0$, and the thickness δ increases as

$$\delta = A\sqrt{x/U}.$$

The momentum thickness is $\theta = B\delta$, and A and B are constants. (a) Find the relation for wall shear stress. (Assume you know nothing about solutions to flat-plate or other boundary-layer examples.) (b) Find the relation for the wall shear-stress coefficient.

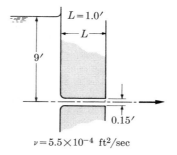

$\nu = 5.5 \times 10^{-4}$ ft^2/sec

FIGURE 10–16

10-16. Consider the steady flow from a slot, as shown in Fig. 10–16. Compute the maximum velocity in the liquid jet at the end of the slot and the discharge per foot of width of the slot. Assume $dp/dx \approx 0$ over the distance L in the flow direction.

10-17. Assume that a steady, laminar boundary layer on a flat plate ($dp/dx = 0$) has velocity profiles given by

$$u/U = \tfrac{3}{2}(y/\delta) - \tfrac{1}{2}(y/\delta)^3.$$

Use the definition and integral momentum equations of Chapter 8 to find (a) θ and c_f as functions of δ and x, and (b) δ as a function of $\mathbf{R}_x$ and x. (c) Compare the results with Blasius' exact solutions given by Eqs. (10–11) through (10–15).

10-18. An assumed velocity distribution for a steady laminar boundary layer with $dp/dx = 0$, such as the one given in Problem 10–17, is formulated to satisfy three conditions:

$$(1) \quad u = U \quad \text{at } y = \delta,$$

$$(2) \quad \frac{\partial u}{\partial y} = 0 \quad \text{at } y = \delta,$$

$$(3) \quad \frac{\partial^2 u}{\partial y^2} = 0 \quad \text{at } y = 0.$$

(a) Discuss the physical significance of (1) and (2) and prove the validity of condition (3) by means of the Prandtl boundary-layer equations. (b) Prove that the velocity distribution given in Problem 10–17 satisfies these conditions. (c) Can a velocity profile of the form

$$u/U = A + B(y/\delta) + C(y/\delta)^2$$

satisfy all of these conditions?

REFERENCES

1. Schlichting, H., *Boundary Layer Theory*, McGraw-Hill Book Co., New. York, 1960, 4th ed. Article VII f provides the complete development.
2. Dhawan, S., "Measurements of Skin Friction," *Nat. Advisory Comm. Aeron*, Tech. Note 2567 (1952).
3. Dhawan, S., *loc. cit. supra.*
4. Janour, *Nat. Advisory Comm. Aeron*, Tech. Memo. 1316.
5. Schlichting, H., *op. cit.* Articles XII b, d give detailed development.
6. Prandtl, L., *Essentials of Fluid Dynamics*, Hafner Publishing Co., Inc., New York, 1952, Article V 10; H. Schlichting, *op. cit.*, Article V b 10.
7. von Kármán, T., "Laminare und Turbulente Reibung," *Z. Angew. Math. Phys.*, **1**, 233 (1921).
8. Cochran, W. G., "The Flow Due to a Rotating Disk," *Proc. Cambridge Phil. Soc.*, **30**, 365 (1934).

CHAPTER 11

Origin of Turbulence
and Turbulent Shear Stress

11-1 SOURCES OF TURBULENCE

A state of turbulence is the result of a breakdown of orderly flow into eddies which spread to "contaminate" a region of the flow with irregular fluctuating motions. If the conditions are right, this turbulence persists as a quasi-steady state of motion.

One source of turbulence generating eddies is found in surfaces of flow discontinuity which occur whenever two fluid streams come together in such a way as to leave a sharp jump in velocity between adjacent layers. Examples are at the tips of sharp projections at the edges of bluff bodies, at the trailing edges of airfoils and guide vanes, and at zones of boundary-layer separation. These are illustrated schematically in Fig. 11–1, and one example obtained with smoke tracers is shown in Fig. 11–2. At surfaces of velocity discontinuity, there is a tendency for waviness to develop, either by accident from external causes or from disturbances transported by the fluid. It is this waviness, in turn, which tends to be unstable and grow in amplitude. This is described quantitatively by a simple classical explanation [1]. Relative to the wave

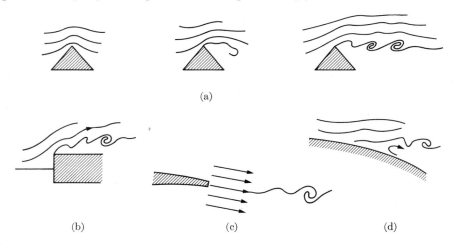

(a)

(b)   (c)   (d)

FIG. 11-1.   Eddy formation at velocity discontinuity surfaces: (a) sharp projection; (b) bluff body; (c) trailing edge; (d) boundary-layer separation.

FIG. 11–2. Turbulence generation at the surface of separation. (Courtesy Prof. F. N. M. Brown, Notre Dame University.)

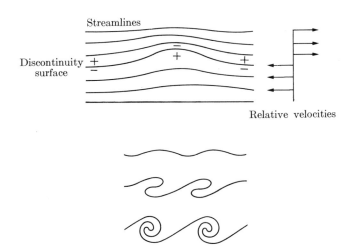

FIG. 11–3. Eddies arising from waves at a surface of discontinuity.

(which advances with the mean of the fluid velocities), the streamlines are as shown in Fig. 11–3. Applying Bernoulli's equation along each stream tube, we conclude that a high pressure occurs on the concave side of each wave crest or valley and a low pressure on the convex side. Consequently, the wavy surface is unstable and tends to amplify and then curl over and break up into separate eddies. Under a limited range of velocity conditions, these vortexlike eddies may retain their individual identities. More generally, however, they will degenerate into random fluctuations.

Turbulence is generated also in shear flows where a velocity gradient occurs without an abrupt discontinuity. Reynolds' experiment with a dyestreak in a glass tube, described in Chapter 8, is an example of shear flow becoming unstable and degenerating into turbulence. Similar to the case of discrete discontinuities, there is a tendency for disturbances to appear in ordinary shear flows. Once they appear, the question remains whether they will amplify or disappear. There have been several mechanisms of instability postulated. One for which experiments and theory have found a large measure of agreement is based on disturbances being composed of oscillations of a range of frequencies which can be selectively amplified by the hydrodynamic flow field. A small-perturbation theory (known as the Tollmien-Schlichting theory) leads to the conclusion that below a critical Reynolds number all disturbances will be damped. Above this critical Reynolds number certain frequencies will be amplified and others damped. The results of computations for a two-dimensional boundary with two-dimensional disturbances are shown by the solid curve in

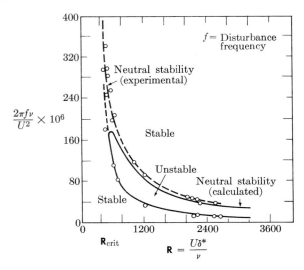

FIG. 11–4. Tollmien-Schlichting stability diagram for a laminar boundary layer. The ordinate is a dimensionless frequency function. Above R_{crit}, disturbances of the frequencies falling within the neutral stability loop are unstable and will amplify. Disturbances of all other frequencies will be stable [2].

Fig. 11–4 [2]. On this diagram, the ordinate is a dimensionless frequency and the abscissa is a boundary-layer Reynolds number based on the displacement thickness δ^*. A looping line which encloses a region above the critical Reynolds number $\mathbf{R}_{crit}$ denotes the conditions for neutral stability. Below $\mathbf{R}_{crit}$ disturbances of all frequencies will be suppressed by viscous damping. Above $\mathbf{R}_{crit}$, disturbances of frequencies within the neutral stability loop will amplify and all others will be damped. This type of behavior has been verified experimentally as shown by the dashed curve in Fig. 11–4. Disturbances were introduced into a laminar boundary layer by a metal ribbon placed on its edge in the boundary layer normal to the flow direction. It was given controlled vibration by applying electric current of a selected frequency in the presence of a magnetic field. By varying the frequencies of vibration, the line of neutral stability and zones of stability and instability were determined. The measured conditions for neutral stability lie very close to the theoretical.

The stability theories have assumed that instability occurs simultaneously over the entire region of the fluid in which the critical conditions are met. Thus, for a two-dimensional boundary layer on a flat surface, instability should be evident along the full length of a line which is perpendicular to the flow. This is shown schematically in Fig. 11–5(a). However, this does not seem to be the case in nature. Disturbances appear in restricted zones or "spots" within the fluid [3]. These spots grow as they are swept downstream, encroaching into the laminar fluid until a merging of separate spots develops general turbulence. The spread and amplification of a spot disturbance into a turbulence patch is

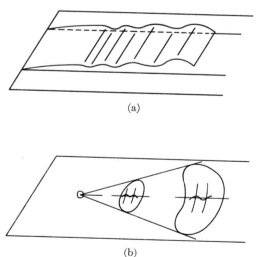

(a)

(b)

FIG. 11–5. Two-dimensional versus localized spot disturbance in a laminar boundary layer: (a) two-dimensional wave amplification by Tollmien-Schlichting hypothesis; (b) observed spread and amplification of a spot disturbance into a turbulence patch.

illustrated in Fig. 11–5(b). It is clear that the onset of turbulence is three-dimensional in character. A similar behavior apparently occurs in circular pipes where Reynolds' transition has been observed to initiate with spots along the pipe walls [4]. Apparently the breakdown of a dyed streamline into turbulence, which Reynolds' experiment shows, is a consequence of disturbances growing from such spots. The mechanism of the initiation of spots of turbulence seems to be related to what happens when the small disturbances, whose amplification is predicted by the small-perturbation theory, become large. One view is that the initially two-dimensional disturbances are distorted into concentrated three-dimensional eddy systems and thence to turbulent spots. Another is that the basic disturbance is three-dimensional and "hairpin" eddies are generated which become unstable and appear as turbulence spots.

11–2 VELOCITIES, ENERGIES, AND CONTINUITY IN TURBULENCE

Turbulent motion, once established, has a random nature making it difficult to describe exactly. Because of this random nature, we assume that it can be described by a set of statistical properties. For this purpose it is convenient to set the instantaneous velocity equal to the sum of a mean value plus a fluctuating component. Thus, for the xyz-coordinate directions, we write

$$u = \bar{u} + u', \qquad v = \bar{v} + v', \qquad w = \bar{w} + w', \qquad (11\text{–}1)$$

where

$$\bar{u} = \frac{1}{T}\int_0^T u\,dt \qquad \text{etc. for } \bar{v}, \bar{w}, \qquad (11\text{–}2)$$

with T being a long time in comparison to the time scale of the turbulence we want to examine. Because the fluctuations are both plus and minus, the mean of u' is

$$\bar{u'} = \frac{1}{T}\int_0^T u'\,dt \equiv 0. \qquad (11\text{–}3)$$

Statistical quantities which are useful include the *root-mean-square* (rms) of the fluctuations. We denote, for the x-component,

$$\text{rms} = \sqrt{\bar{u'^2}} = \left\{\frac{1}{T}\int_0^T u'^2\,dt\right\}^{1/2}, \qquad (11\text{–}4)$$

and similarly for the y- and z-components. This is the average intensity of the turbulence.

The average *kinetic energy of turbulence* per unit mass is

$$\frac{\text{average KE of turbulence}}{\text{mass}} = \tfrac{1}{2}(\bar{u'^2} + \bar{v'^2} + \bar{w'^2}), \qquad (11\text{–}5a)$$

which is proportional to the sum of the squares of the intensities. As for any random process, the kinetic energy of turbulence can be decomposed into an *energy spectrum* of *energy density* versus frequency. We define the energy density $\Phi(f)$ at a point in the flow as the limit of the average kinetic energy per unit mass contained in a particular frequency bandwidth Δf divided by the bandwidth, or

$$\Phi(f) = \lim_{\Delta f \to 0} \frac{\text{average KE/mass contained in } \Delta f}{\Delta f}, \qquad (11\text{--}5\text{b})$$

where

$$f = \text{ordinary frequency in cycles per second}$$
$$= \frac{\omega}{2\pi}.$$

The integral of the energies over the range of frequencies appearing in the turbulence has the average value given by Eq. (11–5a). Thus

$$\frac{\text{average KE of turbulence}}{\text{mass}} = \int_0^\infty \Phi(f) \, df = \tfrac{1}{2}(\overline{u'^2} + \overline{v'^2} + \overline{w'^2}). \qquad (11\text{--}5\text{c})$$

A *correlation* exists between events if they are related systematically in some way. The degree of correlation may vary from exact (one-to-one correspondence) to zero (complete independence). Our primary interest is in the degree to which the velocity fluctuations u', v', and w' are interdependent, as given by expressions of the form

$$\overline{u'v'} = \frac{1}{T} \int_0^T u'v' \, dt. \qquad (11\text{--}6)$$

In a shear flow in an xy-plane, $\overline{u'v'}$ is finite and, as we shall see, is related to the magnitude of the shear stress.

An important instrument for measuring turbulent fluctuations is the hot-wire anemometer. It operates on the principle that a change in temperature affects the current flow or voltage drop through a wire. A short length of very fine platinum wire stretched between two fingers is heated electrically by a circuit that maintains either current or voltage drop constant. When inserted into a stream, the cooling, which is a function of the velocity, can be detected as variations in voltage (for constant current) or current (for constant voltage). The smaller the mass and heat capacity of the wire, the more rapid the response to velocity changes. When we orient the wire in different planes and directions, it will respond to different velocity components. Using two or more wires at one point in the flow, one can make simultaneous measurements of components. The mean value can be subtracted and the rms-values, correlations, and energy spectra computed using the fluctuating remainders. All these operations can be performed electronically.

Continuity must be satisfied for turbulent as for laminar motion. For incompressible fluids, the divergence of the velocity equals zero. Using the relations of Eqs. (11–1), we have

$$\frac{\partial(\overline{u} + u')}{\partial x} + \frac{\partial(\overline{v} + v')}{\partial y} + \frac{\partial(\overline{w} + w')}{\partial z} = 0, \qquad (11-7)$$

or

$$\frac{\partial \overline{u}}{\partial x} + \frac{\partial \overline{v}}{\partial y} + \frac{\partial \overline{w}}{\partial z} + \frac{\partial u'}{\partial x} + \frac{\partial v'}{\partial y} + \frac{\partial w'}{\partial z} = 0.$$

Taking averages of each term, we see that

$$\frac{\overline{\partial u'}}{\partial x} + \frac{\overline{\partial v'}}{\partial y} + \frac{\overline{\partial w'}}{\partial z} = 0.$$

Therefore

$$\frac{\partial \overline{u}}{\partial x} + \frac{\partial \overline{v}}{\partial y} + \frac{\partial \overline{w}}{\partial z} = 0, \qquad (11-8)$$

and also

$$\frac{\partial u'}{\partial x} + \frac{\partial v'}{\partial y} + \frac{\partial w'}{\partial z} = 0. \qquad (11-9)$$

Thus both the mean-motion components and the superposed turbulent-motion components must satisfy the continuity condition.

11–3 TURBULENT SHEAR STRESSES AND EDDY VISCOSITIES

The resistance to motion with laminar flow increases directly with the velocity. We saw this with steady uniform laminar flow through a circular tube, where, by Eq. (6–41),

$$\frac{d(p + \gamma h)}{dz} \propto V_z,$$

in which z is the axis of symmetry of the pipe. This relation is shown as a straight line of 45° slope on the logarithmic plot of Fig. 11–6. When turbulence sets in, the resistance is observed first to jump sharply and then to increase at a greater rate than for laminar flow. The jump may be 100% or more. The subsequent rate of increase is according to

$$\frac{d(p + \gamma h)}{dz} \propto V_z^n,$$

where n is nearly 2. Figure 11–6 also shows the jump and higher rate of increase. Since $d(p + \gamma h/dz$ is proportional to the shear stress resisting the motion, we have the condition that at equal flow rates,

$$\tau|_{\text{turb}} > \tau|_{\text{lam}}.$$

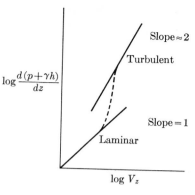

FIG. 11-6. Pressure gradient with laminar and turbulent flow in a conduit.

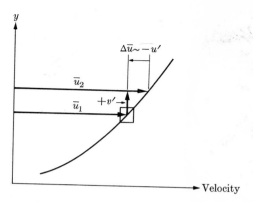

FIG. 11-7. Momentum transport by turbulent velocity fluctuation.

An early French investigator, Boussinesq, expressed the increased shear with turbulence in terms of an *eddy viscosity* that depends on the state of the turbulent motion. He wrote, for the simple two-dimensional case,

$$\tau|_{\text{turb}} = \mu \frac{d\bar{u}}{dy} + \eta \frac{d\bar{u}}{dy}, \tag{11-10}$$

where

$\bar{u}$ = the mean local velocity defined by Eq. (11-2),

μ = the dynamic molecular viscosity, a property of the fluid,

η = a dynamic *eddy viscosity* that depends on the state of the turbulent motion.

The term $\mu(d\bar{u}/dy)$ is an *apparent stress* computed from the velocity gradient of the mean motion. The term $\eta(d\bar{u}/dy)$ is an *additional* apparent stress associated with the turbulence. With laminar flow the eddy viscosity η is zero. With turbulent flow η becomes much larger than the molecular viscosity μ, so that often $\mu(d\bar{u}/dy)$ is neglected. Moreover, η varies across the fluid stream since it depends on the state of motion, while μ is a property solely of the fluid itself. Like the dynamic molecular viscosity, the dynamic eddy viscosity η has the units (lb-sec)/ft². Dividing by density ρ, we get a *kinematic eddy viscosity* $\epsilon = \eta/\rho$ having the units ft²/sec.

A physical model for explaining the extra contribution of turbulence to the flow resistance and effective stress is based on the momentum interchange between fluid elements as they fluctuate. Consider the shear flow shown in Fig. 11-7. Fluid in layers 1 and 2 moves with different mean velocities. If the lower-velocity fluid in layer 1 were to fluctuate with a v'-velocity into layer 2, its velocity in the direction of the stream would be less than the mean velocity of the new environment. In the new environment, it would appear to be deficient by an amount $-u'$. The drag of the faster moving surroundings would accelerate the element and increase its momentum. The flux crossing from 1 to 2 can be expressed as $\rho v'$ mass/sec-ft^2. Multiplying by u' gives the flow-direction momentum change per second for this flux as $-\rho u'v'$, or on the average over a time period, as

$$-\rho \overline{u'v'} \tag{11–11}$$

The negative sign is a consequence of a plus y-direction flux giving a deficiency in u, or a negative u', in the new environment. Fluctuations in the negative y-direction will give a positive u'. The rate of change of momentum represented by Eq. (11–11) is an effective resistance to motion and an effective shearing stress.

Using Eq. (11–11) to replace $\eta(d\overline{u}/dy)$ in Eq. (11–10) gives for two-dimensional uniform flow,

$$\tau = \mu(d\overline{u}/dy) - \rho\overline{u'v'}. \tag{11–12}$$

For fully developed turbulence, this becomes

$$\tau \approx \eta(d\overline{u}/dy) \approx -\rho\overline{u'v'}. \tag{11–13}$$

Equation (11–13) illustrates why the turbulent-motion resistance increases with nearly the square of the velocity. With increases in the fluctuating velocities in proportion to increases in the mean velocity, the additional apparent stresses represented by the momentum relations follow the square.

By a similar argument, an effective addition to the normal pressure intensity acting in the flow direction is

$$-\rho\overline{u'u'} = -\rho\overline{u'^2}. \tag{11–14}$$

11–4 REYNOLDS EQUATIONS FOR INCOMPRESSIBLE FLUIDS

The equations of motion of a viscous fluid, as derived in Chapter 6, are general and apply to turbulent as well as nonturbulent cases. However the complexity of turbulence, even in the simplest cases, has made it impossible to relate the motion to the boundary conditions and obtain any exact solution. A useful, although limited, alternative is to consider the pattern of the mean turbulent motion even though we cannot establish the true details of the fluctuations. Reynolds converted the equations of motion for an incompressible fluid into a form which does this. His form can be obtained in the manner described below.

We begin with the equations of motion in the basic form of Eq. (6–21). For the x-direction, replacing σ_x with $-p_x$, we have

$$\rho a_x = \rho\left(\frac{\partial u}{\partial t} + u\frac{\partial u}{\partial x} + v\frac{\partial u}{\partial y} + w\frac{\partial u}{\partial z}\right) = \rho g_x - \frac{\partial p_x}{\partial x} + \frac{\partial \tau_{yx}}{\partial y} + \frac{\partial \tau_{zx}}{\partial z}.$$
$$\tag{11–15}$$

By use of the continuity equation for an incompressible fluid, we can write

$$u \frac{\partial u}{\partial x} + u \frac{\partial v}{\partial y} + u \frac{\partial w}{\partial z} = 0.$$

Adding to Eq. (11–15), we obtain

$$\rho \left(\frac{\partial u}{\partial t} + \frac{\partial u^2}{\partial x} + \frac{\partial uv}{\partial y} + \frac{\partial uw}{\partial z} \right) = \rho g_x - \frac{\partial p_x}{\partial x} + \frac{\partial \tau_{yx}}{\partial y} + \frac{\partial \tau_{zx}}{\partial z}. \qquad (11–16)$$

Let

$$u = \bar{u} + u', \qquad v = \bar{v} + v', \qquad w = \bar{w} + w', \qquad p_x = \bar{p}_x + p'_x. \qquad (11–17)$$

We then substitute Eqs. (11–17) in Eq. (11–16), average according to the rules of Section 11–2, and rearrange to obtain

$$\rho \left(\frac{\partial \bar{u}}{\partial t} + \bar{u} \frac{\partial \bar{u}}{\partial x} + \bar{v} \frac{\partial \bar{u}}{\partial y} + \bar{w} \frac{\partial \bar{u}}{\partial z} \right)$$

$$= \rho g_x - \frac{\partial \bar{p}_x}{\partial x} + \frac{\partial \bar{\tau}_{yx}}{\partial y} + \frac{\partial \bar{\tau}_{zx}}{\partial z} - \rho \frac{\overline{\partial u'^2}}{\partial x} - \rho \frac{\overline{\partial u'v'}}{\partial y} - \rho \frac{\overline{\partial u'w'}}{\partial z} \qquad (11–18a)$$

and similarly for the y- and z-directions

$$\rho \left(\frac{\partial \bar{v}}{\partial t} + \bar{u} \frac{\partial \bar{v}}{\partial x} + \bar{v} \frac{\partial \bar{v}}{\partial y} + \bar{w} \frac{\partial \bar{v}}{\partial z} \right)$$

$$= \rho g_y + \frac{\partial \bar{\tau}_{xy}}{\partial x} - \frac{\partial \bar{p}_y}{\partial y} + \frac{\partial \bar{\tau}_{zy}}{\partial z} - \rho \frac{\overline{\partial v'u'}}{\partial x} - \rho \frac{\overline{\partial v'^2}}{\partial y} - \rho \frac{\overline{\partial v'w'}}{\partial z}, \qquad (11–18b)$$

$$\rho \left(\frac{\partial \bar{w}}{\partial t} + \bar{u} \frac{\partial \bar{w}}{\partial x} + \bar{v} \frac{\partial \bar{w}}{\partial y} + \bar{w} \frac{\partial \bar{w}}{\partial z} \right)$$

$$= \rho g_z + \frac{\partial \bar{\tau}_{xz}}{\partial x} + \frac{\partial \bar{\tau}_{yz}}{\partial y} - \frac{\partial \bar{p}_z}{\partial z} - \rho \frac{\overline{\partial w'u'}}{\partial x} - \rho \frac{\overline{\partial w'v'}}{\partial y} - \rho \frac{\overline{\partial w'^2}}{\partial z}. \qquad (11–18c)$$

The primed terms on the right-hand side are the result of transferring turbulence-acceleration terms from the left-hand side.

Equations (11–18) are for the mean motion. They are of the same form as Eq. (11–15) if we recognize as stresses the following sums of terms on the right-hand side:

$$(-\bar{p}_x - \rho \overline{u'^2}), \qquad (\bar{\tau}_{yx} - \rho \overline{u'v'}), \qquad (\bar{\tau}_{zx} - \rho \overline{w'v'}), \qquad (11–19)$$

and similarly for the y- and z-direction equations. Each of these is a sum of an "apparent stress" of the mean motion and an "additional apparent stress" due to turbulent fluctuations.

The apparent stresses are obtained by substituting Eqs. (11–17) in the Newtonian stress relations (Eqs. 5–29 and 5–30) and averaging. In the x-

direction, the Newtonian stresses for incompressible fluids are

$$\sigma_x = -p_x = -p + 2\mu\frac{\partial u}{\partial x}, \quad \tau_{yx} = \mu\left(\frac{\partial v}{\partial x} + \frac{\partial u}{\partial y}\right), \quad \tau_{zx} = \mu\left(\frac{\partial u}{\partial z} + \frac{\partial w}{\partial x}\right),$$

which lead to

$$-\bar{p}_x = -\bar{p} + 2\mu\frac{\partial \bar{u}}{\partial x}, \quad \bar{\tau}_{yx} = \mu\left(\frac{\partial \bar{v}}{\partial x} + \frac{\partial \bar{u}}{\partial y}\right), \quad \bar{\tau}_{zx} = \mu\left(\frac{\partial \bar{u}}{\partial z} + \frac{\partial \bar{w}}{\partial x}\right)$$

$$\tag{11–20a}$$

and similarly, for the y- and z-directions,

$$-\bar{p}_y = -\bar{p} + 2\mu\frac{\partial \bar{v}}{\partial y}, \quad \bar{\tau}_{xy} = \mu\left(\frac{\partial \bar{v}}{\partial x} + \frac{\partial \bar{u}}{\partial y}\right), \quad \bar{\tau}_{zy} = \mu\left(\frac{\partial \bar{w}}{\partial y} + \frac{\partial \bar{v}}{\partial z}\right),$$

$$\tag{11–20b}$$

$$-\bar{p}_z = -\bar{p} + 2\mu\frac{\partial \bar{w}}{\partial z}, \quad \bar{\tau}_{xz} = \mu\left(\frac{\partial \bar{u}}{\partial z} + \frac{\partial \bar{w}}{\partial x}\right), \quad \bar{\tau}_{yz} = \mu\left(\frac{\partial \bar{w}}{\partial y} + \frac{\partial \bar{v}}{\partial z}\right).$$

$$\tag{11–20c}$$

The *additional* apparent stresses are

$$\begin{array}{ccc} -\rho\overline{u'^2}, & -\rho\overline{u'v'}, & -\rho\overline{u'w'}, \\ -\rho\overline{v'u'}, & -\rho\overline{v'^2}, & -\rho\overline{v'w'}, \\ -\rho\overline{w'u'}, & -\rho\overline{w'v'}, & -\rho\overline{w'^2}. \end{array} \tag{11–21}$$

These are the momentum stresses already described in Section 11–3. The six (three independent) shear stresses will be finite so long as a correlation exists among the fluctuations in velocity components.

If the equations of motion are written in the Navier-Stokes form and converted for application to turbulence, the results for the incompressible fluid are

$$\rho\left(\frac{\partial \bar{u}}{\partial t} + \bar{u}\frac{\partial \bar{u}}{\partial x} + \bar{v}\frac{\partial \bar{u}}{\partial y} + \bar{w}\frac{\partial \bar{u}}{\partial z}\right)$$

$$= \rho g_x - \frac{\partial \bar{p}}{\partial x} + \mu\nabla^2\bar{u} - \rho\left(\frac{\partial \overline{u'^2}}{\partial x} + \frac{\partial \overline{u'v'}}{\partial y} + \frac{\partial \overline{u'w'}}{\partial z}\right),$$

$$\rho\left(\frac{\partial \bar{v}}{\partial t} + \bar{u}\frac{\partial \bar{v}}{\partial x} + \bar{v}\frac{\partial \bar{v}}{\partial y} + \bar{w}\frac{\partial \bar{v}}{\partial z}\right)$$

$$= \rho g_y - \frac{\partial \bar{p}}{\partial y} + \mu\nabla^2\bar{v} - \rho\left(\frac{\partial \overline{v'u'}}{\partial x} + \frac{\partial \overline{v'^2}}{\partial y} + \frac{\partial \overline{v'w'}}{\partial z}\right), \qquad (11–22)$$

$$\rho\left(\frac{\partial \bar{w}}{\partial t} + \bar{u}\frac{\partial \bar{w}}{\partial x} + \bar{v}\frac{\partial \bar{w}}{\partial y} + \bar{w}\frac{\partial \bar{w}}{\partial z}\right)$$

$$= \rho g_z - \frac{\partial \bar{p}}{\partial z} + \mu\nabla^2\bar{w} - \rho\left(\frac{\partial \overline{w'u'}}{\partial x} + \frac{\partial \overline{w'v'}}{\partial y} + \frac{\partial \overline{w'^2}}{\partial z}\right).$$

Example 11–1: Turbulent flow between parallel plates

As an example of the application of Reynolds equations, consider the steady uniform motion in the x-direction between parallel horizontal walls. For this condition

$$\frac{\partial}{\partial t} = 0, \qquad \frac{\partial(\text{vel})}{\partial x} = 0, \qquad \bar{v} = 0. \tag{11–23}$$

From Eqs. (11–22), the x- and y-component equations become

$$0 = \rho g_x - \frac{\partial \bar{p}}{\partial x} + \mu \frac{\partial^2 \bar{u}}{\partial y^2} - \rho \frac{\partial \overline{u'v'}}{\partial y}, \qquad 0 = \rho g_y - \frac{\partial \bar{p}}{\partial y} - \rho \frac{\partial \overline{v'^2}}{dy}. \tag{11–24}$$

The latter equation gives

$$\frac{\partial}{\partial y}(\bar{p} + \gamma h) + \rho \frac{\partial \overline{v'^2}}{\partial y} = 0, \tag{11–25}$$

or, on integrating, we have

$$(\bar{p} + \gamma h) + \rho \overline{v'^2} = \text{const.} \tag{11–26}$$

Thus the static pressure distribution in planes perpendicular to the flow direction differs from the hydrostatic one by an amount $\rho \overline{v'^2}$. This is small but can be important in special cases.

In the first of Eqs. (11–24), if the turbulence contribution to the shear is over-riding, there remains only

$$\frac{\partial(\bar{p} + \gamma h)}{\partial x} = -\rho \frac{\partial \overline{u'v'}}{\partial y}, \tag{11–27}$$

where $\partial(\bar{p} + \gamma h)/\partial x$ is independent of y. Integrating with respect to y, we have

$$\frac{d(\bar{p} + \gamma h)}{dx} y = -\rho \overline{u'v'} = \tau, \tag{11–28}$$

which is a linear relation for shear stress. Thus for uniform flows, laminar and turbulent, the shear-stress distribution is linear with distance from the wall.

Example 11–2: Equations for a turbulent boundary layer

As another example, the Reynolds form of the Navier-Stokes equations can be specialized for two-dimensional turbulent boundary layers along plane or slightly curved boundaries using order-of-magnitude arguments as was done in Chapter 8 for laminar flow. Neglecting body forces, we obtain

$$\rho\left(\frac{\partial \bar{u}}{\partial t} + \bar{u}\frac{\partial \bar{u}}{\partial x} + \bar{v}\frac{\partial \bar{u}}{\partial y}\right) = -\frac{\partial \bar{p}}{\partial x} - \rho\frac{\partial \overline{u'^2}}{\partial x} + \mu\frac{\partial^2 \bar{u}}{\partial y^2} - \rho\frac{\partial \overline{u'v'}}{\partial y},$$

$$0 = -\frac{\partial(\bar{p} + \rho\overline{v'^2})}{\partial y}, \tag{11–29a}$$

and continuity is

$$\frac{\partial \bar{u}}{\partial x} + \frac{\partial \bar{v}}{\partial y} = 0. \tag{11–29b}$$

Verification of these equations will be left to the reader.
Integrating the second equation of (11–29a)

$$\bar{p} + \rho \overline{v'^2} = \text{const.} \tag{11–30}$$

Thus, like the flow between plates, the pressure intensity in the boundary layer depends on the mean-square value of the v'-turbulence. Assuming that gravity acts in the negative h-direction, we find that Eq. (11–26) results for this case also.

11–5 MIXING LENGTH AND SIMILARITY HYPOTHESES IN SHEAR FLOW

The concept of a mixing length was introduced by Prandtl [5] in an attempt to express the momentum shear stresses in terms of mean velocities. It has been applied primarily to simple cases in which only one shear component is important. He assumed that the average distance traversed by a fluctuating fluid element before it acquired the velocity of the new region is related to an average magnitude of the fluctuating velocity. He also assumed that two orthogonal fluctuating velocities are proportional to each other. For the shear flow in the xy-plane pictured in Fig. 11–7, we can write

$$\overline{|v'|} \propto l \left| \frac{d\bar{u}}{dy} \right|, \qquad \overline{|u'|} \propto \overline{|v'|}, \tag{11–31}$$

where $l = l(y)$ is a mixing length, $\|$ denotes absolute magnitudes, and $\overline{|u'|}$, $\overline{|v'|}$ are average values. The mixing length is a function of position because it depends on the state of turbulence. Using these proportionalities in Eq. (11–13) and allowing l to absorb the coefficients of proportionality, we have

$$\tau = \rho l^2 \left| \frac{d\bar{u}}{dy} \right| \frac{d\bar{u}}{dy}, \tag{11–32}$$

and the dynamic eddy viscosity is represented by

$$\eta = \rho l^2 \left| \frac{d\bar{u}}{dy} \right|. \tag{11–33}$$

Equation (11–32) can be used to replace $-\rho \overline{u'v'}$ if $l = l(y)$ can be established. Working assumptions have been made for $l(y)$ in a few particular cases. However, for flows in general, it has not been possible to predict the mixing-length function *a priori*. Hence this formulation has a restricted usefulness.

Kármán [6] used a different model of turbulence in a shear field to obtain an expression that eliminates the mixing length. He assumed that the turbulent motion differs from point to point only by length and time scales. With these assumptions he obtained

$$\tau = \rho\kappa^2 \, \frac{(d\bar{u}/dy)^4}{(d^2\bar{u}/dy^2)^2}\,, \tag{11-34}$$

where $\kappa = $ a universal constant (kappa). Kármán's result is equivalent to setting

$$l = \kappa \, \frac{d\bar{u}/dy}{d^2\bar{u}/dy^2}\,. \tag{11-35}$$

Experiments show that κ is not a universal constant. On the other hand, it falls within the same range of magnitudes for a variety of shear flows. The use of Prandtl's mixing length and the Kármán kappa will be illustrated in later chapters.

PROBLEMS

11–1. A disturbance having a fundamental frequency of 20 cps occurs in the laminar boundary layer of 40°F water flowing over a hydrofoil. The free-stream velocity is 5 ft/sec. In what range of streamwise locations along the hydrofoil must the disturbance occur to cause turbulence? Assume that the information for zero-pressure gradient shown in Fig. 11–4 is applicable.

11–2. The velocity data listed in Table 11–1 were obtained at a point in a turbulent flow of sea water. Compute the energy of turbulence per unit volume.

11–3. Use the velocity data given in Problem 11–2 for a one-second interval. Determine the mean velocity in the x-direction $\bar{u}$ and verify that $\overline{u'} = 0$.

11–4. Use the velocity data given in Problem 11–2 for a one-second interval. Determine the magnitude of the three independent turbulent shear stresses in Eq. (11–21).

TABLE 11–1

time, t, sec	u, ft/sec	u', ft/sec	v', ft/sec	w', ft/sec
0	2.95	−0.15	+0.05	+0.03
0.1	3.12	+0.02	0.0	−0.01
0.2	3.38	+0.28	−0.12	−0.07
0.3	3.27	+0.17	−0.04	−0.02
0.4	3.02	−0.08	−0.02	+0.01
0.5	2.88	−0.22	+0.08	+0.03
0.6	3.05	−0.05	+0.03	−0.02
0.7	2.98	−0.12	+0.06	+0.02
0.8	3.15	+0.05	+0.02	+0.03
0.9	3.07	−0.03	+0.01	−0.02
1.0	3.23	+0.13	−0.05	−0.04

11–5. If a stagnation tube responds only to the instantaneous longitudinal velocity u, derive an expression for the time-average stagnation pressure $\bar{p}_{stg}$ in terms of the time-average free-stream pressure $\bar{p}_0$, the time-average longitudinal velocity $\bar{u}$, and the mean square value of the fluctuating velocity component $\overline{u'^2}$.

11–6. Steady uniform flow discharge through a horizontal water main 4 ft in diameter is at the rate of 64 cfs when the temperature is 50°F. Measurements of velocity give the data listed below (radius $= r$, velocity $= \bar{u}$). The pressure-head drop in the flow direction is 0.00257 ft/ft of pipe. (a) Compute and plot the eddy viscosity, η, as a function of y/R, where y is distance from the wall and R is the pipe radius. Show on the plot the value of the dynamic viscosity μ. [*Hint:* Verify first that $\mu \ll \eta$.] (b) Compare the shear stress at a radius of 1.8 ft with the value that would exist if the flow were laminar at the same discharge.

r, ft	0*	0.3	0.6	0.9	1.2	1.5	1.8
$\bar{u}$, ft/sec	6.6	6.5	6.25	6.1	5.7	5.2	4.4

* Pipe centerline

11–7. Using an order-of-magnitude argument verify the three equations in (11–29a) and (11–29b).

11–8. Discuss *boundary-layer transition* and *boundary-layer separation*. Illustrate your discussion with carefully drawn sketches and descriptions of one example of each of these phenomena.

REFERENCES

1. PRANDTL, L., *Essentials of Fluid Dynamics*, Hafner Publishing Co., Inc., New York, 1952, Article II b.
2. SCHUBAUER, G. B., and H. K. SKRAMSTAD, "Laminar Boundary Layer Oscillations on a Flat Plate," *J. Aeron. Sci.*, **14**, 69 (1947). Also see *Nat. Advisory Comm. Aeron.*, Rept. 909 (1948).
3. EMMONS, H. W., "The Laminar Transition in a Boundary Layer—Part I," *J. Aeron. Sci.*, **18**, 7 (July, 1951) is a good example.
4. LINDGREN, E. R., "Some Aspects of the Change Between Laminar and Turbulent Flow of Liquids in Turbulent Tubes," *Arkiv Fysik*, **7**, 293 (1954).
5. PRANDTL, L., *op. cit.*, pp. 117–118.
6. VON KÁRMÁN, T., "Turbulence and Skin Friction," *J. Aeron. Sci.*, **1**, 1 (1934).

CHAPTER 12

Wall Turbulence. Boundary-Layer Flows

12–1 INTRODUCTION

Turbulence occurs most commonly in various forms of shear flow, i.e., flows in which there is a spatial variation of the mean velocity. Shear flows can be classified according to whether they occur along solid surfaces or at the interface between fluid zones having different velocities, such as at the boundaries of a jet. Turbulence generated in the shear flow near a solid boundary is called wall turbulence; that due to the velocity differential between fluid zones is called free turbulence. Turbulent motion in shear flows tends to be self-sustaining in the sense that turbulence arises as a consequence of the shear and the shear persists as a consequence of the turbulent fluctuations. Turbulence can neither arise nor persist without shear. Turbulent fluctuations can only decay in a shear-free field.

Wall turbulence may be the consequence of the shear flow in nonuniform boundary layers as found on immersed bodies, or in uniform layers found with fully developed flow in uniform conduits. This chapter will discuss nonuniform boundary layers for incompressible flow along walls both with and without pressure gradients.

12–2 STRUCTURE OF A TURBULENT BOUNDARY LAYER

Consider a fluid stream flowing past a solid boundary. Shear at the boundary retards the fluid and establishes a boundary-layer zone of viscous influence. For smooth bodies, this layer is initially laminar (as in Chapter 10) and is described by $u = u(y)$. If the Reynolds number $\mathbf{R}_x$ exceeds a certain critical value $\mathbf{R}_{crit}$, the retarded flow becomes unstable and turbulence may set in. With turbulence, the boundary layer is defined in terms of the temporal mean velocity $\bar{u} = \bar{u}(y)$. The turbulence quickly engulfs the original laminar boundary layer and reaches out into the free stream to entrain and mix more fluid to form a thicker boundary layer. At the same time, the mean velocity very near the boundary is increased, giving a velocity profile which is fuller than in the laminar case. If the body is rough, the turbulent boundary layer may be established very near the leading edge of the body without a preceding stretch of laminar flow.

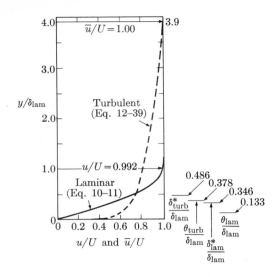

FIG. 12–1.   Comparison of laminar and turbulent boundary-layer profiles on a smooth wall: $dp/dx = 0$, $\mathbf{R}_x = 500,000$.

Figure 12–1 compares laminar and turbulent boundary-layer profiles for flows of the same Reynolds number, $\mathbf{R}_x = 500{,}000$, past a flat plate. This Reynolds number is in the critical range where a transition to turbulence sometimes occurs, and where either a laminar or turbulent boundary layer might be expected. The pronounced effect of turbulence in increasing the thickness is illustrated. In this example,

$$\frac{\delta_{\text{turb}}}{\delta_{\text{lam}}} = 3.9.$$

Also shown are the displacement thickness δ^* (Eq. 8–9b) and the momentum thickness θ (Eq. 8–10b). Neither of these increases by as large a ratio as δ because of the higher flux of mass and momentum through the zone nearest the wall. They are, respectively,

$$\frac{\delta_{\text{turb}}^*}{\delta_{\text{lam}}^*} = 1.41 \quad \text{and} \quad \frac{\theta_{\text{turb}}}{\theta_{\text{lam}}} = 2.84.$$

Outside a boundary layer, the free-stream flow is nearly shearless and the field of mean velocities conforms closely to potential flow. This field usually is slightly turbulent but is considered to be nonturbulent relative to the much higher turbulence inside a turbulent boundary layer. A close examination of the interior of the turbulent layer shows it to be composed of regions of different types of flow. Although there is a smooth and continuous transition of the mean velocity from the wall outward into the nonturbulent flow, the instantaneous border between turbulent and nonturbulent fluid is irregular and changing. At a particular instant, the border consists of fingers of turbulence extending into the nonturbulent fluid and fingers of nonturbulent fluid extending deep into the turbulent region. If a hot-wire anemometer is used for measuring the velocity in the border region, the record caused by the passing stream will show turbulent and nonturbulent conditions intermittently. Deep in the boundary layer, the intermittency factor Ω, the fraction of time during which the flow is turbulent, is 1.0. In the free stream it is zero. The eddies in the zone of intermittency are similar to the eddies in the wake of an immersed body. Figure 12–2 is a diagram constructed from hot-wire measurements showing a typical instantaneous composition of a turbulent boundary layer. The figure shows turbulent fluid overreaching δ to distances as great as $1.2\,\delta$. Nonturbulent fluid protrudes into the boundary layer as far as $0.4\,\delta$. The average position of the turbulent-nonturbulent interface is $0.78\,\delta$, 22% short of the distance over which a mean shear stress exists.

The nonturbulent region well beyond δ has the free-stream velocity. In the intermittent zone, the mean velocity is reduced. However, recordings taken with a fixed hot wire show the nonturbulent and the turbulent fluid passing through this intermediate zone to have different velocities. The nonturbulent is higher with a step to a lower velocity as a turbulent patch passes.

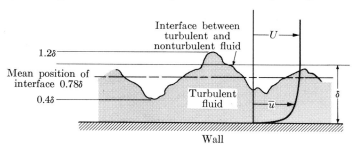

FIG. 12–2. Schematic diagram of a turbulent boundary layer.

The turbulent energy in a boundary layer is measured in terms of the mean square values of the turbulent fluctuations, as defined by Eq. (11–5a). A useful dimensionless energy expression is

$$\frac{\overline{u'^2} + \overline{v'^2} + \overline{w'^2}}{u_*^2},\tag{12–1}$$

where

$$u_* = \sqrt{\tau_0/\rho} = \text{shear velocity.}\tag{12–2}$$

Figure 12–3 [1] is a dimensionless plot of a measured distribution of energy across a boundary layer on a smooth wall. The data are for $\mathbf{R}_\delta = U\delta/\nu$ of 73,000. For a turbulent layer, this corresponds approximately to $\mathbf{R}_x = 4 \times 10^6$. The plotted values are temporal mean squares from records including both turbulent

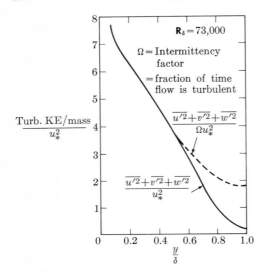

FIG. 12–3.   Turbulent energy in a boundary layer [1].

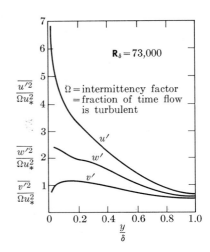

FIG. 12–4.   Turbulence intensities in a boundary layer [1].

and nonturbulent regions. Assuming no fluctuations in the nonturbulent regions and no effect of the step in velocity between regions, division by the intermittency factor should give the turbulent energy average for the turbulent regions only. This is shown by the dashed line. Figure 12–4 [1] shows the distributions for each of the components computed for the turbulent regions only. These records show the important features of peaks very near the wall with decreases toward the wall and toward the free stream. At a smooth wall the fluctuations are effectively zero. In fact, the flow is laminar in the limit. However, at $y = \delta$, the turbulent energy remains finite in keeping with the observation that the turbulent fluid extends beyond δ into the nonturbulent free stream. With a rough boundary, turbulent fluctuations may exist clear to the wall depending on the magnitude of the roughness.

12–3   MEAN-FLOW CHARACTERISTICS

It is useful to have relations describing the mean-flow characteristics which will permit the prediction of velocity magnitudes throughout the boundary layer as well as their relation to wall shear and pressure gradient forces in different situations. Furthermore, it is desirable that these relations should not require knowledge of the turbulence details. In principle, it should be possible to obtain a solution by integrating the equations of the mean motion in analogy to the laminar boundary-layer solution of Chapter 10. It should only be necessary to replace the momentum stresses by a suitable stress-strain relation for the turbulent mean flow. However, as we have just noted in the previous paragraphs, the turbulent boundary layer is composed of zones of different types of flow. Thus the effective viscosity, which is the sum of the dynamic molecular and dynamic eddy viscosities in Eq. (11–10), varies from the wall out through the layer. Consequently, a theoretical solution is not practical for the general nonuniform boundary layer. Instead a semiempirical procedure is followed in which the form of the relation between wall shear and velocity distribution is predicted.

We will consider first the case of zero or negligible pressure gradients. More generally, the pressure gradient influences the velocity profile, and hence the wall shear, but we will postpone a discussion of this effect.

12–3.1   Universal velocity and friction laws: smooth walls

Velocity-profile regions. As for the instantaneous boundary layer, distinct regions are shown in the mean velocity profile along a smooth boundary. These are illustrated in Fig. 12–5(a). At the wall there is a zone in which $d\bar{u}/dy$ is very nearly linear. This is a zone in which the mean shear stress is controlled by the dynamic molecular viscosity μ. Recent investigations [2] have shown that the flow structure in this region is highly three-dimensional with intermittent formation of swirling eddies which streak in the flow direction. However, the energy of the fluctuations is practically zero, and the mean flow is laminar. This region has been named the laminar sublayer.

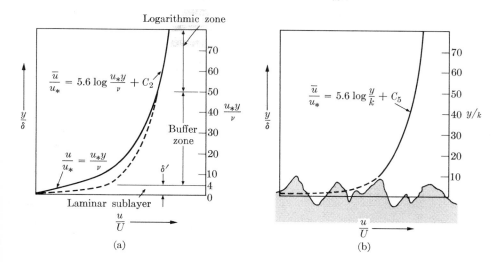

FIG. 12–5.   Velocity-profile regions. (a) Close to a smooth wall the velocity profile (solid line) deviates from the logarithmic equation (dashed line). (b) In the case of a rough wall the mean velocity $\bar{u}$ follows a logarithmic law all the way to the roughness projections of the wall.

Proceeding beyond the thin sublayer, one finds a second narrow zone in which the laminar motion undergoes a change to intense turbulence. This is sometimes called the *buffer* zone. It is in this region that the sharp peak in turbulent energy occurs (Fig. 12–3). Farther out into the boundary layer, but inside $y = 0.4\delta$ where the intermittent behavior (Fig. 12–2) begins, the flow is fully turbulent. The intensity of turbulence decreases from its peak value, as we saw in Fig. 12–4. In the range of $y = 0.4\delta$ to 1.2δ, the flow is intermittently turbulent and nonturbulent with continual decrease in turbulent energy. Of course, all zones merge smoothly into a continuous mean velocity profile.

Large roughness will disrupt the laminar sublayer as illustrated in Fig. 12–5(b). Under such conditions, turbulence extends to the wall.

The wall law and velocity-defect law. Unfortunately no single equation will describe the velocity profile over its entire thickness δ. It is found, however, that equations which tend to be universal can be developed for the several zones of a boundary layer beginning with two different functional relations or laws. A *law of the wall* applies close to smooth boundaries, and a *velocity-defect law* applies to the outer reaches of boundary layers for both smooth and rough walls.

Close to a smooth boundary, molecular viscosity has a dominant influence in producing the velocity profile and in generating turbulence. The law of the wall assumes that the relation between wall shear stress and the velocity $\bar{u}$ at distance y from the wall depends only on fluid density and viscosity. Expressed nondimensionally, the functional relation is

$$\frac{\bar{u}}{u_*} = f\left(\frac{u_* y}{\nu}\right). \tag{12–3}$$

The existence of a unique law of the wall is shown clearly in a semilog plot of $\bar{u}/u_*$ versus $u_* y/\nu$ such as Fig. 12–6. This figure, which is based on experimental data from many sources, shows $\bar{u}/u_*$ as a single line up to values of $(u_* y)/\nu$ ranging from 500 to 2000. At higher values of $(u_* y)/\nu$, a single velocity law no longer holds because viscosity is no longer a major factor. Instead, the upper limit depends primarily on y/δ, the relative distance from the wall. In the range of the law of the wall distinct regions can be defined as follows:

Laminar sublayer: $0 < \dfrac{u_* y}{\nu} < 4,$

Buffer zone: $4 < \dfrac{u_* y}{\nu} < 30 \text{ to } 70,$ (12–4)

Turbulent zone: $\dfrac{u_* y}{\nu} > 30 \text{ to } 70, \quad y/\delta < 0.2.$

The sublayer and turbulent zones have special significance. In the laminar sublayer the velocity is $\bar{u} \equiv u$ and the velocity gradient $\partial u/\partial y$ is nearly constant.

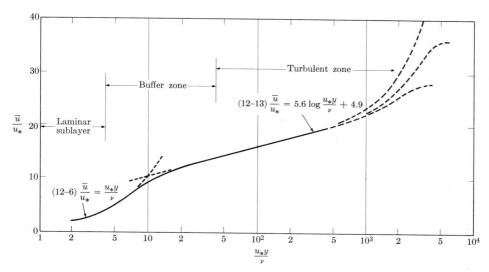

FIG. 12–6. Velocity-curves profile illustrating the law of the wall.

Hence, the shear stress may be assumed constant and equal to τ_0 as follows

$$\tau \approx \tau_0 = \mu \left.\frac{\partial u}{\partial y}\right|_{y=0} \equiv \mu \frac{u}{y}. \tag{12–5}$$

Then Eq. (12–3) becomes

$$\frac{u}{u_*} = \frac{u_* y}{\nu}, \tag{12–6}$$

which can be shown by rearranging Eq. (12–5). We can define a thickness of the laminar sublayer as the value of y at $u_* y/\nu = 4$ or

$$\delta' = \frac{4\nu}{u_*} = \frac{4\nu}{U\sqrt{c_f/2}}, \tag{12–7}$$

since

$$\tau_0 = c_f \rho U^2/2,$$

and c_f is the local shear-stress coefficient. Thus δ' decreases as the shear stress increases. We will see later that c_f decreases slowly with increasing Reynolds number $\mathbf{R}_x$ (as was true for the laminar boundary layer), so that δ' increases with distance along the surface if U/ν is constant.

In the turbulent zone of this close-to-the-wall region, the mean shear remains nearly equal to the wall shear. With the approximation $\tau \approx \tau_0$, the first of Eqs. (11–29) for the turbulent boundary layer reduces to

$$\rho \left(\frac{\partial \bar{u}}{\partial t} + \bar{u} \frac{\partial \bar{u}}{\partial x} + \bar{v} \frac{\partial \bar{u}}{\partial y} \right) + \frac{\partial(\bar{p} + \overline{\rho u'^2})}{\partial x} \approx \frac{\partial \tau_0}{\partial y} = 0, \tag{12–8}$$

where $\tau_0 \approx \tau = \mu(d\bar{u}/dy) - \rho\overline{u'v'}$. Hence a solution is obtained if we have a relation for τ_0. Let us assume that $\tau = \tau_0$ is given by Eq. (11–32) in terms of Prandtl's mixing length, namely

$$\tau_0 \approx \tau = \rho l^2 \left|\frac{d\bar{u}}{dy}\right| \frac{d\bar{u}}{dy}.$$

Integrating this with the assumption that the mixing length depends on distance from the wall by

$$l = \kappa y \tag{12–9}$$

yields

$$\frac{\bar{u}}{u_*} = \frac{1}{\kappa} \ln y + C_1. \tag{12–10}$$

Extrapolated toward the wall, the logarithm gives a definite value $y = y'$ at $\bar{u} = 0$. This "positioning" of the profile depends on ν and τ_0. Putting y' proportional to ν/u_*, which is a characteristic length, at $\bar{u} = 0$ we have

$$C_1 = -\frac{1}{\kappa} \ln y' = C_2 - \frac{1}{\kappa} \ln \frac{\nu}{u_*}. \tag{12–11}$$

Introducing C_1 into Eq. (12–10) and changing the logarithm to base 10 gives

$$\frac{\bar{u}}{u_*} = \frac{2.3}{\kappa} \log \left(\frac{u_* y}{\nu}\right) + C_2, \tag{12–12}$$

where κ and C_2 are dimensionless constants. Because of the conditions of the derivation, this equation is expected to hold only over a limited portion of the boundary layer.

Alternatively, the Kármán relation, Eq. (11–34), can be used to derive a logarithmic equation. If we make the same approximation that $\tau \approx \tau_0$, Eq. (12–12) is obtained, again with κ and C_2 as dimensionless constants. This suggests that κ should be a universal constant as Kármán's development concluded. Experiments show that κ falls within a definite range of values but does not have a universal single value. For the inner region of the boundary layer, where Eq. (12–12) is meant to apply, empirical values [3] for κ and C_2 which hold up to approximately $y/\delta = 0.15$ are $\kappa = 0.41$ and $C_2 = 4.9$. Hence

$$\frac{\bar{u}}{u_*} = 5.6 \log \left(\frac{u_* y}{\nu}\right) + 4.9, \quad \frac{u_* y}{\nu} > 30 \text{ to } 70, \quad \frac{y}{\delta} < 0.15. \tag{12–13}$$

In the outer reaches of the turbulent boundary layer, Reynolds stresses dominate the viscous stresses to produce the velocity profile. Evidence of this appears in the velocity-defect law which stems from early observations of boundary layers in pipes and channels and later along walls. It was observed that the velocity reduction, or defect, $(U - \bar{u})$ at y-values well out into the boundary layer was almost solely dependent on the magnitude of the wall shear stress and independent of how the stress arose. Thus, for both rough and

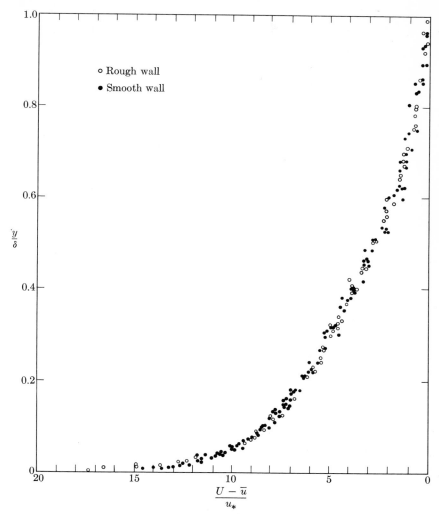

FIG. 12–7.   Velocity-defect profiles for rough and smooth walls.

smooth walls, a dimensionless relation similar to the following should hold:

$$\frac{U - \bar{u}}{u_*} = g\left(\frac{y}{\delta}\right). \tag{12–14}$$

An example is shown in Fig. 12–7, where data are plotted for constant-pressure boundary layers due to different roughnesses and Reynolds numbers. The general correlation is surprisingly good over most of the turbulent zones of the boundary layer. It cannot hold in the laminar sublayer but apparently will overlap into the turbulent region where Eq. (12–3) applies.

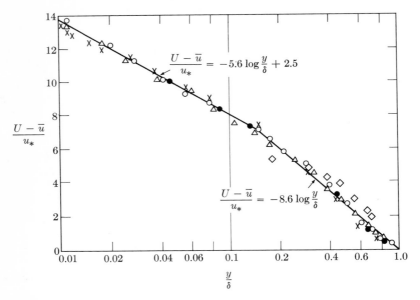

FIG. 12–8. Velocity-defect law: ✕ $R_\delta = 10.1 \times 10^4$ — Freeman [4]; ○ 4.8 $\times 10^4$ — Klebanoff and Diehl [5]; △ 15.2 $\times 10^4$ — Klebanoff and Diehl [5]; ● 2.7 $\times 10^4$ — Schultz-Grunow [6]; ◇ — Moore [7].

A logarithmic relation for the function g can be obtained by assuming that Eq. (12–12) will give $\bar{u} = U$ if we put $y = \delta$ and modify C_2. Then, by subtracting, we get

$$\frac{U - \bar{u}}{u_*} = -\frac{2.3}{\kappa} \log\left(\frac{y}{\delta}\right) + C_3, \qquad (12\text{–}15)$$

where κ and C_3 are empirical constants.

Figure 12–8 [4–7] shows a semilog plot of velocity-defect data for zero-pressure gradient flow along flat plates. As in Fig. 12–7, the experiments correlate well on the velocity-defect basis. However, a single log-equation does not fit the data over the entire boundary layer. Instead one equation will fit an inner region overlapping with Eq. (12–13). A second will approximate the outer region. We thus have for the inner region, where $\kappa = 0.41$ and $C_3 = 2.5$,

$$\frac{U - \bar{u}}{u_*} = -5.6 \log\left(\frac{y}{\delta}\right) + 2.5, \qquad \frac{y}{\delta} < 0.15; \qquad (12\text{–}16)$$

and for the outer region, where $\kappa = 0.267$ and $C_3 = 0$,

$$\frac{U - \bar{u}}{u_*} = -8.6 \log\left(\frac{y}{\delta}\right), \qquad \frac{y}{\delta} > 0.15. \qquad (12\text{–}17)$$

These two equations are shown in Fig. 12–8. Let us emphasize that Eqs. (12–16) and (12–17) apply to both smooth and rough surfaces. Moreover, Eq. (12–16) overlaps with Eq. (12–13).

The above equations which stem from the law of the wall and the velocity-defect law seem to be universal in that they apply over a wide range of Reynolds numbers. The relations are summarized in Table 12–1.

Surface-resistance formulas: smooth walls. The velocity-profile equations are also shear-stress relations, as we can see by noting that

$$u_* = \sqrt{\tau_0/\rho} = U\sqrt{c_f/2}, \tag{12–18}$$

in which c_f is the local shear-stress coefficient as defined by Eq. (8–19). Using this, and again assuming that Eq. (12–12) will give $\bar{u} = U$ if we put $y = \delta$ and modify C_2, we obtain

$$\sqrt{2/c_f} = \frac{2.3}{\kappa} \log\left(\frac{U\delta}{\nu} \sqrt{c_f/2}\right) + C_4. \tag{12–19}$$

Here κ and C_4 are best evaluated directly from friction-loss data in view of the approximation introduced by extrapolating Eq. (12–12) to $y = \delta$. We see from Eq. (12–19) that c_f is a function only of Reynolds number for smooth boundaries.

Equation (12–19) has the drawback that c_f is not given explicitly. Moreover, δ is not easily established so that $\mathbf{R}_\delta$ lacks precision. It would be better to use the displacement thickness δ^* or the momentum thickness θ. Approximate relations that include converting to δ^* or θ and evaluating constants from flat-plate friction data are

$$\text{Clauser [3]:} \quad 1/\sqrt{c_f} = 3.96 \log \mathbf{R}_{\delta^*} + 3.04. \tag{12–20}$$

$$\text{Squire and Young [8]:} \quad 1/\sqrt{c_f} = 4.17 \log \mathbf{R}_\theta + 2.54, \tag{12–21}$$

where

$$\mathbf{R}_{\delta^*} = \frac{U\delta^*}{\nu},$$
$$\tag{12–22}$$
$$\mathbf{R}_\theta = \frac{U\theta}{\nu}.$$

The boundary-layer thickness depends on the distance x measured from the leading edge of the surface, and $\mathbf{R}_{\delta^*}$ and $\mathbf{R}_\theta$ are functions of $\mathbf{R}_x$. Assuming turbulence all the way from the leading edge (i.e., no preceding stretch of laminar boundary layer), Kármán [9] converted Eq. (12–19) and evaluated the constants from direct measurements of local shear stresses to obtain

$$1/\sqrt{c_f} = 4.15 \log (\mathbf{R}_x c_f) + 1.7. \tag{12–23}$$

TABLE 12–1

VELOCITY-PROFILE EQUATIONS FOR BOUNDARY LAYERS WITH $d\bar{p}/dx = 0$

	Zone	Smooth walls	Rough walls
		UNIVERSAL EQUATIONS	
Law of the wall:			
Laminar sublayer	$\dfrac{u_* y}{\nu} < 4$	(12-6) $\dfrac{u}{u_*} = \dfrac{u_* y}{\nu}$	—
Buffer zone	$4 < \dfrac{u_* y}{\nu} < 30$ to 70	—	—
Logarithmic zone	$\left.\begin{array}{c}\dfrac{u_* y}{\nu} > 30 \text{ to } 70 \\[4pt] \dfrac{y}{\delta} < 0.15\end{array}\right\}$	(12-13) $\dfrac{\bar{u}}{u_*} = 5.6 \log \dfrac{u_* y}{\nu} + 4.9$	(12-43) $\dfrac{\bar{u}}{u_*} = -5.6 \log \dfrac{k}{y} + C_5$ $C_5 = f$ (roughness size, shape, and distribution)
Velocity-defect law:			
Inner region (overlaps with logarithmic wall law)	$\dfrac{y}{\delta} < 0.15$	(12-16) $\dfrac{U - \bar{u}}{u_*} = -5.6 \log \dfrac{y}{\delta} + 2.5$	
Outer region (approximate formula)	$\dfrac{y}{\delta} > 0.15$	(12-17) $\dfrac{U - \bar{u}}{u_*} = -8.6 \log \dfrac{y}{\delta}$	
		POWER LAW	
(3,000 < $\mathbf{R}_\delta$ < 70,000) Outer region	—	(12-35) $\dfrac{u}{u_*} = 8.74 \left(\dfrac{u_* y}{\nu}\right)^{1/7}$	—

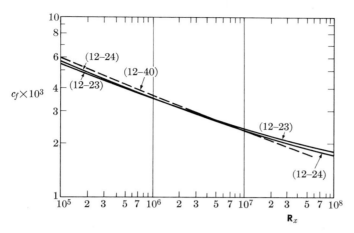

FIG. 12–9.　Local coefficient of resistance.

This equation is useful for ships and aircraft wings where the laminar boundary layer extends only an insignificant distance before transition to turbulence. An extrapolation formula developed by Schultz-Grunow [6] gives c_f explicitly as

$$c_f = \frac{0.370}{(\log \mathbf{R}_x)^{2.58}}.\qquad(12\text{–}24)$$

Equations (12–23) and (12–24) are compared graphically in Fig. 12–9.

The average shear coefficient over a distance l along a flat plate of width b is defined by

$$C_f = \frac{\text{total surface drag}}{bl\rho U^2/2},\qquad(12\text{–}25)$$

where the total surface drag is the integral effect of the local wall-shear stresses. Using measurements of the total surface drag, Schoenherr [10] derived the following widely used equation for boundary layers that are turbulent from the leading edge,

$$1/\sqrt{C_f} = 4.13 \log (\mathbf{R}_l C_f),\qquad(12\text{–}26)$$

where

$$\mathbf{R}_l = Ul/\nu.$$

This expresses C_f in terms of $\mathbf{R}_l$ in parallel to Kármán's relation for c_f. Schultz-Grunow's extrapolation formula for average shear is

$$C_f = \frac{0.427}{(\log \mathbf{R}_l - 0.407)^{2.64}},\qquad(12\text{–}27)$$

which applies when $10^6 < \mathbf{R}_l < 10^9$. The last two equations are compared in Fig. 12–10.

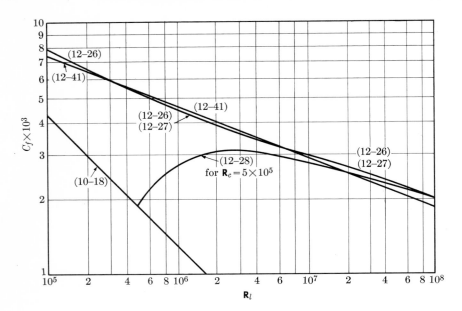

FIG. 12–10. Average coefficient of resistance for flat plates.

When there is a significant stretch of laminar boundary layer preceding the turbulent layer, the total friction is the laminar portion up to x_{crit} plus the turbulent portion from x_{crit} to l. Therefore, the average shear coefficient is lower than Eqs. (12–26) and (12–27) predict, falling between the laminar and turbulent curves in Fig. 12–10. A transition formula for C_f is

$$C_f = \frac{0.427}{(\log \mathbf{R}_l - 0.407)^{2.64}} - \frac{A}{\mathbf{R}_l}, \qquad (12\text{–}28)$$

where the correction $A/\mathbf{R}_l$ is computed by proportioning the laminar and turbulent friction. Following Prandtl it is assumed that the turbulent portion of the boundary layer behaves as though it were turbulent from the leading edge. Computed magnitudes of A for different $\mathbf{R}_{\text{crit}}$ values are given in Table 12–2.

TABLE 12–2*

VALUES OF TRANSITION CONSTANT A IN EQ. (12–28)

$\mathbf{R}_{\text{crit}} = \dfrac{Ux_{\text{crit}}}{\nu}$	3×10^5	4×10^5	5×10^5	6×10^5	10^6
A	1060	1400	1740	2080	3340

* Values in this table were computed using Eqs. (10–18) and (12–41).

TABLE 12-3

SURFACE RESISTANCE FORMULAS FOR BOUNDARY LAYERS WITH $d\bar{p}/dx = 0$

	Smooth walls	Rough walls
LOCAL SHEAR		
Universal equations		
Clauser (12-20)	$1/\sqrt{c_f} = 3.96 \log \mathbf{R}_{\delta^*} + 3.04$	(12-46) $\dfrac{1}{\sqrt{c_f}} = 3.96 \log \dfrac{\delta}{k} + C_8$ $C_8 = f$ (size, shape, and distribution of roughness)
Squire and Young (12-21)	$1/\sqrt{c_f} = 4.17 \log \mathbf{R}_\theta + 2.54$	
von Kármán (12-23)	$1/\sqrt{c_f} = 4.15 \log (\mathbf{R}_x c_f) + 1.7$	
Schultz-Grunow (12-24)	$c_f = \dfrac{0.370}{(\log \mathbf{R}_x)^{2.58}}$	
Power law (12-40)	$c_f = \dfrac{0.0466}{\mathbf{R}_\delta^{1/4}} = \dfrac{0.059}{\mathbf{R}_x^{1/5}}$	
AVERAGE SHEAR		
Universal equations		
Schoenherr (12-26)	$1/\sqrt{C_f} = 4.13 \log (\mathbf{R}_l C_f)$	
Schultz-Grunow (12-27)	$C_f = \dfrac{0.427}{(\log \mathbf{R}_l - 0.407)^{2.64}}$	
Power law (12-41)	$C_f = \dfrac{0.074}{\mathbf{R}_l^{1/5}}$	
Transition formula		
Schultz-Grunow-Prandtl (12-28)	$C_f = \dfrac{0.427}{(\log \mathbf{R}_l - 0.407)^{2.64}} - \dfrac{A}{\mathbf{R}_l}$ $A = f(\mathbf{R}_{\mathrm{crit}})$ as given in Table 12-2	

A typical transition for $\mathbf{R}_{crit} = 5 \times 10^5$ is shown in Fig. 12–10. For flow along smooth boundaries with $d\bar{p}/dx \approx 0$, $\mathbf{R}_{crit}$ falls in the approximate range

$$300{,}000 < \mathbf{R}_{crit} < 600{,}000,$$

depending on the free-stream turbulence. High turbulence causes lower $\mathbf{R}_{crit}$. Roughness projections will also cause earlier transition.

When $d\bar{p}/dx < 0$, transition is delayed and $\mathbf{R}_{crit}$ can reach 10^6 or more. This feature has been used to design *laminar-flow airfoils* which have low drag because x_{crit}/l approaches 0.7 to 0.8.

Table 12–3 summarizes the several surface-resistance formulas for smooth walls with $d\bar{p}/dx \approx 0$.

Example 12–1: Turbulent boundary-layer velocities and thickness

An aircraft flies at 25,000 ft with a speed of 410 mph (600 ft/sec). Assuming standard air, compute the following items for the boundary layer at a distance 10 ft from the leading edge of the wing of the craft. (Treat the wing surface as a smooth flat plate). (a) Thickness δ' of the laminar sublayer. (b) Velocity $\bar{u}$ at $y = \delta'$. (c) Velocity $\bar{u}$ at $y/\delta = 0.15$. (d) The distance y at $y/\delta = 0.15$ and thickness δ.

From Table 1–7 for standard air at 25,000 ft elevation, we have

$$\nu = 3 \times 10^{-4} \text{ ft}^2/\text{sec}, \qquad \rho = 1.07 \times 10^{-3} \text{ slug/ft}^3, \qquad \mathbf{R}_x = \frac{600 \times 10}{3 \times 10^{-4}} = 2 \times 10^7.$$

Since $\mathbf{R}_{crit}$ is in the range of 5×10^4, the transition to turbulence in the boundary layer occurs within 3 in. of the leading edge of the wing. Hence the laminar portion may be neglected and the boundary layer computed as though it were turbulent from the leading edge. Then, using Eq. (12–24), we have

$$c_f = \frac{0.37}{(\log \mathbf{R}_x)^{2.58}} = \frac{0.37}{(7.30)^{2.58}} = 2.18 \times 10^{-3}.$$

Hence

$$\tau_0 = c_f \rho \frac{U^2}{2} = 2.18 \times 10^{-3} \times 1.07 \times 10^{-3} \frac{(600)^2}{2} = 0.42 \text{ lb/ft}^2,$$

$$u_* = \sqrt{\tau_0/\rho} = U\sqrt{c_f/2} = 19.8 \text{ ft/sec}.$$

(a) The laminar sublayer thickness is given by Eq. (12–7),

$$\delta' = \frac{4\nu}{u_*} = \frac{4\nu}{U\sqrt{c_f/2}} = \frac{4 \times 3 \times 10^{-4}}{19.8} = 0.000060 \text{ ft} = 0.00072 \text{ in.}$$

(b) If we use Eq. (12–5), $\bar{u}$ at $y = \delta'$ is given as

$$\tau \approx \tau_0 \approx \mu \frac{\bar{u}}{y},$$

or

$$\bar{u} \approx \frac{\tau_0 y}{\rho \nu} \approx \frac{0.42 \times 6.0 \times 10^{-5}}{1.07 \times 10^{-3} \times 3 \times 10^{-4}} = 79 \frac{\text{ft}}{\text{sec}}.$$

(c) Using Eq. (12–16), we can obtain $\bar{u}$ at $y/\delta = 0.15$:

$$\frac{U - \bar{u}}{u_*} = -5.6 \log \frac{y}{\delta} + 2.5$$

$$\frac{600 - \bar{u}}{19.8} = -5.6 \log 0.15 + 2.5 = 7.11,$$

which gives $\bar{u} = 459$ ft/sec.

(d) We can obtain y at $y/\delta = 0.15$, using the previous result in Eq. (12–13),

$$\frac{\bar{u}}{u_*} = 5.6 \log \frac{u_* y}{\nu} + 4.9 = \frac{459}{19.8},$$

from which

$$y = 0.028 \text{ ft} = 0.33 \text{ in.}$$

Then the thickness is

$$\delta = \frac{y}{0.15} = \frac{0.028}{0.15} = 0.186 \text{ ft} = 2.24 \text{ in.}$$

Example 12–2: Surface resistance on a smooth boundary

For the conditions of Example 12–1 compute the displacement thickness δ^* and determine the local surface-resistance coefficient c_f using Eq. (12–20).

From Eq. (8–9b) we can obtain

$$\frac{\delta^*}{\delta} = \int_0^{h/\delta} \left(1 - \frac{\bar{u}}{U}\right) d\left(\frac{y}{\delta}\right),$$

where $h/\delta \geq 1$. In the turbulent zone of the boundary layer the velocity defect is given by Eqs. (12–16) and (12–17). In dimensionless terms,

$$1 - \frac{\bar{u}}{U} = -5.6 \frac{u_*}{U} \log \frac{y}{\delta} + 2.5 \frac{u_*}{U}, \qquad \frac{y}{\delta} < 0.15 \qquad (12\text{–}16)$$

$$1 - \frac{\bar{u}}{U} = -8.6 \frac{u_*}{U} \log \frac{y}{\delta}, \qquad \frac{y}{\delta} > 0.15 \qquad (12\text{–}17)$$

Neglecting the contribution of the laminar sublayer to the velocity defect and approximating the contribution of buffer zone with Eq. (12–16), we can evaluate δ^*/δ as follows:

$$\frac{\delta^*}{\delta} = \frac{u_*}{U} \int_{\delta'/\delta}^{0.15} \left(-2.43 \ln \frac{y}{\delta} + 2.5\right) d\left(\frac{y}{\delta}\right) + \frac{u_*}{U} \int_{0.15}^{1.0} \left(-3.74 \ln \frac{y}{\delta}\right) d\left(\frac{y}{\delta}\right),$$

which gives $\delta^*/\delta \approx 3.74(u^*/U)$. Then

$$\delta^* = 3.75 \frac{19.8}{600} 0.186 = 0.023 \text{ ft.}$$

Using Eq. (12–20) for c_f, we have

$$\frac{1}{\sqrt{c_f}} = 3.96 \log \mathbf{R}_{\delta^*} + 3.04, \qquad \mathbf{R}_{\delta^*} = \frac{600 \times 0.023}{3 \times 10^{-4}} = 46,000.$$

Then

$$c_f = \frac{1}{(3.96 \times 4.663 + 3.04)^2} = 21.6 \times 10^{-3}.$$

12–3.2   Power-law formulas: smooth walls

The equations introduced in the previous sections for the velocity profile and the shear-stress coefficient have the important property of being practically universal; that is, they apply over almost the entire range of Reynolds numbers for which experimental data exist.   Historically, the first attempt to relate turbulent-flow velocity profiles to wall shears produced *power-law* formulas rather than the logarithmic equations.   These power laws are limited in the Reynolds-number range over which they apply.   However, they are important and useful because the relations are simple and give $\bar{u}/U$ and c_f in explicit terms. Moreover, they yield an explicit relation for boundary-layer thickness δ in terms of Reynolds number and distance x which the universal relations do not produce.

The power laws stem from two facts that hold for turbulent boundary layers with negligible pressure gradients when $\mathbf{R}_\delta < 5 \times 10^5$. First, except very near the wall, the mean velocity is closely proportional to a root of the distance y from the wall.   Second, the shear coefficient c_f is inversely proportional to a root of $\mathbf{R}_\delta$.   Let us begin with the latter and write

$$c_f = \frac{A}{(U\delta/\nu)^m}, \tag{12–29}$$

where m and A are constants.   Note that this is similar to Eq. (10–15a) for laminar flow, where $m = 1$.   Using this equation and Eq. (12–18), we can derive

$$U/u_* = B(u_*\delta/\nu)^{m/(2-m)}, \tag{12–30}$$

where B is another constant. Assume now that the local mean velocity $\bar{u}$ depends on y by the same relation, namely

$$\bar{u}/u_* = B(u_*y/\nu)^{m/(2-m)}. \tag{12–31}$$

The ratio of the last two equations is

$$\bar{u}/U = (y/\delta)^{m/(2-m)}. \tag{12–32}$$

By this equation, all profiles are *similar* and can be represented by a single dimensionless curve. Laminar boundary-layer profiles also possess this similarity (Section 10–2 and Fig. 10–4). However, turbulent profiles are not truly similar, and Eq. (12–32) will apply for different Reynolds-number ranges only if the constant m is varied.

Empirical information, first from pipe flows and later from boundary-layer measurements, shows that in the range of $3000 < \mathbf{R}_\delta < 70{,}000$, a value of $m = \frac{1}{4}$ expresses c_f in Eq. (12–29) with good accuracy. Then

$$\bar{u}/U = (y/\delta)^{1/7}. \tag{12–33}$$

Evaluating constants A and B from experimental data results in

$$U/u_* = 8.74(u_*\delta/\nu)^{1/7}, \tag{12–34}$$

$$\bar{u}/u_* = 8.74(u_*y/\nu)^{1/7}, \tag{12–35}$$

and

$$c_f = 0.0466/\mathbf{R}_\delta^{1/4}. \tag{12–36}$$

To convert the above relations to the variable x measured from the leading edge of a body, we make use of the integral-momentum equation from Chapter 8 which, as pointed out there, is applicable to turbulent flow. Using Eq. (8–18a) for the case of $d\bar{p}/dx = 0$ and steady motion, we write

$$\frac{\tau_0}{\rho} = c_f \frac{U^2}{2} = U^2 \frac{d\theta}{dx} = U^2 \frac{d\theta}{d\delta} \frac{d\delta}{dx}, \tag{12–37}$$

where θ is the momentum thickness defined by Eq. (8–10b), namely

$$\theta = \int_0^h \frac{\bar{u}}{U}\left(1 - \frac{\bar{u}}{U}\right) dy.$$

Since the velocity profiles are similar, θ varies linearly with δ. Using Eq. (12–33) and integrating gives

$$\theta = \tfrac{7}{72}\, \delta. \tag{12–38}$$

Introducing Eq. (12–36) and (12–38) into Eq. (12–37) and integrating with the boundary condition that $\delta = 0$ at $x = 0$, we get the explicit expression

$$\delta = \frac{0.38x}{(Ux/\nu)^{1/5}}, \qquad \mathbf{R}_x < 10^7. \tag{12–39}$$

Then Eq. (12–36) is transformed to

$$c_f = 0.059/\mathbf{R}_x^{1/5}, \qquad \mathbf{R}_x < 10^7. \tag{12–40}$$

The average shear coefficient for a distance $x = l$, as obtained by integrating, is

$$C_f = 0.074/\mathbf{R}_l^{1/5}, \qquad \mathbf{R}_l < 10^7. \tag{12–41}$$

This equation is compared with universal log-law equations in Fig. 12–10. Equation (12–41) was used in computing the transition constants in Table 12–2.

12–3.3 Laws for rough walls

Roughness and its basic effects. In the preceding developments for smooth walls the Reynolds number was the only characteristic parameter for velocity distribution and resistance. For rough surfaces, the roughness magnitude, form, and distribution also become parameters. For a given form and distribution, the magnitude k of the roughness projection should characterize the surface condition. For the natural roughnesses of most surfaces, k is a statistical quantity as illustrated in Fig. 12–11. For artificial roughnesses, more definite measures can be established from the geometry, as Fig. 12–11 also shows. One standard scale for comparing roughness effects derives from experiments with sand grains cemented to smooth surfaces. The roughness value is taken to be the height k_s which uniformly graded sand grains will project above the smooth surface. By comparing the hydrodynamic behavior with other types and magnitudes of roughness, an equivalent sand-grain roughness can be assigned. Whether or not a geometric or sand-grain equivalent is used to define k, both the form and distribution of roughness are relevant parameters. However it has not yet been possible to put both of these into a general formulation. Consequently, in the following, the general correlation of roughness effects is on the basis of magnitude only, and comparisons between roughness types are made by adjusting an empirical constant.

Roughness disrupts the laminar sublayer in amounts depending on the relative heights of the roughness projection k and sublayer thickness δ'. Using the sand-grain scale $k = k_s$ and δ' as defined by Eq. (12–7), we find that for $k_s/\delta' < 1$, roughness has negligible effect on the wall shear. Such surfaces are hydrodynamically smooth. Above this value, the roughness effects appear, the smooth-wall relations for velocity and c_f no longer hold, and the surface becomes hydrodynamically rough. When k_s/δ' exceeds 15 to 25, the friction and velocity distribution become dependent only on the roughness, and we speak of "fully rough" flow conditions.

Let us define k_{crit} as the critical roughness obtained when $k_s = \delta'$. According to Eq. (12–7), δ' increases with decreasing c_f, and hence with increasing downstream distance, for a constant U. Consequently, a given roughness will have a diminishing effect on the laminar sublayer along an immersed surface. Therefore, it is possible for a surface to be hydrodynamically rough upstream, yet hydrodynamically smooth downstream.

Rough-wall velocity profiles. For fully rough flow conditions, if we assume that the height k accounts for form and distribution of the roughness as well as magnitude, the functional relation that should apply is

$$\bar{u}/u_* = f(y/k). \tag{12–42}$$

We have seen that the velocity-defect law applies for both rough and smooth boundaries and can be represented by semilog-equations (Eqs. 12–16 and 12–17). Therefore, if Eq. (12–42) is to overlap the velocity-defect law, it too

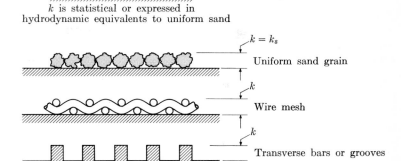

FIG. 12–11. Example of roughness types and definitions of roughness magnitude *k*.

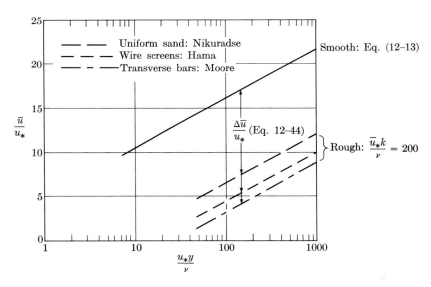

FIG. 12–12. Boundary-layer velocity-profile data illustrating effect of roughness.

will be a logarithmic function. An equation for rough walls that applies in the wall region is

$$\bar{u}/u_* = -5.6 \log (k/y) + C_5 \quad \begin{cases} y/\delta < 0.15 \\ u_* y/\nu > 50 \text{ to } 100, \end{cases} \quad (12\text{–}43)$$

where $5.6 = 2.3/\kappa$, $\kappa = 0.41$, and C_5 is a constant which depends on the size, shape, and distribution of the roughness. The equivalent relation for smooth walls is Eq. (12–13), namely

$$\bar{u}/u_* = 5.6 \log (u_* y/\nu) + C_2 \quad \begin{cases} y/\delta < 0.15 \\ u_* y/\nu > 30 \text{ to } 70, \end{cases}$$

TABLE 12-4

VALUES OF CONSTANTS IN ROUGH-WALL EQUATIONS FOR THE WALL REGION

$(y/\delta < 0.15; u_*k/\nu > 50$ to $100)$

Roughness type	Source of data	C_5, Eq. (12-43)	C_6, Eq. (12-44)	C_8, Eq. (12-46)
Uniform sand grains	Nikuradse [11] (pipes)	8.2	−3.3	7.55
Wire screens	Hama [12] (plates)	6.1	−1.2	6.1
Transverse bars	Moore [7] (plates)	4.9	0	5.25

Eq. (12-43): $\bar{u}/u_* = -5.6 \log (k/y) + C_5,$

Eq. (12-44): $\Delta\bar{u}/u_* = 5.6 \log (u_*k/\nu) + C_6,$

Eq. (12-46): $1/\sqrt{c_f} = 3.96 \log (\delta/k) + C_8.$

(Constants in this table were evaluated graphically from Fig. 12-12.)

where $C_2 = 4.9$. A comparison of the smooth and rough profiles is illustrated in Fig. 12-5. Figure 12-12 is a plot of $\bar{u}/u_*$ versus $\log (u_*y/\nu)$ for hydrodynamically smooth and fully rough walls which shows that roughness reduces the local mean turbulent velocity $\bar{u}$ near the wall for the same wall friction. Taking the difference between the last two equations gives this effect as

$$\frac{\Delta\bar{u}}{u_*} = \frac{\bar{u}_{\text{smooth}} - \bar{u}_{\text{rough}}}{u_*} = 5.6 \log \left(\frac{u_*k}{\nu}\right) + C_6, \qquad (12\text{-}44)$$

where C_6 depends empirically on the type of roughness. It can be seen that $\Delta\bar{u}/u_*$ is a function of u_*k/ν which is proportional to the ratio k/δ'. Values of C_5 and C_6 were evaluated graphically from Fig. 12-12 for the three types of roughness shown and are given in Table 12-4.

Surface-resistance formulas: rough walls. Combining Eqs. (12-43) with Eq. (12-16) gives for the completely rough regime,

$$U/u_* = 5.6 \log (\delta/k) + C_7, \qquad (12\text{-}45)$$

and

$$1/\sqrt{c_f} = 3.96 \log (\delta/k) + C_8, \qquad (12\text{-}46)$$

where $C_8 = (2.5 + C_5)/\sqrt{2}$ depends on the type of roughness. Values are given in Table 12-4.

If in Eq. (12-43) we allow C_5 to be evaluated from the empirical data of $[\bar{u}/u_* + 5.6 \log (k/y)]$ versus u_*k/ν, the relation will apply over the entire range from smooth to fully rough conditions. Doing this and then using Eq.

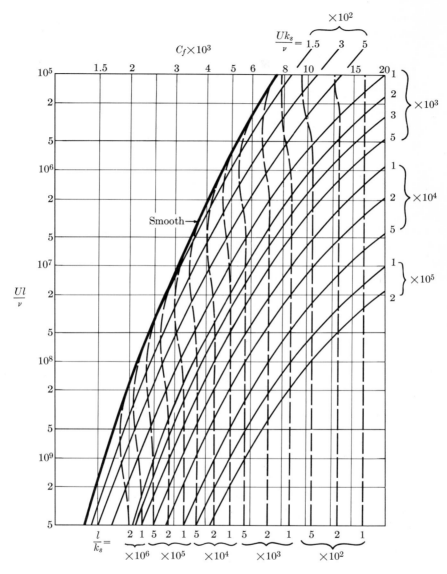

FIG. 12–13. Average shear coefficient C_f for sand-roughened plates [14].

(12–43) in the integral momentum equation (Eq. 8–21), Prandtl and Schlichting [13] converted the relation for the local coefficient c_f into a function of the parameters $x/k = f(\delta/k)$, Uk/ν and Ux/ν. A similar procedure gave the mean coefficient C_f in terms of l/k, Uk/ν, and Ul/ν. A diagram of the latter for sand-roughened plates (computed using sand-roughened pipe data) is given in Fig. 12–13 [14]. This figure uses the sand-grain roughness scale $k = k_s$.

The roughness that is *admissible* without increasing the resistance over that of a smooth plate is the critical roughness k_{crit}. With the condition

$$k_{\text{adm}} = k_{\text{crit}} \approx \delta',$$

we can use Eq. (12–7) to obtain

$$\frac{k_{\text{adm}}}{\delta'} \approx 1 \approx \frac{u_* k_{\text{adm}}}{4\nu} = \left(\frac{U k_{\text{adm}}}{\nu}\right) \frac{\sqrt{c_f}}{4\sqrt{2}},$$

or   (12–47)

$$k_{\text{adm}} \leq \frac{\text{const}}{\sqrt{c_f}\, U/\nu}.$$

Thus k_{adm} increases slightly with increasing distance along the surface if U/ν is constant. Therefore, along a surface with a given roughness, the most critical condition occurs at the place where the boundary layer first becomes turbulent. Using c_f at $\mathbf{R}_x = 10^5$ (which is an early, and hence conservative, transition Reynolds number), we get as a theoretical limit

$$(k_{\text{adm}})_{\mathbf{R}_{\text{crit}}=10^5} \leq \frac{75}{U/\nu}.$$ (12–47a)

For a plate of any length l, this limiting value of k_{adm} can be written

$$\frac{k_{\text{adm}}}{l} \leq \frac{75}{\mathbf{R}_l}.$$

Values of k_s/l at which C_f departs from the hydrodynamically smooth curve in Fig. 12–13 agree closely with the equation

$$\frac{k_{\text{adm}}}{l} \leq \frac{100}{\mathbf{R}_l}.$$

For practical purposes, we have approximately

$$k_{\text{adm}} \leq \frac{100}{U/\nu}.$$ (12–47b)

Note that k_{adm} by these relations does not depend on the length of the surface being considered.

Example 12–3: Rough wall velocity distribution and local skin friction coefficient

Examine the effect of roughness by comparing the boundary layers on a smooth plate and on a plate roughened by sand grains. The boundary layers result from the flow of water past the plates. Make the comparison for the condition of the same local shear stress τ_0.

Given: $\tau_0 = 0.485 \text{ lb/ft}^2$ on both plates,

$U = 10 \text{ ft/sec}$ past the rough plate,

$k_s = 0.001 \text{ ft}$,

Water temperature $= 58°\text{F}$ on both plates.

Find: (a) The velocity reduction Δu due to roughness, (b) the velocity $\bar{u}$ on each plate at $y = 0.007 \text{ ft}$, (c) the boundary layer thickness δ on the rough plate.

Solution: (a) Using Eq. (12–44), we have

$$\frac{\Delta u}{u_*} = 5.6 \log \frac{u_* k_s}{\nu} + C_6,$$

with $C_6 = -3.3$ from Table 12–4. Then

$$\frac{u_* k_s}{\nu} = \frac{0.5 \times 0.001}{1.25 \times 10^{-5}} = 40,$$

$$\Delta u = 0.5(5.6 \times 1.602 - 3.3)$$

$$= 2.83 \text{ ft/sec.}$$

(b) Rough plate. Using Eq. (12–43), we have

$$\frac{\bar{u}}{u_*} = 5.6 \log \frac{y}{k} + C_5,$$

with $C_5 = 8.2$ from Table 12–4. Then

$$\bar{u} = 0.5(5.6 \log 7 + 8.2) = 6.47 \text{ ft/sec.}$$

Smooth plate. From Eq. (12–13), we have

$$\frac{\bar{u}}{u_*} = 5.6 \log \frac{u_* y}{\nu} + 4.9, \qquad \frac{u_* y}{\nu} = \frac{0.5 \times 0.007}{1.25 \times 10^{-5}} = 280,$$

$$\bar{u} = 0.5(5.6 \log 280 + 4.9) = 9.3 \text{ ft/sec.}$$

Check. $\Delta \bar{u} = 9.3 - 6.47 = 2.83 \text{ ft/sec}$ as found in (a).

(c) The thickness δ for the rough plate is found by means of Eq. (12–46):

$$\frac{1}{\sqrt{c_f}} = 3.96 \log \frac{\delta}{k_s} + C_8$$

with $C_8 = 7.55$ from Table 12–4. Then

$$\log \frac{\delta}{k_s} = \frac{1}{3.96} \left(\frac{1}{\sqrt{0.005}} - 7.75 \right) = 1.66,$$

$$\frac{\delta}{k_s} = 46 \qquad \text{and} \qquad \delta = 0.046 \text{ ft.}$$

12–4 TURBULENT BOUNDARY LAYERS WITH PRESSURE GRADIENTS

12–4.1 Effects of pressure gradients

As for laminar boundary layers (Section 10–3), the effects of boundary curvature in the flow direction are primarily the consequence of a finite pressure gradient, $d\bar{p}/dx$. The profile of mean velocity departs from the zero-gradient forms presented in the previous section. It tends to become increasingly fuller for $d\bar{p}/dx < 0$ and to develop increasingly more pronounced inflections for $d\bar{p}/dx > 0$. The fundamental problem is to describe the rate of change of boundary-layer shape and thickness along a surface and to predict whether and where separation will occur.

A relation between boundary-layer growth, wall shear stress, and pressure gradient in terms of velocity-profile shape can be represented by Kármán's integral momentum equation (Eq. 8–21). For steady flow this can be written

$$\frac{d\theta}{dx} = \frac{c_f}{2} + \left(\frac{H+2}{2}\right)\left(\frac{\theta}{q}\right)\frac{d\bar{p}}{dx}, \tag{12–48}$$

where

$H = \delta^*/\theta =$ profile-shape factor,

$\quad \delta^*, \theta =$ displacement and momentum thickness by the usual definitions (Eqs. 8–9b and 8–10b),

$\quad\quad q = \rho U^2/2.$

For $d\bar{p}/dx = 0$, this becomes Eq. (8–18a), and the local shear coefficient c_f is directly proportional to the rate of increase in thickness. In general, the shape factor H grows larger with both distance x and increasing $d\bar{p}/dx$. Thus large H is associated with $d\bar{p}/dx > 0$ and with inflected profiles and ultimate flow separation. Low H corresponds to accelerated flow and to full, stable profiles.

Experimental dimensionless velocity data for a range of H-values along a smooth wall are shown in Fig. 12–14.* An orderly family of profiles results which corresponds to the conditions

$$\begin{aligned}
d\bar{p}/dx &< 0, & H &< 1.4, & \text{accelerating flow,} \\
d\bar{p}/dx &= 0, & H &= 1.4, & \\
d\bar{p}/dx &> 0, & H &> 1.4, & \text{decelerating flow.}
\end{aligned}$$

Separation is likely to occur when H is 2.6 to 2.7. This condition is realized by a combination of both position x and gradient $d\bar{p}/dx$.

---

* Figure 12–14 [15] applies for the nonequilibrium profiles discussed here. Clauser showed the existence of a special class of equilibrium profiles (with H constant for increasing x) which does not fit the same family but has the same general features otherwise.

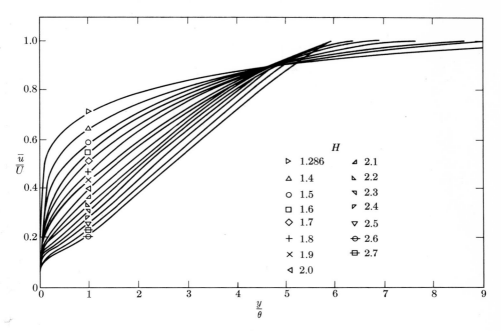

FIG. 12–14. Nonequilibrium profiles as functions of shape parameter H [15].

In Fig. 12–14, two features are worth special note. The first, of course, is the large defect near the wall with increasing $d\overline{p}/dx$. The second is the zone of steep rise in $\overline{u}/u_*$ at the wall even after large defects and pronounced inflections have developed. The explanation lies in the fact that the wall region is merely the zone of predominance of viscous stresses over which the law of the wall applies even with adverse pressure gradients. Thus, although $\overline{u}$ is reduced near the wall, the shear stress remains related to $\overline{u}$ by $\overline{u}/u_* = f(u_*y/\nu)$. Thus, as $d\overline{p}/dx$ becomes positive, the logarithmic part of the wall law (Eq. 12–13) continues to apply but over a decreasing range of u_*y/ν. Figure 12–15 (after Coles [16]) shows this. Landweber [17] has shown that the logarithmic portion does not exist for $U\delta^*/\nu < 725$.

It follows from Eq. (12–13) that with the reduction of $\overline{u}$ near the wall when $d\overline{p}/dx > 0$, τ_0 is also reduced. Measurements confirm that τ_0 decreases steadily to zero at separation. A formula [18] which gives the reduced skin-friction coefficient closely so long as Eq. (12–13) applies is

$$c_f = \frac{0.246}{10^{0.678H}\mathbf{R}_\theta{}^{0.268}} \tag{12–49}$$

When $d\overline{p}/dx$ is very large (positively), the second right-hand term of Eq. (12–48) predominates, and the growth of θ depends primarily on the internal momentum losses in the zone of large-velocity defect. Flow along a boundary

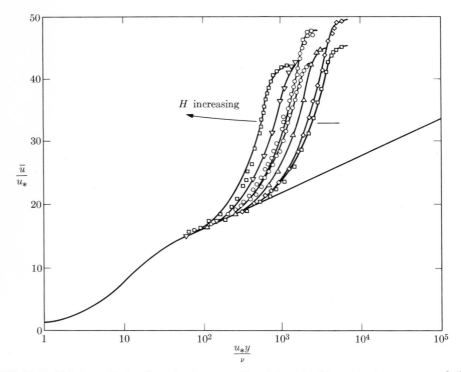

FIG. 12–15. Velocity profile data illustrating departure of law of the wall with increasing shape parameter [16].

progresses against a pressure rise only by giving up its momentum, with a maximum exchange in the absence of a boundary layer. With a boundary layer in an adverse pressure gradient, the efficiency of pressure recovery is impaired. The maximum shear stress, which now occurs some place in the boundary layer away from the wall, is increased and the turbulence production and energy dissipation is greater. Thus the total work of resistance forces is increased. Although a reduced skin friction results as evidenced by a reduced c_f, the total of both skin friction and pressure drag is higher.

12–4.2 Boundary-layer development computations

In the computation of rate of growth and the location of flow separation along a boundary for an arbitrary pressure distribution, the successful methods employ the Kármán integral momentum equation. The widely used method of von Doenhoff and Tetervin [15] is based on the profile correlation of Fig. 12–14 for which $H = \delta^*/\theta$ is the shape factor. Then, for thin two-dimensional layers, Eq. (12–48) may be applied.

For thin layers, it is assumed that the pressure across the layer is constant. Moreover, it is assumed that the mean velocity and pressure outside the boundary layer can be computed using Bernoulli's equation, as for potential flow.

Then $d\bar{p}/dx$ along the surface is given by

$$\rho U \frac{dU}{dx} \approx -\frac{dp}{dx}. \qquad (12\text{-}50)$$

Equation (12–48) becomes

$$\frac{d\theta}{dx} = \frac{c_f}{2} - (H+2)\frac{\theta}{U}\frac{dU}{dx} = \frac{c_f}{2} - (H+2)\frac{\theta}{\rho U^2}\frac{dq}{dx}. \qquad (12\text{-}51)$$

For an arbitrary potential flow $U(x)$, Eq. (12–51) can be integrated if we have data on how θ, δ^*, and c_f are related. For the nonequilibrium profiles of Fig. 12–14, von Doenhoff and Tetervin used Eq. (12–21) (thus ignoring the effect of pressure gradient on c_f) and developed the empirical relation

$$\theta \frac{dH}{dx} = e^{4.680(H-2.975)}\left[-\frac{\theta}{q}\frac{dq}{dx}\frac{2q}{\tau_0} - 2.035(H - 1.286)\right]. \qquad (12\text{-}52)$$

Then, with dq/dx specified by the potential flow, Eqs. (12–51) and (12–52) may be solved by a step-by-step procedure.

PROBLEMS

12–1. A large stabilizing fin projects from the hull of an ocean liner traveling at a speed of 30 knots. Assume the fin may be treated as a smooth flat plate and determine the following boundary-layer items at a distance 10 ft from the leading edge of the fin: (a) thickness δ' of the laminar sublayer, (b) velocity $\bar{u}$ at $y = \delta'$, (c) velocity $\bar{u}$ at $y/\delta = 0.15$, (d) the distance y at $y/\delta = 0.15$ and thickness δ. Compare with the results for the aircraft wing in Example 12–1.

12–2. Measurements of wind speed near the ground give the following results.

Distance above ground, ft	Wind velocity, ft/sec
6	7.0
12	8.0

(a) The Kármán constant κ is 0.41. Determine the value of the constant C_1 in Eq. (12–10). (b) Compute the laminar sublayer thickness δ' and compare it with the parameter y'. (c) Compute the wind velocity that would be expected 25 ft above the ground.

12–3. If a total-pressure (stagnation-pressure) tube is placed in contact with the fixed smooth boundary, it will measure a pressure that is dependent on the magnitude of the boundary shear stress. Discuss the reasons for this relationship. Discuss the factors that must be considered in calibrating the probe.

12–4. A smooth flat plate is held in water (60°F) moving at 4 ft/sec. The plate is 10 ft long and 4 ft wide. (a) Estimate the point on the plate where the boundary layer might be expected to become turbulent. (b) Compute the total drag of the plate. (c) Compute the total drag of the same plate if a strip of sandpaper is glued to the leading edge.

12–5.   The skin friction of a boat can be approximated by assuming that it is the same as that of a flat plate of the same area and length as the wetted surface of the boat. Assume that the skin friction of a canoe is the same as that for one side of a flat plate 1.5 ft wide and 12 ft long. Given that the canoe is moving at 1.5 ft/sec: (a) estimate the total skin friction force; (b) calculate the shear stress at the mid length of the canoe; (c) calculate the boundary-layer thickness at the mid length and end of the canoe.

12–6.   Determine the ratios of the displacement thickness and momentum thickness to the boundary-layer thickness (δ^*/δ and θ/δ) for the following velocity distributions.

(a) $\bar{u}/U = y/\delta$,   (b) $\bar{u}/U = (y/\delta)^{1/7}$,   (c) $\bar{u}/U = \sin(\pi/2)(y/\delta)$.

12–7.   A new streamlined train will travel between Tokyo and Osaka, Japan, a distance of 515 km, at an average speed of 172 km/hr. Assume that the exterior surfaces are smooth and compute the power required to overcome surface resistance along the top and sides of a train of 10 cars. Assume that the air temperature is 5°C, and the cars are 25m long, 3.4m wide, and 4.5m high.

12–8.   A submarine is 200 ft long and has a cigar shape with a diameter of 20 ft. It travels submerged in 40°F sea water at 15 knots (25.3 ft/sec). The exterior surface is smooth. Estimate the surface-friction drag and power, assuming that the body-surface area can be approximated by a circular cylinder of the same diameter and overall length.

12–9.   A harpoon is a smooth spear 6 ft long and 1 in. in diameter. It is launched from a gun and enters 60°F sea water at 50 ft/sec. Find the surface-friction drag. What will be the distance from the tip back to the point of boundary-layer transition to turbulence? What will be the thickness of the boundary layer at the end of the spear?

12–10.   A rocket is 15 ft in diameter and 120 ft tall. As it is launched into the atmosphere it accelerates so that at an elevation of 25,000 ft it is traveling at 1000 mph. Compute the thrust necessary to overcome surface-friction drag at that elevation.

12–11.   A 14-ft model of a 685-ft aircraft carrier is tested at prototype Froude number in a towing tank in fresh water at 60°F. The design speed of the prototype is 30 knots at 68°F in sea water. The model surface-friction drag is 6.0 lb. The wetted area of the model is 40 ft², and the boundary layer behaves like that on a flat plate of the equivalent length and area. Determine the prototype friction drag.

12–12.   A smooth flat plate is used as a flow divider in a water stream at 60°F. Assume that the boundary layer is turbulent from the leading edge. At a distance x from the leading edge the wall shear stress is 0.066 lb/in². (a) Determine the velocity $\bar{u}$ at a distance $y = 0.024$ in. from the wall. (b) Given that the velocity outside the boundary layer is $U = 75$ ft/sec, compute the boundary-layer thickness δ at the distance x from the leading edge of the plate and compute the distance x. (c) If the plate were roughened to the equivalent of uniform sand grains of diameter 0.01 in., what would be the velocity at $y = 0.024$ in. for the same wall shear?

12–13.   Consider a turbulent boundary layer of atmospheric air at 60°F on a plate. The plate is 20 ft long and the velocity outside the boundary layer is 100 ft/sec. Assume that the plate is artificially roughened with sand grains and that the transition to turbulence occurs at $R_x = 10^5$. (a) Find the maximum size of the sand grains which can be used if the wall shear stress is to be the same as for a smooth wall. (b) Explain why a certain degree of roughness on a wall will not necessarily affect the wall shear stress. Use *carefully drawn* sketches.

12–14. Consider the two-dimensional flow from a nozzle (Fig. 12–16) discharging a jet of water (68°F) onto a smooth glass plate having a slope such that the water velocity at the free surface remains constant. (a) The initial water velocity is 5 ft/sec, and the water depth at the beginning of the glass plate ($x = 0$) is 0.2 ft. Estimate the distance ($x = L$) to the point at which turbulent fluctuations should first appear on the free surface of the flow. (Neglect boundary-layer development within the nozzle.) (b) Sketch the vertical velocity distributions at the points $x = L/2$, $x = L$, and $x = 4L$. (c) Would you expect the water depth to change in the distance 0 to L? If so, explain whether it should increase or decrease.

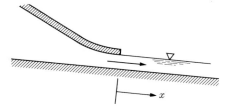

x **FIGURE 12–16**

12–15. A 6-ft model of a ship 600 ft long is tested in a fresh-water towing tank (68°F) at the Froude number corresponding to the prototype ship speed of 25 knots in 68°F sea water. Determine the admissible roughness for model and prototype. Assume the boundary layer behaves like that on a flat plate and is turbulent from the leading edge.

REFERENCES

1. Schubauer, G. B., "Turbulent Processes as Observed in Boundary Layer and Pipe," *J. Appl. Phys.*, **25**, 2 (Feb., 1954).

2. Runstadler, P. W., S. J. Kline, and W. C. Reynolds, "An Experimental Investigation of the Flow Structure of the Turbulent Boundary Layer," *Stanford University, Mechanical Engineering Department*, Rept. MD-8 (June, 1963).

3. Clauser, F. H., "The Turbulent Boundary Layer," *Advances in Applied Mechanics*, Vol. IV, Academic Press, Inc., New York, 1956, pp. 2–51.

4. Freeman, H. B., "Force Measurements on a 1/40-Scale Model of the U.S. Airship Akron," *Nat. Advisory Comm. Aeron.*, Rept. 432 (1932).

5. Klebanoff, P. S., and Z. W. Diehl, "Some Features of Artificially Thickened Fully Developed Turbulent Boundary Layers with Zero-Pressure Gradient," *Nat. Advisory Comm. Aeron.*, Rept. 1110 (1952).

6. Schultz-Grunow, F., "Neues Widerstandsgesetz für glatte Platten," *Luftfahrtforschung*, **17**, 239 (1940). See also *Nat. Advisory Comm. Aeron.*, Tech. Memo. 986 (1941).

7. Moore, W. F., *An Experimental Investigation of the Boundary Layer Development Along a Rough Surface*, Ph.D. Dissertation, State University of Iowa, 1951.

8. Squire, H. B., and A. D. Young, "The Calculation of the Profile Drag of Airfoils," *Brit. ARC*, Rept. and Memo. 1838 (1938).

9. VON KÁRMÁN, T., "Turbulence and Skin Friction," *J. Aeron. Sci.*, **1**, 1 (Jan., 1934).

10. SCHOENHERR, K. E., "Resistance of Flat Surface Moving Through a Fluid," *Trans. SNAME*, **40**, 279 (1932).

11. NIKURADSE, J., Strömungsgesetze in rauhen Rohren, *Ver. deut. Ing. Forschungsheft*, **361**, issue (1933).

12. HAMA, F. R., Boundary Layer Characteristics for Smooth and Rough Surfaces, *Trans. SNAME*, **62**, 333 (1954).

13. PRANDTL, L., and H. SCHLICHTING, "Das Widerstandsgesetz rauher Platten," *Werft, Reederei, Hafen*, **15**, 1 (1934). See also SCHLICHTING, H., *Boundary Layer Theory*, McGraw-Hill Book Co., New York, 4th ed., 1960, Chapter XXI.

14. SCHLICHTING, H., *Boundary Layer Theory*, McGraw-Hill Book Co., New York, 4th ed., 1960.

15. DOENHOFF, A. E., and N. TETERVIN, "Determination of General Relations for the Behavior of Turbulent Boundary Layers," *Nat. Advisory Comm. Aeron.*, Rept. 772 (1943).

16. COLES, D., "The Law of the Wake in the Turbulent Boundary Layer," *J. Fluid Mech.*, **1**, Part 2 (1956).

17. LANDWEBER, L., "The Frictional Resistance of Flat Plates in Zero Pressure Gradient," *Trans. SNAME*, **61**, 5 (1953).

18. LUDWIEG, H., and W. TILLMAN, "Investigations of the Wall-Shearing Stress in Turbulent Boundary Layers," *Nat. Advisory Comm. Aeron.*, Tech. Memo. 1285 (1950); Transl. from *Z. Angew. Math. Mech.*, **29** (1949).

CHAPTER 13

Wall Turbulence. Flows in Uniform Conduits

13–1 INLET-REGION FLOW VERSUS FULLY DEVELOPED FLOW

The flow entering a conduit is nonuniform over an *entrance length*. Figure 13–1 illustrates two examples of steady flow into uniform ducts. Initially, the entering fluid is largely without shear. Boundary layers form, leaving an unsheared *core* which, in turn, shrinks with distance as the growing layers envelop the core flow. The pressure gradient $d\overline{p}/dx$ varies over this distance. If the conduit is long relative to width or depth, a uniform state is reached with no further change of velocity profile or other mean flow quantities. The pressure gradient assumes a constant value. To preserve continuity, the core velocity U increases in the flow direction to a maximum value as the core vanishes. The final uniform condition is called *fully developed* flow.

The distance l_E to the fully developed profile is a momentum entrance length because the boundary-layer growth represents a changing momentum flux. The longitudinal sum of pressure and gravity forces has components due both to wall friction and to momentum flux change. Both the entrance length and

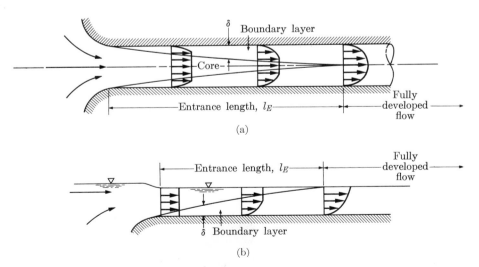

FIG. 13–1. Development of uniform boundary layers: (a) circular tube; (b) two-dimensional open channel.

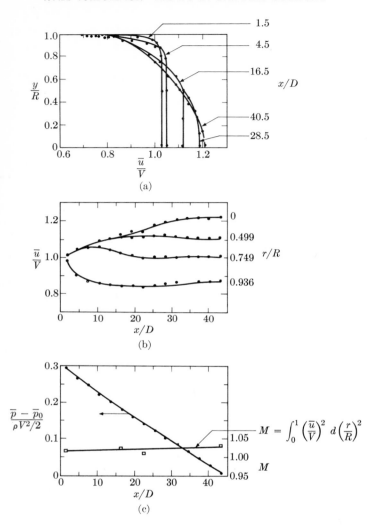

FIG. 13–2. Velocity profiles, pressure drop, and momentum flux for turbulent flow in the inlet region of a pipe [1].

the resistance coefficient in this region depend on whether or not the boundary layer remains laminar. For the laminar case in a pipe, a theoretical entrance length due to Boussinesq which agrees well with experiment is

$$l_E/D = 0.065\mathbf{R}, \tag{13–1}$$

where $\mathbf{R} = VD/\nu$ and $D =$ pipe diameter. Thus, for $\mathbf{R} = 2000$, l_E is 130 diameters. If the boundary layer becomes turbulent, the entrance length is shorter. Experimentally, l_E is in the range of 50 through 100 diameters. The exact length depends on the location of the transition to turbulence, and a general relation for l_E/D does not exist. Figure 13–2 [1] is an example of velocity profiles and the change in velocity in the inlet region when the transition to turbulence is artifically stimulated at $x/D = 0.25$. The fully developed conditions are approached at $x/D \approx 45$ for the Reynolds number of the experiment. Figure 13–2 also shows the variation of the dimensionless drop in static pressure, $(\bar{p} - \bar{p}_0)/(\rho V^2/2)$ (where $\bar{p}_0$ is a reference wall pressure) and of the dimensionless momentum flux parameter,

$$M = \int_0^1 \left(\frac{\bar{u}}{V}\right)^2 d\left(\frac{r}{R}\right)^2. \tag{13-2}$$

The wall shear stress may be calculated from the gradients of these two components by the equation

$$\frac{\tau_0}{\rho V^2/2} = -\frac{D}{\rho V^2/2}\frac{d(\bar{p} - \bar{p}_0)}{dx} + \frac{D}{2}\frac{dM}{dx}. \tag{13-3}$$

It can be seen from the figure that the contribution of the momentum flux is small, and in this case, the wall shear is essentially the terminal value of the flux after $x/D = 15$.

The fully developed velocity profiles in ducts are special cases of boundary-layer flow. As for turbulent nonuniform boundary layers, the distribution of mean velocity is the consequence of turbulence generated through wall shear. In general, there is one notable difference from the nonuniform layer. Turbulence extends without interruption clear across the conduit. There is no intermittent passage of nonturbulent fluid in the region of maximum mean velocity. On the other hand, mean properties of boundary layers in conduits possess many similarities to the nonuniform layers discussed in Chapter 12.

Circular pipe flow is the most common case of fully developed motion. This is a special two-dimensional case having axial symmetry. Here the mean flow is composed of parallel streamlines. The axially symmetric velocity profile has the form of two-mirrored boundary layers which are coplaner on every radius. Truly two-dimensional flow is seldom realized in conduits. However, in rectangular ducts which are very wide relative to their height, two-dimensional conditions are approximated. Here also the uniform flow is composed of mirrored coplanar boundary layers. For both axial symmetry and two-dimensional cases, the boundary-layer mean properties are very nearly the same as those of the nonuniform layer.

Most other conduit cross sections, even though uniform in shape and size along the duct, produce turbulent flows which depart from the two-dimensionality of the circular and wide rectangular cross sections. Profiles of axial velocities tend to become three-dimensional, especially near corners. As shown by the velocity contours for two examples in Fig. 13–3 [2], the flow-direction wall shear tends to be less in corners than the average over the perimeter. In

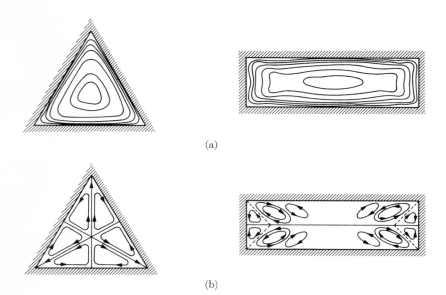

(a)

(b)

FIG. 13–3. Velocity contours and diagrams of secondary motions for fully developed flow in noncircular ducts: (a) velocity contours; (b) secondary circulation patterns [2].

addition, secondary circulations in planes normal to the duct axis are also shown by the diagrams in Fig. 13–3.

The remainder of this chapter is concerned with fully developed flow in both closed and open conduits, and with relations among mass discharge rate, fluid properties, wall shear, and distributions of velocity and turbulence.

13-2 INCOMPRESSIBLE FLOW IN PIPES

13-2.1 Reynolds equations for axially symmetric flows

For cylindrical coordinates (x, r, θ) with velocity components $(\overline{u}, \overline{v}, \overline{w})$, the Reynolds equations (11–22) become*

$$\rho \left(\frac{\partial \overline{u}}{\partial t} + \overline{u} \frac{\partial \overline{u}}{\partial x} + \overline{v} \frac{\partial \overline{u}}{\partial r} + \frac{\overline{w}}{r} \frac{\partial \overline{u}}{\partial \theta} \right) = \rho g_x - \frac{\partial \overline{p}}{\partial x} + \mu \nabla^2 \overline{u}$$

$$- \rho \left(\frac{\partial \overline{u'^2}}{\partial x} + \frac{1}{r} \frac{\partial r \overline{u'v'}}{\partial r} + \frac{1}{r} \frac{\partial \overline{u'w'}}{\partial \theta} \right),$$

---

* The coordinate designations (x, r, θ) and (u, v, w) are introduced here instead of (z, r, θ) and (v_z, v_r, v_θ) used in earlier chapters, to conform to the widely used notation for pipe flow.

(continued)

$$\rho \left(\frac{\partial \bar{v}}{\partial t} + \bar{u} \frac{\partial \bar{v}}{\partial x} + \bar{v} \frac{d\bar{v}}{\partial r} + \frac{\bar{w}}{r} \frac{\partial \bar{v}}{\partial \theta} - \frac{\bar{w}^2}{r} \right)$$

$$= \rho g_r - \frac{\partial \bar{p}}{\partial r} + \mu \left(\nabla^2 \bar{v} - \frac{\bar{v}}{r^2} - \frac{2}{r^2} \frac{\partial \bar{w}}{\partial \theta} \right)$$

$$- \rho \left(\frac{\partial \overline{u'v'}}{\partial x} + \frac{1}{r} \frac{\partial r \overline{v'^2}}{\partial r} + \frac{1}{r} \frac{\partial \overline{v'w'}}{\partial \theta} - \frac{\overline{w'^2}}{r} \right),$$

$$\rho \left(\frac{\partial \bar{w}}{\partial t} + \bar{u} \frac{\partial \bar{w}}{\partial x} + \bar{v} \frac{\partial \bar{w}}{\partial r} + \frac{\bar{w}}{r} \frac{\partial \bar{w}}{\partial \theta} + \frac{\bar{v}\bar{w}}{r} \right)$$

$$= \rho g_\theta - \frac{1}{r} \frac{\partial \bar{p}}{\partial \theta} + \mu \left(\nabla^2 \bar{w} + \frac{2}{r^2} \frac{\partial \bar{w}}{\partial \theta} - \frac{\bar{w}}{r^2} \right)$$

$$- \rho \left(\frac{\partial \overline{w'u'}}{\partial x} + \frac{\partial \overline{v'w'}}{\partial r} + \frac{1}{r} \frac{\partial \overline{w'^2}}{\partial \theta} - \frac{2 \overline{v'w'}}{r} \right),$$

where   (13–4)

$$\nabla^2 = \frac{\partial^2}{\partial x^2} + \frac{\partial^2}{\partial r^2} + \frac{1}{r} \frac{\partial}{\partial r} + \frac{1}{r^2} \frac{\partial^2}{\partial \theta^2}.$$

13–2.2 Equations of motion and energy for pipe flow

Consider steady fully developed turbulent pipe flow. Using the notation of Fig. 13–4, we find that the condition of axial symmetry gives

$$\partial/\partial\theta = 0, \qquad \bar{p} = \bar{p}(x, r) \text{ only,} \tag{13–5a}$$

and flow uniformity requires

$$\partial(\text{vel})/\partial x = 0, \qquad \bar{v} = \bar{w} = 0, \qquad \overline{v'w'} = \overline{w'u'} = 0. \tag{13–5b}$$

With $\rho = $ const, Eqs. (13–4) reduce to

$$\frac{\partial \bar{p}}{\partial x} = -\frac{1}{r} \frac{\partial}{\partial r} (r\rho \overline{u'v'}) + \mu \left(\frac{\partial^2 \bar{u}}{\partial r^2} + \frac{1}{r} \frac{\partial \bar{u}}{\partial r} \right), \qquad \frac{\partial \bar{p}}{\partial r} = -\frac{1}{r} \frac{\partial}{\partial r} (r\rho \overline{v'^2}) + \rho \frac{\overline{w'^2}}{r},$$

$$\tag{13–6}$$

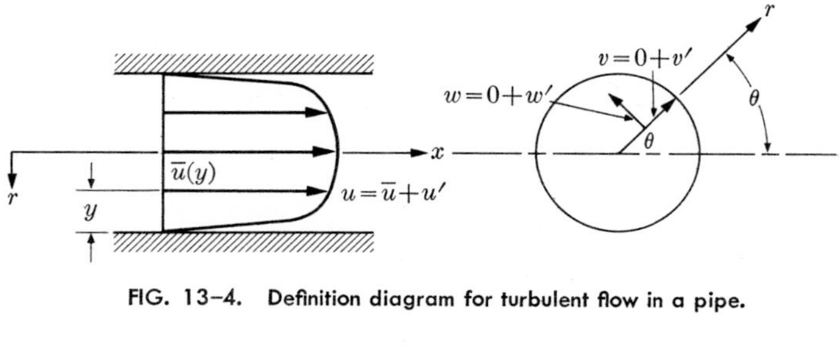

FIG. 13–4. Definition diagram for turbulent flow in a pipe.

where we have omitted the gravity-force term. From the second equation,

$$\frac{\partial^2 \overline{p}}{\partial x \partial r} = 0 = \frac{\partial^2 \overline{p}}{\partial r \partial x},$$

so that $\partial \overline{p}/\partial x$ is independent of r. Now, integrating both equations with respect to r, we have

$$\frac{r^2}{2} \frac{\partial \overline{p}}{\partial x} = -r \left(\rho \overline{u'v'} - \mu \frac{\partial \overline{u}}{\partial r} \right) + A(x),$$

$$\overline{p}(x, r) - \rho \overline{v'^2} - \rho \int_r^R \frac{\overline{v'^2} - \overline{w'^2}}{r} \, dr = \overline{p}_0(x),$$

$$(13\text{--}7)$$

where $\overline{p}_0(x)$ is the static wall pressure at $r = R$. Evaluating the first of Eqs. (13–7) at $r = 0$, we see that $A(x)$ must be zero. Then, since the mean conditions are independent of x, $\partial \overline{p}/\partial x = \partial \overline{p}_0/\partial x$ is constant. Now, evaluating at $r = R$ and integrating, we obtain

$$\overline{p}_0(x) - \overline{p}_0(0) = - \frac{2}{R} \tau_0 x. \qquad (13\text{--}8)$$

With this result, Eqs. (13–7) can be written

$$\frac{r}{R} \tau_0 = \rho \overline{u'v'} - \mu \frac{\partial \overline{u}}{\partial r} = \tau,$$

$$\overline{p}(x, r) - \overline{p}_0(0) = - \frac{2}{R} \tau_0 x + \rho \overline{v'^2} + \rho \int_r^R \frac{\overline{v'^2} - \overline{w'^2}}{r} \, dr.$$

$$(13\text{--}9)$$

Note that $dr = -dy$ in Fig. 13–4, so that $\partial \overline{u}/\partial r = -\partial \overline{u}/\partial y$. Also, then, the sense of positive v' is changed so that *positive* v' correlates with *positive* u' in the product $\overline{u'v'}$.

By the first of Eqs. (13–9), we see that the shear τ is linearly distributed across the tube. At the pipe centerline where $r = 0$, $\overline{u'v'}$ and $d\overline{u}/dy$ are zero and $\tau = 0$. At the wall where $r = R$,

$$\overline{u'v'} = 0$$

and

$$\tau = \tau_0 = \mu (d\overline{u}/dy)_{r=R}.$$

Like the two-dimensional cases of Examples 11–1 and 11–2, the second of Eqs. (13–9) show that the local pressure $\overline{p}$ differs from the wall pressure $\overline{p}_0$ by a small amount depending on the turbulence.

Let us define a shear stress or friction coefficient for pipe flow by

$$\tau_0 = \frac{f}{4} \rho \frac{V^2}{2}, \qquad (13\text{--}10)$$

where $V = Q/A$, f = pipe friction factor. Using Eq. (13–8) with $D = 2R$ and $x = L$, we have

$$\bar{p}_0(0) - \bar{p}_0(L) = f \frac{L}{D} \rho \frac{V^2}{2},\tag{13–11}$$

or

$$h_f = \frac{\bar{p}_0(0) - \bar{p}_0(L)}{\gamma} = f \frac{L}{D} \frac{V^2}{2g}.\tag{13–12}$$

The preceding developments omitted the gravity force term from the equations of motion, so that Eq. (13–12) applies to a horizontal pipe. For an inclined pipe having elevations $h(0)$ and $h(L)$ at $x = 0$ and L, we have

$$h_f = \left[\frac{\bar{p}_0(0)}{\gamma} + h(0)\right] - \left[\frac{\bar{p}_0(L)}{\gamma} + h(L)\right] = f \frac{L}{D} \frac{V^2}{2g}.\tag{13–12a}$$

Equation (13–12) is the Darcy equation for head loss. For steady uniform flow and for constant temperature, h_f is the "loss" of mechanical energy per pound of fluid due to conversion to heat by friction. Its units are (ft-lb)/lb, or merely feet measured as the equivalent height of a column of the fluid in question. The Darcy equation is used for noncircular as well as circular conduits. The friction factor f depends on duct shape, wall roughness, and Reynolds number **R**, where

$$\mathbf{R} = VD/\nu.\tag{13–13}$$

If we multiply Eq. (13–12) by the flow rate in lb/sec, we have the total rate of energy conversion to heat, or the power loss, in (ft-lb)/sec. Thus the total energy loss rate is

$$\gamma Q h_f = \gamma Q f \frac{L}{D} \frac{V^2}{2g}.\tag{13–14}$$

The rate of energy conversion varies over the cross section of a shear field. If we multiply the first of Eqs. (13–9) by the mean velocity gradient $d\bar{u}/dr$, we obtain an equation for the rate per unit volume at which mechanical energy is given up by the mean flow at points across a radius, namely

$$\frac{r}{R} \tau_0 \frac{d\bar{u}}{dr} = \overline{\rho u'v'} \frac{d\bar{u}}{dr} - \mu \left(\frac{d\bar{u}}{dr}\right)^2,\tag{13–15}$$

with the units (ft-lb)/(ft³-sec). It expresses the balance

$$\begin{bmatrix} \text{rate of energy} \\ \text{availability because} \\ \text{of pressure drop} \end{bmatrix} = \begin{bmatrix} \text{rate of partial} \\ \text{conversion into} \\ \text{turbulent energy} \end{bmatrix} + \begin{bmatrix} \text{rate of partial} \\ \text{conversion to heat} \\ \text{by direct dissipation} \end{bmatrix}$$

The integral of Eq. (13–15) over the volume of the pipe must equal the total given by Eq. (13–14).

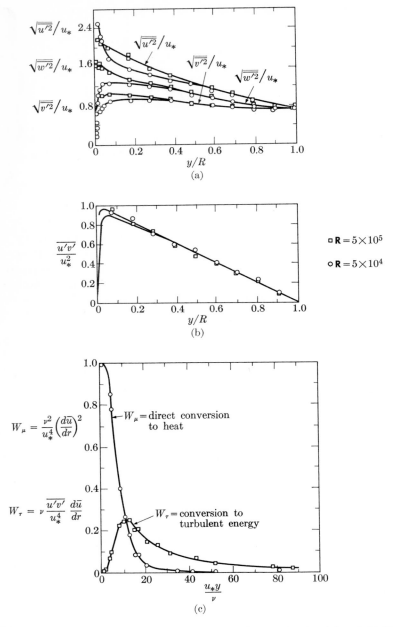

FIG. 13–5. Turbulence intensities and turbulent shear stresses for fully developed flow: (a) turbulence intensities; (b) turbulent shear stresses (Reynolds stresses); (c) distribution of energy given up by mean flow [3].

13–2.3   Structure of uniform shear flow in pipes

As already noted, turbulence extends without interruption throughout the cross section of a fully developed flow. The distribution of total turbulent energy for a pipe agrees closely with the energy found in the fully turbulent regions of a boundary layer and shown by the dashed line in Fig. 12–3. Turbulence intensities measured for two Reynolds numbers are shown in Fig. 13–5(a) [3]. Peaks near the wall are observed as found for nonuniform boundary layers (Fig. 12–4). The intensities of the v'- and w'-components are again less than the u', but are closer to each other in magnitude than for the nonuniform boundary layer. Figure 13–5(b) [3] gives the turbulent stresses for the same two Reynolds numbers.

Figure 13–5(c) [3] is a dimensionless plot of the distribution of the energy given up by the mean flow as computed, using Eq. (13–15), from the same measurements. It can be seen that the production of turbulence is a maximum at $u_*y/\nu \approx 12$ which, as we will observe in a later paragraph, is very near the edge of the laminar sublayer. The viscous dissipation occurs primarily in the laminar sublayer. Note that the energy is not completely dissipated in the zone of its production. Energy given up by the mean flow in the center of the pipe is transported by turbulent mixing to the region of highest shear where the bulk of the dissipation occurs. These observations for uniform pipe-boundary layers also hold qualitatively for the nonuniform layers discussed in Chapter 12.

13–2.4   Velocity and friction laws: smooth pipes

The profiles of mean velocity in pipes are similar to those of nonuniform boundary layers, differing only slightly in their details. We find, as shown in Figs. 13–6 [4] and 13–7 [4], the same inner region of the law of the wall and outer region of the velocity-defect law which were discussed in Chapter 12. Moreover, the same zones of the law of the wall appear, namely

Laminar sublayer $\quad 0 < u_*y/\nu < 4,$

Buffer zone $\qquad\quad 4 < u_*y/\nu < 30$ to 70, $\hfill$ (13–16)

Turbulent zone $\qquad u_*y/\nu > 30$ to 100, $\dfrac{y}{\delta} < 0.2.$

In the laminar sublayer, $\bar{u} = u$, and

$$\tau \approx \tau_0 \approx \mu(u/y), \tag{13–17}$$

and

$$u/u_* = u_*y/\nu. \tag{13–18}$$

The thickness δ' is

$$\delta' = 4\nu/u_* = 4\nu/V\sqrt{f/8} = 4\sqrt{8}\, D/\mathbf{R}\sqrt{f}. \tag{13–19}$$

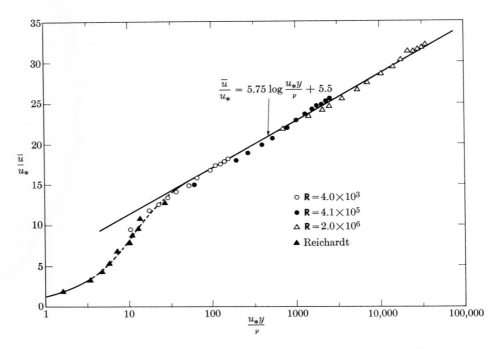

FIG. 13–6. Fully developed velocity profiles in a smooth pipe: inner region [4].

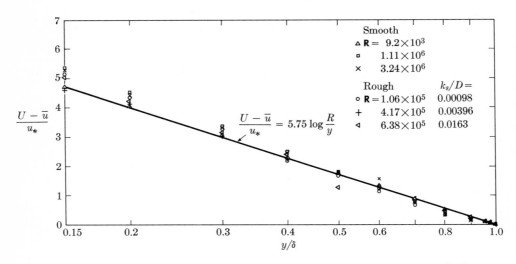

FIG. 13–7. Velocity-defect profiles in the outer region of both smooth and rough pipes [4,5].

The friction factor f decreases only slowly with Reynolds number **R**, so that δ'/D decreases as **R** increases.

In the turbulent zone, the same procedures as followed in Chapter 12 lead to universal logarithmic relations for velocity distribution. However, for pipes, it is possible and convenient to write an equation for $\bar{u}$ which holds with good accuracy across almost the entire pipe radius. Measurements in smooth pipes give

$$\bar{u}/u_* = 5.75 \log (u_* y/\nu) + 5.5. \tag{13–20}$$

This is shown in Fig. 13–6. The velocity defect for the outer portion of the profile is represented by

$$(U - \bar{u})/u_* = 5.75 \log (R/y). \tag{13–21}$$

As shown in Fig. 13–7, this applies to both smooth and rough pipes. In both Eqs. (13–20) and (13–21), the Kármán constant κ is 0.4.

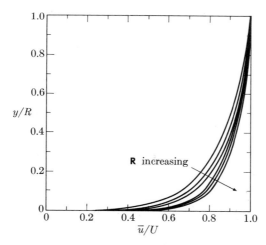

FIG. 13–8.   Effect of Reynolds number on velocity profiles in smooth pipes [4].

One feature of the universal profile is its variation with Reynolds number. Figure 13–8 [4] plots $\bar{u}/U$ versus y/R and shows increasing fullness of the profile with increasing **R**. With respect to the nonuniform boundary layers, it is possible to represent the pipe-velocity profiles by a power law. Thus, in analogy to Eq. (12–32), we write

$$\frac{\bar{u}}{U} = \left(\frac{y}{R}\right)^{m/(2-m)} = \left(\frac{y}{R}\right)^{1/n}. \tag{13–22}$$

This leads to the ratio of the average to maximum velocity,

$$\frac{V}{U} = \frac{2n^2}{(n+1)(2n+1)}. \tag{13–23}$$

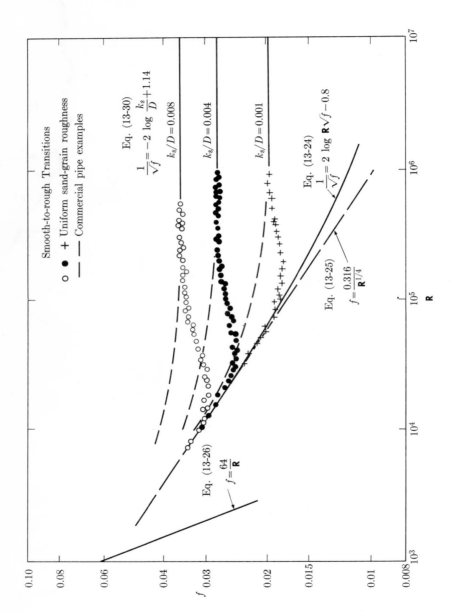

FIG. 13-9.   Friction factors for pipe flow.

TABLE 13–1

EMPIRICAL EXPONENTS FOR POWER-LAW EQUATIONS

R	4×10^3	2.3×10^4	1.1×10^5	1.1×10^6	2×10^6	3.2×10^6
n	6.0	6.6	7.0	8.8	10	10
V/U	0.791	0.806	0.817	0.853	0.865	0.865

Empirically good fits, using the Nikuradse data, are provided by the value of n listed in Table 13–1. For **R** in the range of 10^4 to 10^5, an average value of 7 can be used.

Integrating Eq. (13–20) over the radius and dividing by the cross-sectional area πR^2 will give a relation for the average velocity $V = Q/\pi R^2$. Dividing V by $u_* = \sqrt{f/8}\ V$ [from Eq. (13–10)] will result in the friction law for smooth pipes in the form

$$1/\sqrt{f} = A \log \mathbf{R}\sqrt{f} + B,$$

where A and B are constants.

Experiments of many investigators show that $A = 2.0$ and $B = -0.8$, so that

$$1/\sqrt{f} = 2.0 \log \mathbf{R}\sqrt{f} - 0.8. \tag{13–24}$$

An empirical relation due to Blasius that holds well up to $\mathbf{R} = 10^5$ is

$$f = 0.316/\mathbf{R}^{1/4}. \tag{13–25}$$

Below $\mathbf{R} = 2000$, the flow is laminar, and the f computed from Eq. (6–42) is

$$f = 64/\mathbf{R}. \tag{13–26}$$

Equations (13–24), (13–25), and (13–26) are shown in Fig. 13–9.

If we use $V/U = 0.8$, which corresponds approximately to $n = 7$ in Table 13–1, Eqs. (13–10) and (13–25) yield

$$U/u_* = 8.74[u_* R/\nu]^{1/7}.$$

Assuming that the maximum values, U and R, can be replaced by the local values, $\bar{u}$ and y, we again have the power law for the velocity profile,

$$\bar{u}/u_* = 8.74[u_* y/\nu]^{1/7}. \tag{13–27}$$

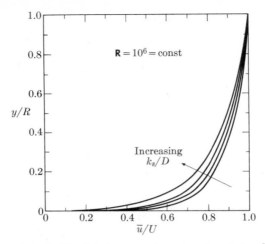

FIG. 13–10.   Velocity profiles in sand-roughened pipes [5].

13–2.5   Roughness effects

Wall roughness disrupts the laminar sublayer with effects similar to those for nonuniform boundary layers.   For "fully rough" conditions, Fig. 13–10 [5] shows how sand-grain roughness k_s causes a peaking of the velocity profiles for the same Reynolds number.   These curves are described by

$$\bar{u}/u_* = 5.75 \log (y/k_s) + 8.5. \tag{13–28}$$

Using Eq. (13–28) at $\bar{u} = U$ when $y = R$ will lead to the velocity-defect equation (13–21).   Subtracting Eq. (13–28) from Eq. (13–20) yields

$$\frac{\Delta\bar{u}}{u_*} = \frac{\bar{u}_{\text{smooth}} - \bar{u}_{\text{rough}}}{u_*} = 5.75 \log \frac{u_* k_s}{\nu} + C. \tag{13–29}$$

For sand-grain roughness, $C = -3$.   Figure 13–11 is a plot of $\bar{u}/u_*$ versus log $(u_* y/\nu)$ for smooth and sand-roughened walls showing the increase in $\Delta\bar{u}/u^*$ with increase in $u_* k_s/\nu$.

The friction law for fully rough pipe flow can be derived from Eq. (13–28) by integrating over the radius and dividing by the pipe area and by $u_* = \sqrt{f/8}\, V$. The result is $1/\sqrt{f} = C \log (R/k_s) + E$, where C and E are constants.   Thus, when the laminar sublayer is completely disrupted by the roughness projections, the pipe flow f will be independent of the Reynolds number.   For Nikuradse's sand-grain roughness, the constants are $C = 2$ and $E = 1.74$.   Then, using $D = 2R$, we have

$$1/\sqrt{f} = -2 \log (k_s/D) + 1.14. \tag{13–30}$$

The constant values of f given by this equation are shown in the right-hand region of Fig. 13–9.

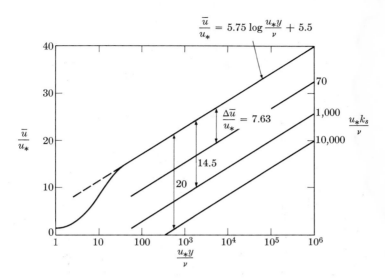

FIG. 13–11. Effect of roughness on pipe-velocity profiles.

Transition values of the friction factor f between "smooth" and "fully rough" conditions depend on the kind of roughness. The effects with uniform sand grains and with the random roughness of commercial pipe are compared in Fig. 13–9. The commercial pipe roughness is expressed as an equivalent sand roughness. Curves are shown for different relative roughnesses k_s/D. With sand grains, the departure of the friction factor f from the smooth wall curve is approximately at the same value as the ultimate rough wall value of f. Commercial pipe roughness causes an earlier departure of f from the smooth-pipe curve and a continuous drop to the ultimate full rough magnitude.

The commercial-roughness transition is described by the Colebrook-White semiempirical formula

$$\frac{1}{\sqrt{f}} + 2 \log \frac{k_s}{D} = 1.14 - 2 \log \left(1 + 9.35 \frac{D/k_s}{\mathbf{R}\sqrt{f}} \right). \qquad (13\text{–}30a)$$

This equation is asymptotic to both the smooth-pipe equation (13–24) and the rough-pipe equation (13–30) as indicated in Fig. 13–9.

It should be emphasized again that the above discussion of velocity and friction is for fully developed uniform flow, and Fig. 13–9 applies only to this condition. In the pipe-entrance length, l_E, the pressure drop is higher than for uniform flow because it includes both a higher wall shear and a pressure drop due to changing momentum flux. Thus an entrance length f computed from entrance-length pressure drop using Eq. (13–11) will be higher than the uniform flow f at the same value of $\mathbf{R} = VD/\nu$. For long ducts, entrance-length effects usually may be neglected.

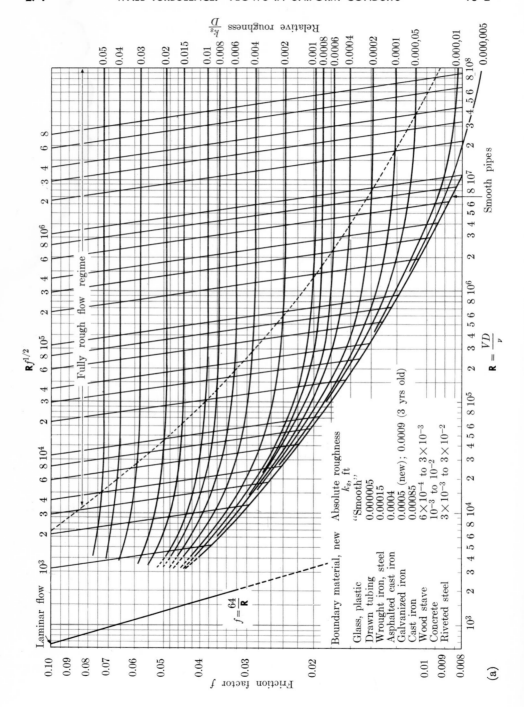

(a)

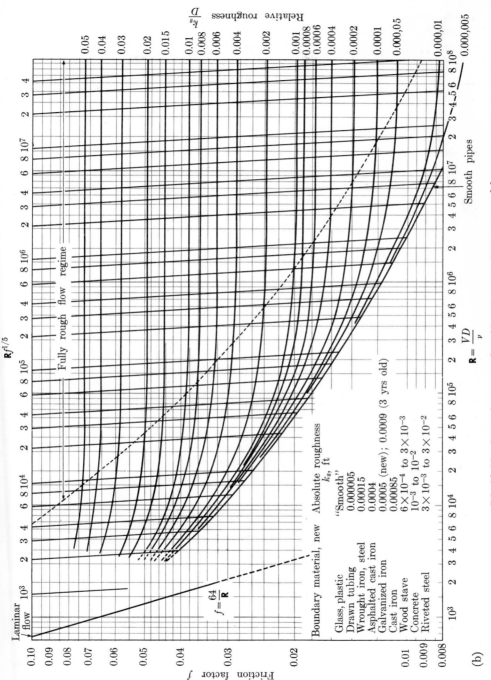

FIG. 13–12.   Friction factor versus Reynolds number and relative roughness [6].

(b)

The preceding results for smooth, rough, and smooth-to-rough transition friction factors were used by Moody to develop a general resistance diagram for uniform flow in conduits. Figure 13–12 is adapted from Moody's [6] diagram and will be used for the computations in the following sections. Curves are shown for different relative roughness k_s/D. The values of k_s for particular pipe surfaces are tabulated in Fig. 13–12.

13-2.6   Pipe-flow computations

The computation of steady flow of constant-density fluids through pipes involves the simultaneous solution of the two equations below.

$$\text{Continuity:} \quad Q = AV,$$

$$\text{Darcy:} \quad h_f = f \frac{L}{D} \frac{V^2}{2g},$$

with $f = f(\mathbf{R}, k/D)$, $\mathbf{R} = VD/\nu$.

There are three basic types of problems, namely,

(a) Head loss. Given: Q, L, D, ν, k;   find: h_f.
(b) Flow rate. Given: h_f, L, D, ν, k;   find: Q.
(c) Diameter. Given: h_f, Q, L, ν, k;   find: D.

In all three problems, the fundamental step is the identification of f as a function of the fluid properties, the pipe size and roughness, and the flow variables. In general, f is known as a function of $\mathbf{R}$ and k/D and is usually given in the form of Figs. 13–9 and 13–12.

The first problem of head loss is solved readily by selection of f directly from Fig. 13–12, using values of $\mathbf{R}$ and k/D computed from the given data. As shown schematically in Fig. 13–13(a), we move upward on the diagram along the line of given $\mathbf{R}$ to the intersection with the curve for the specified roughness condition. Then if we move horizontally to the left-hand scale, we read f. The loss h_f is then computed from the Darcy equation.

For a direct solution of flow rate Q, we rearrange the Darcy equation

$$V = \left[\frac{2gh_f}{L/D}\right]^{1/2} \frac{1}{f^{1/2}}$$

$$= \frac{M}{f^{1/2}},$$

and

$$\mathbf{R}f^{1/2} = \frac{VD}{\nu} f^{1/2} = \frac{D^{3/2}}{\nu} \left[\frac{2gh_f}{L}\right]^{1/2}.$$

(13–31)

(13–31a)

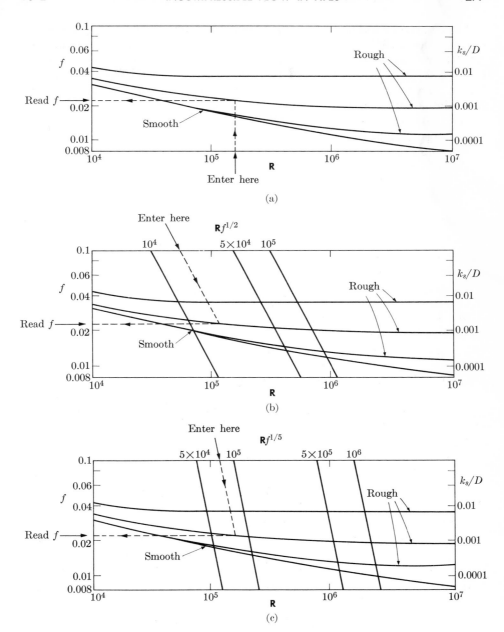

FIG. 13–13.   Diagrams illustrating direct determination of f for the three pipe-flow problems: (a) solution for h_f; (b) solution for Q; (c) solution for D.

Thus $\mathbf{R}f^{1/2}$ can be evaluated in terms of given quantities. Then, if f is presented as a function of $\mathbf{R}f^{1/2}$, its direct determination is possible. This is accomplished with an auxiliary scale of $\mathbf{R}f^{1/2}$ on the f versus $\mathbf{R}$ diagram, as shown schematically in Fig. 13–13(b). To use it, move down the diagram as indicated by the dashed line and arrows for the value of $\mathbf{R}f^{1/2}$ computed from Eq. (13–31a). At the intersection with the curve for the given roughness condition, read horizontally to the left-hand scale for f. With f known, compute V from Eq. (13–31) and Q from the continuity equation.

For a direct solution for pipe diameter, we use continuity and the Darcy equation to write

$$D = \left[\frac{8LQ^2}{\pi^2 g h_f}\right]^{1/5} f^{1/5} = Nf^{1/5}, \tag{13–32}$$

and

$$\mathbf{R}f^{1/5} = \frac{4}{\nu}\left[\frac{g h_f Q^3}{8\pi^3 L}\right]^{1/5}. \tag{13–32a}$$

Here $\mathbf{R}f^{1/5}$ can be computed by means of given quantities. In this case, we can determine f directly, using an auxiliary scale of $\mathbf{R}f^{1/5}$ as illustrated in Fig. 13–13(c), and D can be computed from Eq. (13–32).

The friction-factor diagram in Fig. 13–12 has the scale for $\mathbf{R}f^{1/2}$ superposed on part (a) and the scale for $\mathbf{R}f^{1/5}$ superposed on part (b). Hence the one figure can be used for the three different basic flow computations.

Example 13–1: Head loss in a tube

Ethyl alcohol at 68°F flows at the rate of 20 gal/min (gpm) through 20 ft of $\frac{1}{2}$-in. stainless-steel drawn tubing. Compute the head loss and power consumption.

Given:

$$Q = 20 \text{ gal/min} = 0.0445 \text{ cfs*}, \qquad L = 20 \text{ ft},$$
$$D = \tfrac{1}{2} \text{ in.} = \tfrac{1}{24} \text{ ft}, \qquad A = 0.196 \text{ in}^2,$$
$$k_s = 5 \times 10^{-6} \text{ ft} \quad \text{(Fig. 13–12)}, \qquad k_s/D = 1.2 \times 10^{-4},$$
$$\nu_{68} = 1.64 \times 10^{-5} \text{ ft}^2/\text{sec} \quad \text{(Table 1–2)},$$
$$\gamma = 49.3 \text{ lb/ft}^3 \quad \text{(Table 1–2)},$$
$$V = Q/A = 0.0445(144/0.196) = 32.6 \text{ ft/sec},$$
$$\mathbf{R} = VD/\nu = (32.6 \times 1/24)/(1.64 \times 10^{-5}) = 8.3 \times 10^4.$$

Solution. From Fig. 13–12, we have $f = 0.0195$ for $\mathbf{R} = 8.3 \times 10^4$, $k_s/D = 1.2 \times 10^{-4}$. The head loss is

$$h_f = f\frac{L}{D}\frac{V^2}{2g} = 0.0195\,\frac{20}{1/24}\,\frac{(32.6)^2}{2g},$$

$$= 154 \text{ ft of alcohol}.$$

---

* The conversion between gpm and cfs is 450 gpm = 1 cfs.

The power loss is $\qquad \gamma Q h_f = 49.3 \times 0.0445 \times 154,$

$$= 339 \text{ (ft-lb)/sec,}$$

$$= 0.615 \text{ hp.}$$

Example 13–2: Direct computation of Q in a rough pipe

Gasoline flows in a horizontal commercial steel pipe of 3-in. diameter a distance of 1000 ft with a pressure drop of 25 lb/in² (psi). Determine the flow rate Q when the specific gravity is 0.68 and temperature is 100°F.

Given:

$$L = 1000 \text{ ft,} \qquad D = 3 \text{ in.} = \tfrac{1}{4} \text{ ft,}$$

$$k_s = 1.5 \times 10^{-4} \text{ ft} \quad \text{(Fig. 13–12),} \qquad k_s/D = 6 \times 10^{-4},$$

$$h_f = 25(144/42.4) = 85 \text{ ft of gasoline,}$$

$$\nu_{100} = 3.9 \times 10^{-6} \text{ ft}^2/\text{sec} \quad \text{(Fig. 1–4).}$$

Solution. Using Eq. (13–31a), we have

$$\mathbf{R}f^{1/2} = \frac{D^{3/2}}{\nu} \left[\frac{2gh_f}{L} \right]^{1/2} = 7.5 \times 10^4.$$

From the $\mathbf{R}f^{1/2}$-scale in Fig. 13–12, we find that the value of 7.5×10^4 intersects the line of $k_s/D = 6 \times 10^{-4}$ at $f \approx 0.018$. From Eq. (13–31), we have

$$M = \left[\frac{2gh_f D}{L} \right]^{1/2} = 1.17, \qquad V = \frac{M}{f^{1/2}} = \frac{1.17}{0.135} = 8.7 \text{ ft/sec,}$$

$$Q = \frac{\pi D^2}{4} V = 0.428 \text{ cfs}$$

$$= 193 \text{ gpm.}$$

Alternatively, we can readily obtain solutions for both Q and D by trial, using a successive approximation procedure. For example, to find the flow rate an assumed f is used with the given data in Eq. (13–31) to obtain a trial value of V. With this V, a trial Reynolds number is computed, and a new f determined from the Moody diagram (Fig. 13–12). The cycle is repeated until successive values of f agree. Usually, agreement to two significant figures is sufficient. For the initial assumption, *any* value may be chosen arbitrarily for f, but convergence occurs more quickly if a value is chosen in the practical range.

In solving for diameter, the given data and an assumed f are used in Eq. (13–32) to obtain a trial D. A trial Reynolds number is found by means of

$$\mathbf{R} = VD/\nu = (4Q/\pi\nu)1/D, \qquad (13\text{–}33)$$

and a trial relative roughness k_s/D is computed. A new f is found from the Moody diagram and the cycle repeated to convergence. In this case, the computed diameter is finally adjusted to the next larger even dimension to conform to the possible standard commercial pipe sizes.

Example 13-3: Calculation of Q for a smooth pipe

Water at 60°F flows through 200 ft of smooth pipe 1 in. in diameter with a pressure drop of 20 lb/in^2 (psi). Compute the flow rate Q.

Given:

$$L = 200 \text{ ft}, \qquad D = 1 \text{ in.} = \tfrac{1}{12} \text{ ft}, \qquad k_s = 0 \quad \text{(smooth)},$$

$$h_f = 20(144/62.4) = 20 \times 2.31 = 46.2 \text{ ft of water},$$

$$\nu_{60} = 1.22 \times 10^{-5} \text{ ft}^2/\text{sec} \quad \text{(Table 1-3)}.$$

Solution.

Assume: $f = 0.015$ (a mid-value in turbulent range for smooth pipe).
From Eq. (13-31), we have

$$M = \left[\frac{2gh_f D}{L}\right]^{1/2} = 1.11.$$

First trial:

$$V = \frac{M}{f^{1/2}} = \frac{1.11}{0.122} = 9.1 \frac{\text{ft}}{\text{sec}}, \qquad \mathbf{R} = \frac{VD}{\nu} = \frac{9.1 \times 1/12}{1.22 \times 10^{-5}} = 6.2 \times 10^4.$$

From Fig. 13-12, we have $f = 0.0198$, $f^{1/2} = 0.141$.

Second trial:

$$V = \frac{1.11}{0.141} = 7.9 \text{ ft/sec}, \qquad \mathbf{R} = \frac{7.9 \times 1/12}{1.22 \times 10^{-5}} = 5.4 \times 10^4.$$

From Fig. 13-12, we have $f = 0.0204$, which agrees with the previous value to the accuracy of the diagram, and

$$Q = \frac{\pi D^2}{4} V = \frac{\pi}{4 \times 144} 7.9 = 0.0431 \text{ cfs}$$

$$= 19.4 \text{ gpm}.$$

The tabulation given in Table 13-2 is useful in systemizing the computations.

TABLE 13-2

$\dfrac{k_s}{D}$	$M = \left[\dfrac{2gh_f}{L/D}\right]^{1/2}$	D/ν	Assumed f	$V = \dfrac{M}{f^{1/2}}$	$\mathbf{R} = \dfrac{VD}{\nu}$	f (Fig. 13-12)
0	1.11	6830	0.015	9.1	6.2×10^4	0.0198
			0.0198	7.9	5.4×10^4	0.0204
			0.0204	7.8	5.3×10^4	0.0205

Example 13-4: Calculation of D for a rough pipe

Water at 60° is to flow through 5000 ft of pipe under a 50-ft head loss. A riveted steel pipe with smooth walls is to be used. Although the wall material may be smooth, the rivets cause a flow resistance equivalent to that produced by a degree of wall roughness. Compute the necessary diameter for a flow of 100 cfs, assuming the "smoothest" riveted pipe.

Given: $L = 5000$ ft, $Q = 100$ cfs, $h_f = 50$ ft of water,

         $\nu_{60} = 1.22 \times 10^{-5}$ ft^2/sec   (Table 1-3).

         $k_s = 3 \times 10^{-3}$ ft   (Fig. 13-12; for smoothest riveted pipe).

Any value of f may be assumed. However, it is sometimes useful to start with the f for fully rough flow. This Q is a large flow and will require a pipe diameter of 4 to 6 ft (an estimate based on judgment which the reader will acquire after a few computations). Assume

$$D = 5 \text{ ft} \quad \text{and} \quad k_s/D = 6 \times 10^{-4}.$$

From Fig. 13-12, f for fully rough flow with this k_s/D is 0.0175. Starting with this value, we proceed with the step-by-step calculation using Eqs. (13-32) and (13-33). The results appear in the Table 13-3. The computed diameter is 3.43 ft = 41.2 in. The next commercial size is 42 in.

TABLE 13-3

$N = \left[\dfrac{8LQ^2}{\pi^2 g h_f}\right]^{1/5}$	$\dfrac{4Q}{\pi \nu}$	Assumed f	$D = Nf^{1/5}$	$R = \dfrac{4Q}{\pi \nu}\dfrac{1}{D}$	k_s/D	f (Fig. 13-12)
7.6	1.04×10^7	0.0175	3.38	3.25×10^6	8.9×10^{-4}	0.0192
		0.0192	3.43	3.03×10^6	8.8×10^{-4}	0.0192

13-3   INCOMPRESSIBLE FLOW IN NONCIRCULAR DUCTS

13-3.1   Friction losses in closed conduits

As we noted in the introduction to this chapter, the wall shear in the flow direction is less in corners of noncircular ducts than along the exposed walls. The secondary circulations shown in Fig. 13-2 are a consequence of fluid from the high shear zone being propelled toward the center of the flow while a return flow occurs toward the corners. The momentum transport produced by this circulation tends to even out the wall shear, and if the cross section has a ratio of area to wetted perimeter close to that of a circumscribing circle or semicircle, the head loss per unit length will be nearly the same as for a pipe. This is the case for sections like squares, equilateral triangles, and ovals, and the friction-loss data for circular pipes may be used.

We employ the Darcy equation in a slightly different form for computations of flow in noncircular sections. Consider the balance of forces on the element of

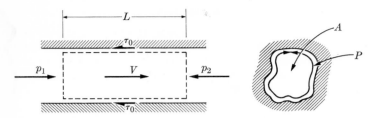

FIG. 13–14.   Free-body diagram for steady flow in a constant-area conduit.

fluid in steady flow in Fig. 13–14. Summing all forces in the flow direction, we obtain $(p_1 - p_2)A = \tau_0 PL$, where P is the wetted perimeter of the conduit. Using Eq. (13–10) and dividing by γ, we have, for a horizontal conduit,

$$h_f = \frac{p_1 - p_2}{\gamma} = f \frac{L}{4R_h} \frac{V^2}{2g}, \tag{13–34}$$

where

$$R_h = A/P = hydraulic\ radius,$$

$$h_f = friction\ head\ loss.$$

The hydraulic radius is a parameter which depends on the cross-section shape. For pipes, $R_h = D/4$, and Eq. (13–34) reduces to Eq. (13–12). Assuming that variations in R_h for noncircular sections have the same significance as variations in D for circles, we write

$$\mathbf{R} = \frac{4V R_h}{\nu} \quad \text{and} \quad \frac{k_s}{D} = \frac{k_s}{4R_h}, \tag{13–35}$$

and we use the pipe-flow friction data and diagrams.

An example of the effect of departing from "open" sections is illustrated by the measurements [7] for isosceles triangles with varying apex angles. The results are compared with the laminar and turbulent friction factors for circular pipes in Fig. 13–15 [7] using diagrams of $f\mathbf{R}$ and $f\mathbf{R}^{1/4}$, both with $\mathbf{R}$ as defined

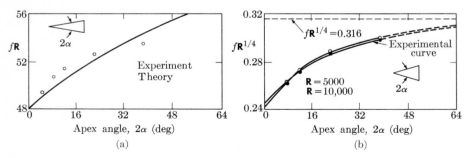

FIG. 13–15.   Friction factors for fully developed flow in triangular ducts: (a) laminar flow ($f\mathbf{R} = 64$ for circular pipes); (b) turbulent flow ($f\mathbf{R}^{1/4} = 0.316$ for circular pipes [7]).

by Eq. (13–35). At the same Reynolds number, f is reduced from the pipe value at all apex angles, and is reduced by increasing amounts as the apex angle becomes more acute. For turbulent flow, the friction factor for a 38.8° apex angle is within 5% of the circular-pipe value of f. The laminar flow results are within 2% of theoretical solutions [8] for a circular sector wedge-shaped passage with the same apex angle.

Example 13–5: Flow in a square duct

Carbon dioxide flows isothermally through 40 ft of drawn copper ducting having a 1-in² cross section. The exit pressure is 14.7 psia, and the temperature is 110°F. Determine the head loss when the flow rate is 10 cfm. Verify that compressibility of the gas may be neglected in this case.

Given:

$$Q = 10 \text{ cfm} = 0.167 \text{ cfs},$$

$$L = 40 \text{ ft}, \qquad k_s = 5 \times 10^{-6} \text{ ft} \quad \text{(Fig. 13–12)},$$

$$P = 4 \text{ in}, \qquad A = 1 \text{ in}^2,$$

$$p_2 = 14.7 \text{ psia}, \qquad T = 110°F, \qquad \nu_{110} = 2 \times 10^{-4} \text{ ft}^2/\text{sec} \quad \text{(Fig. 1–4)},$$

$$V = \frac{Q}{A} = \frac{0.167}{1/144} = 24 \text{ ft/sec}, \qquad R_h = \frac{A}{P} = \frac{1}{4} \text{ in} = \frac{1}{48} \text{ ft},$$

$$\frac{k_s}{4R_h} = \frac{5 \times 10^{-6}}{4/48} = 6 \times 10^{-5}, \qquad \mathbf{R} = \frac{4V R_h}{\nu} = \frac{4 \times 24/48}{2 \times 10^{-4}} = 10^4.$$

From Fig. 13–12, $f = 0.0305$. [Note: At this low $\mathbf{R}$, the copper is "smooth."] Furthermore,

$$h_f = 0.0305 \frac{40}{4/48} \frac{(24)^2}{2g} = 131 \text{ ft of carbon dioxide.}$$

The head loss causes a pressure drop of $h_f\gamma$. To compute γ, we find the gas constant for carbon dioxide from Table 1–6.

$$R = 35.1 \frac{\text{ft-lb}}{\text{lbm-°R}} = 35.1 \text{ g}_c \frac{\text{ft-lb}}{\text{slug-°R}}.$$

Then, using Eq. (1–17), we have

$$\rho_2 = \frac{p_2}{RT} = \frac{14.7 \times 144}{35.1 \text{g}_c \, 570} = \frac{0.106}{\text{g}_c},$$

$$\gamma_2 = \rho_2 g = \frac{0.106 g}{g_c} = 0.106 \frac{\text{lb}}{\text{ft}^3},$$

$$h_f \gamma_2 = 131 \times 0.106 = 13.9 \text{ psf}$$

$$= 0.096 \text{ psi.}$$

Therefore, the inlet pressure is $p_1 \approx 0.096 + 14.7 = 14.8$ psia, and the pressure ratio is $14.7/14.8 \approx 0.99$, which will produce a negligible compressibility effect.

13–3.2   Two-dimensional flows

Turbulent Poiseuille flows.   Laminar flow under pressure drop between two parallel fixed plates has a parabolic velocity distribution as described in Section 6–5. Two-dimensional turbulent flow has been investigated in wide rectangular channels where the secondary circulation associated with the corners is concentrated toward the side walls as indicated schematically in Fig. 13–3.   For fully developed flow in smooth channels of 60:1 and 12:1 width-to-depth ratios, Laufer [9] found that the velocity could be represented by

$$\bar{u}/u_* = 6.9 \log (u_* y/\nu) + 5.5, \tag{13–36}$$

for the range $5 \times 10^4 < 2VB/\nu < 2.5 \times 10^5$ of his experiments.   This equation corresponds to a value of Kármán's constant, $\kappa = 0.333$.

The friction laws for rough and smooth channels are similar to the pipe laws, but the changes in cross-sectional form cause the constants in the equations to be different.   The friction factor may be expressed [10] for smooth channels as

$$\frac{1}{\sqrt{f}} = 2.03 \log \frac{2BV}{\nu} \sqrt{f} - 0.47, \tag{13–36a}$$

and for rough channels as

$$\frac{1}{\sqrt{f}} = 2.03 \log \frac{B/2}{k_s} + 2.11. \tag{13–36b}$$

Turbulent Couette flows.   The flow through a duct with a pressure drop has a varying shear stress across the stream. For pipe flow, τ varies linearly as given by Eq. (13–9). There are important cases where the pressure drop is zero and the shear stress is constant, or nearly constant, because the passage walls move relative to each other. Although such passages are not conduits in the same sense as pipes are, we will discuss the basic two-dimensional model here to compare it with pressure flow.

Consider plane parallel boundaries moving in opposite directions as shown in Fig. 13–16(a). If we superpose a constant velocity $-U_2$ on the system, the same situation appears in Fig. 13–16(b) with one moving and one stationary wall. For $\mathbf{R} = (UB/2)/\nu < 1500$, the flow is laminar, which is the situation treated in Section 6–5. With zero pressure drop, the motion is due solely to the shear field created by the relative boundary motion. This is called *Couette flow.* The shear stress is constant, and the velocity is linear by Eq. (6–35).

For $\mathbf{R} > 1500$, the flow is turbulent. As for Poiseuille flow, Reynolds equations (11–22) reduce to Eq. (11–24), and integration shows that the total mean shear

$$\tau = \mu(d\bar{u}/dy) - \rho\overline{u'v'}$$

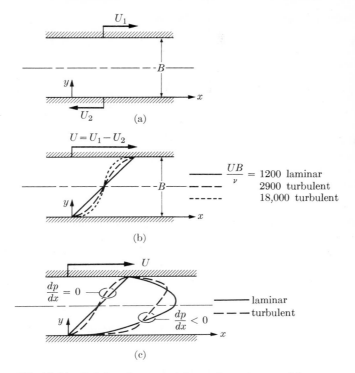

FIG. 13-16. Turbulent Couette and Couette-Poiseuille flows: (a) parallel walls in relative motion; (b) Couette; (c) Couette-Poiseuille.

remains constant over the gap B. However, the local velocity $\bar{u}$ departs from the linear as shown. Using Eq. (11-10) for the mean shear, we see that

$$\tau_0 = \tau = \mu(d\bar{u}/dy) + \eta(d\bar{u}/dy) = \text{const.}$$

It can be seen that $d\bar{u}/dy$ cannot be constant in this gap if the boundary layers are fully developed and $\tau = \tau_0 = $ constant.

Reichardt [11] deduced relations for the turbulent velocity profile and shear stress coefficient. In the notation of Fig. 13-16, they are for smooth surfaces,

$$\bar{u}/u_* = 5.75\,[\log\,(u_* y/\nu) - \log\,(1 - y/B)] + 5.5, \qquad (13\text{-}37)$$

and

$$1/\sqrt{c_f} = 4.06 \log\,(UB/\nu)\sqrt{c_f} - 0.83, \qquad (13\text{-}37\text{a})$$

where $u_* = \sqrt{\tau_0/\rho} = (U/2)\sqrt{c_f/2}$ and c_f is the local wall shear stress coefficient.

If a pressure drop is superposed, the flow is the sum of Couette and Poiseuille velocity distributions as indicated schematically in Fig. 13-16(c).

13–4 COMPRESSIBLE UNIFORM FLOW IN PIPES

Friction, through its internal heat generation and the effect on density and pressure changes, is a major factor in compressible flow over bodies and through ducts. The surface heating on reentry of space craft is such a surface-resistance problem. In pipes, heat transfer and the limits to flow and to pressure drop or pressure recovery are strongly dependent on flow resistance.

The following paragraphs will discuss the compressible-flow relations among flow variables and friction losses, using the one-dimensional point of view. We shall consider the isothermal case typified by pipe lines which have reached equilibrium with the external temperature and the adiabatic case, which is approached in insulated ducts.

13–4.1 One-dimensional equations for compressible flow

Four relations will be useful in this and later chapters for treating one-dimensional compressible-flow problems. They are an equation of state for the fluid, the first law of thermodynamics in the form of an energy equation, the continuity equation, and a momentum equation.

The *equation of state* will depend upon the type of fluid. In the following, we will restrict ourselves to gases and vapors which are highly superheated, and we will apply Eq. (1–17), the perfect gas law, namely

$$p/\rho = RT, \tag{13–38}$$

or

$$dp/p = d\rho/\rho + dT/T. \tag{13–38a}$$

Let us emphasize that in Eq. (13–38), and in all the following equations, consistent units must be used. Particularly with mass in slugs, absolute pressure in lb/ft^2, and temperature in °R, the gas constant R must be in units of (ft-lb)/slug-°R.

The general one-dimensional *energy equation* for steady flow was given by Eq. (4–21). If we omit shaft work and the elevation term (which is small compared to other terms), this equation becomes

$$\frac{\text{heat transfer}}{\text{slug}} = \left[u + \frac{p}{\rho} + \frac{V^2}{2} \right]_2 - \left[u + \frac{p}{\rho} + \frac{V^2}{2} \right]_1. \tag{13–39}$$

The heat transfer is positive when passing to the fluid from the surroundings. For a gas obeying $p/\rho = RT$, the internal energy and the gas constant R can be expressed by Eqs. (1–10a) and (1–19) as

$$du = c_v \, dT$$

and

$$R = (k - 1)c_v = [(k - 1)/k]c_p,$$

where k = ratio of specific heats = c_p/c_v. The energy equation becomes

$$\frac{V_2^2 - V_1^2}{2} = \frac{k}{k-1}\left(\frac{p_1}{\rho_1} - \frac{p_2}{\rho_2}\right) + \frac{\text{heat transfer}}{\text{slug}}. \tag{13–40}$$

This equation applies to all flows with or without friction. In differential form,

$$\frac{k}{k-1}\,d\left(\frac{p}{\rho}\right) + d\left(\frac{V^2}{2}\right) = \Delta\left(\frac{\text{heat transfer}}{\text{slug}}\right). \tag{13–40a}$$

Using Eq. (13–38) and $c_p = kR/(k-1)$, we have

$$c_p\,dT + d\left(\frac{V^2}{2}\right) = \Delta\left(\frac{\text{heat transfer}}{\text{slug}}\right). \tag{13–40b}$$

For adiabatic flows, with or without friction, the heat-transfer term is zero in the above equations.

The one-dimensional *continuity equation* for compressible fluids is written

$$GA = \rho V A, \tag{13–41}$$

or

$$\frac{d\rho}{\rho} + \frac{dV}{V} + \frac{dA}{A} = 0, \tag{13–41a}$$

where G is the mass flow rate per unit area.

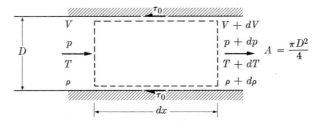

FIG. 13–17.   Control surface for compressible flow in a pipe.

The *momentum equation* is written with the notation of Fig. 13–17 for a straight duct as

$$-A\,dp - \tau_0 P\,dx = GA\,dV = \rho A V\,dV, \tag{13–42}$$

where P is the wetted wall perimeter and the gravity (body) force does not appear. Integrating for $G = \rho V$ = constant and τ_0 = constant, we have

$$(p_1 - p_2)A - \tau_0 PL = GA(V_2 - V_1), \tag{13–43}$$

or

$$F_p + F_s = (\rho V Q)_2 - (\rho V Q)_1, \tag{13–43a}$$

which is the first of Eqs. (4–33) with body force neglected. These three momentum equations apply to all fluids and flow processes. We will replace the wall shear stress τ_0 with the Darcy friction coefficient defined by Eq. (13–10). Then using continuity, we find that Eq. (13–42) becomes

$$dp + \rho V \, dV + f \frac{\rho V^2}{2} \frac{dx}{D} = 0. \tag{13–44}$$

Note that f must always be positive. Introducing the Mach number, we can express Eq. (13–44) as

$$\frac{dp}{p} + \frac{k \mathbf{M}^2}{2} f \frac{dx}{D} + k \mathbf{M}^2 \frac{dV}{V} = 0, \tag{13–44a}$$

where

$$\mathbf{M} = \frac{V}{c} = \frac{V}{\sqrt{kRT}} = \frac{V}{\sqrt{k(p/\rho)}}, \tag{13–45}$$

in which

$$c = \text{velocity of sound,}$$
$$= \sqrt{kRT} \text{ for a gas} \quad \text{(Example 1–3),}$$
$$k = \text{ratio of specific heats} = c_p/c_v.$$

The above equations may be solved if the thermodynamic process is specified and the friction factor is known. Experiments [12] with subsonic flow, that is, $0 < \mathbf{M} < 1$, show no significant compressibility effects when the velocity profile is fully developed. Consequently, the incompressible friction coefficients as given in Fig. 13–12 may be used. Friction tends to decrease the pressure along a pipe for subsonic flow but causes a pressure rise for supersonic velocities. Consequently, the trend is always toward a critical velocity, and continuous transitions from subsonic to supersonic, or vice versa, are impossible. A critical pipe length for continuous flow of either subsonic or supersonic velocity is reached when $\mathbf{M} = 1$ for adiabatic conditions and $\mathbf{M} = 1/\sqrt{k}$ for isothermal conditions. For supersonic flow, this limiting length is so short that fully developed flow is seldom established. Shapiro [13] summarizes experimental data for the apparent friction factors with supersonic flow in the entrance length.

13–4.2   Isothermal gas flow with friction

By means of Eq. (13–41), the Reynolds number can be written as

$$\mathbf{R} = \rho V D/\mu = GD/\mu. \tag{13–46}$$

Thus, since the dynamic viscosity depends only on temperature, $\mathbf{R}$ is a constant along a constant-diameter pipe for steady flow at constant temperature. Combining the equation of state and continuity as given by Eqs. (13–38) and (13–41) together with the definition of the Mach number gives *for isothermal flow* in a pipe,

$$\frac{p_1}{p} = \frac{\rho_1}{\rho} = \frac{V}{V_1} = \frac{V}{\mathbf{M}_1\sqrt{kp_1/\rho_1}}, \quad \text{and} \quad \frac{dp}{p} = -\frac{dV}{V}. \quad (13\text{–}47)$$

Using these relations, we find that Eq. (13–44) can be rearranged to give

$$\frac{1}{k\mathbf{M}_1^2}\frac{2p\,dp}{p_1^2} - 2\frac{dp}{p} + f\frac{dx}{D} = 0. \quad (13\text{–}48)$$

Integrating over $(x_2 - x_1) = L$, we get the equation relating pressure drop and friction for isothermal flow,

$$\frac{p_1^2 - p_2^2}{p_1^2} = k\mathbf{M}_1^2\left[2\ln\frac{p_1}{p_2} + f\frac{L}{D}\right], \quad (13\text{–}49)$$

or, since $k\mathbf{M}_1^2 = G^2/p_1\rho_1$, we have

$$\frac{p_1^2 - p_2^2}{p_1^2} = \frac{G^2}{p_1\rho_1}\left[2\ln\frac{p_1}{p_2} + f\frac{L}{D}\right]. \quad (13\text{–}49\text{a})$$

For long pipes,

$$\frac{fL}{D} \gg 2\ln\frac{p_1}{p_2}, \quad (13\text{–}50)$$

and, approximately,

$$\frac{p_1^2 - p_2^2}{p_1^2} = k\mathbf{M}_1^2 f\frac{L}{D}. \quad (13\text{–}51)$$

Equation (13–44a) can be rewritten for isothermal flow:

$$\frac{dp}{p} = -\frac{dV}{V} = \frac{-k\mathbf{M}^2}{(1 - k\mathbf{M}^2)}f\frac{dx}{2D}. \quad (13\text{–}52)$$

This relation shows that as x increases, friction causes the following:

$$\begin{array}{ll} \text{For } k\mathbf{M}^2 < 1, & p \text{ decreases, } V \text{ increases.} \\ \text{For } k\mathbf{M}^2 > 1, & p \text{ increases, } V \text{ decreases.} \end{array} \quad (13\text{–}53)$$

The rates of change increase with increased friction. Thus we find a critical length of pipe at that x where $\mathbf{M} = 1/\sqrt{k}$. This is the maximum length for which the particular isothermal flow will proceed continuously. If the pipe exceeds this limiting length, either a shock discontinuity occurs or a back-pressure adjustment modifies the inlet pressure-flow conditions. Referring to Eq. (13–40b), we see that with $T = $ const, the heat flux is to the fluid when $\mathbf{M} > 1/\sqrt{k}$ and V decreases in the flow direction.

13–4.3   Adiabatic gas flow with friction in pipes

An approximate solution for adiabatic flow with friction can be obtained if we assume that the pressure-density relation is the same as for an isentropic (frictionless) adiabatic change, and that the friction factor is constant over the pipe length. Then using $p/\rho^k = $ const in Eq. (13–41), we have

$$V = \frac{G}{\rho_1}\left(\frac{p_1}{p}\right)^{1/k}, \qquad dV = -\frac{G}{k\rho_1}\,p_1^{1/k}\,\frac{dp}{p^{[(1+k)/k]}}. \tag{13–54}$$

Introducing these relations into Eq. (13–44), assuming that $f = $ const, and integrating gives

$$\frac{k}{k+1}\frac{p_1^{1/k}}{\rho_1}(p_2^{(k+1)/k} - p_1^{(k+1)/k}) - \frac{G^2}{\rho_1^2}\frac{p_1^{2/k}}{k}\left(\ln\frac{p_2}{p_1}\right) + \frac{G^2}{\rho_1^2}p_1^{2/k}\frac{fL}{2D} = 0. \tag{13–55}$$

Except in the special case of isothermal compressible flow, f varies along the duct, and it is customary to define an average value

$$\bar{f} = \left(1/L\right)\int_0^L f\,dx. \tag{13–56}$$

In practice, the incompressible-flow friction factors are used, and f is estimated from the given data.

Dropping the heat-transfer term in Eq. (13–40a), dividing by T, and using Eqs. (13–38), (13–41a) and (13–45), we can transform the energy equation for adiabatic flow into

$$\frac{dV}{V} = -\frac{1}{1 + (k-1)\mathbf{M}^2}\frac{dp}{p}. \tag{13–57}$$

Substitution into the momentum equation (13–44a) gives

$$\frac{dp}{p} + \frac{k\mathbf{M}^2[1 + (k-1)\mathbf{M}^2]}{2(1 - \mathbf{M}^2)}f\frac{dx}{D} = 0 \tag{13–58}$$

and

$$\frac{dV}{V} = \frac{kM^2}{2(1 - \mathbf{M}^2)}f\frac{dx}{D}. \tag{13–59}$$

Thus, as x increases with adiabatic flow, friction causes the following trends:

$$\begin{array}{ll}\text{For } \mathbf{M} < 1, & p \text{ decreases, } V \text{ increases.} \\ \text{For } \mathbf{M} > 1, & p \text{ increases, } V \text{ decreases.}\end{array} \tag{13–60}$$

The limiting pipe length for continuous adiabatic flow is reached at that x where $\mathbf{M} = 1$. These are the same trends that we observed for isothermal flow, where the limiting condition occurred at $\mathbf{M} = 1/\sqrt{k}$.

Example 13–6: Isothermal gas flow in a pipe

Methane gas is to be pumped through a steel pipe (I.D. 24 in.), connecting two compressor stations 30 miles apart. At the upstream station the pressure is not to exceed 70 psia, and at the downstream station it is to be at least 10 psia. Calculate the maximum allowable rate of flow (in ft³/day at 60°F and 1 atm), assuming that there is sufficient heat transfer through the pipe wall to maintain the gas at 60°F.

Given:

$$p_1 = 70 \text{ psia}, \qquad p_2 = 10 \text{ psia}, \qquad T = 60°F, \qquad D = 24 \text{ in.,}$$
$$L = 30 \text{ miles} = 158{,}400 \text{ ft.}$$

For Methane: (Table 1–6)

$$R = 96.4 \frac{\text{ft-lb}}{\text{lbm-°R}} \quad \mu = 2.4 \times 10^{-7}, \qquad k = 1.26.$$

From Fig. 13–12:

$$k_s = 0.00015 \text{ ft.}$$

Solution. The dynamic viscosity is independent of pressure for a wide range of pressures. Then

$$\nu_1 = \mu/\rho_1 = \frac{\mu}{p_1/RT_1} = \frac{2.4 \times 10^{-7}}{(70 \times 144)/(g_c \times 96.4 \times 520)} = 3.84 \times 10^{-5} \text{ ft}^2/\text{sec.}$$

For isothermal flow in a long pipe, we have from Eq. (13–51),

$$\frac{p_1^2 - p_2^2}{p_1^2} = k\mathbf{M}_1^2 \frac{fL}{D}.$$

From the Moody diagram, Fig. 13–12, for fully developed turbulent flow and for $k_s/D = 0.000075$, we have $f = 0.0115$. Hence

$$\mathbf{M}_1^2 = \frac{(70^2 - 10^2)}{70^2} \left[\frac{2}{1.26 \times 158{,}400} \right] \frac{1}{f} = \frac{9.81 \times 10^{-6}}{f}, \qquad \mathbf{M}_1 = 0.0292.$$

From the definition of **M**, we have

$$\mathbf{M}_1 = \frac{V_1}{c_1} = \frac{V_1}{\sqrt{kp_1/\rho_1}} = \frac{V_1}{\sqrt{kRT_1}}, \qquad \text{or} \qquad V_1 = \sqrt{kRT_1}\,\mathbf{M}_1.$$

For this case,

$$V_1 = \sqrt{1.26 \times g_c \times 96.4 \times 520}\,\mathbf{M}_1$$
$$= 1425\mathbf{M}_1 = 41.7 \text{ ft/sec}$$
$$\mathbf{R}_1 = (41.7 \times 2)/(3.84 \times 10^{-5}) = 2.17 \times 10^6.$$

Returning to Fig. 13–12, we see that for this **R** and for the above (k_s/D)-value, $f = 0.013$. Repeating the calculation with $f = 0.013$, we obtain

$$\mathbf{M}_1 = 0.0275, \qquad V_1 = 39.1 \text{ ft/sec}, \qquad \mathbf{R}_1 = 2.04 \times 10^6,$$

and from Fig. 13-12, we have $f = 0.013$. Checking the assumption that

$$fL/D \gg 2 \ln(p_1/p_2),$$

we find that

$$\frac{fL}{D} = \frac{0.013 \times 158{,}400}{2} = 1030, \quad \text{while} \quad 2 \ln \frac{p_1}{p_2} = 2 \times 1.95 = 3.90.$$

The maximum discharge at 70 psia and 60°F is

$$Q_1 = V_1 A_1 = 39.1 \times \pi \times 1^2 \times 3600 \times 24$$
$$= 1.06 \times 10^7 \text{ cfs/day.}$$

Converted to 14.7 psia and 60°F the maximum discharge is

$$Q = A_1 V_1 \times \frac{70}{14.7} = 5.05 \times 10^7 \text{ cfs/day.}$$

Example 13-7: Adiabatic gas flow with friction in pipes

Air flows into an insulated steel pipe (I.D. 4 in.) at 14.7 psia and at 60°F. The Mach number at the upstream end is 0.3, and the pressure ratio $p_1/p_2 = 3.0$. Determine the length of the pipe. Is this length the maximum possible?

Given:

$$\mathbf{M}_1 = 0.3, \quad k = 1.40 \quad \text{(Table 1-6)}, \quad \rho_1 = 0.00237 \text{ slug/ft}^3,$$
$$R = 53.3 \text{ ft-lb/lbm-°R}, \quad T_1 = 60°F, \quad \nu_1 = 1.6 \times 10^{-4} \text{ ft}^2/\text{sec.}$$

Solution. From

$$\mathbf{M}_1 = \frac{G/\rho_1}{\sqrt{kRT_1}}, \quad G = 0.798$$

we have

$$V_1 = G/\rho_1 = 336 \text{ fps.}$$

For a steel pipe $k_s = 0.00015$ ft and with $D = 0.333$ ft, we have $\mathbf{R}_1 = V_1 D/\nu_1 = 7 \times 10^5$. Using Fig. 13-12 we obtain $f = 0.017$. To obtain the length we substitute into Eq. (13-55) and obtain

$$\left(\frac{1.40}{2.40}\right) \frac{(14.7 \times 144)^{1/1.4}}{0.00237} [(4.9 \times 144)^{2.4/1.4} - (14.7 \times 144)^{2.4/1.4}]$$
$$+ \left(\frac{0.798}{0.00237}\right)^2 \frac{(14.7 \times 144)^{2/1.4}}{1.4} \ln 3$$
$$+ \left(\frac{0.798}{0.00237}\right)^2 (14.7 \times 144)^{2/1.4} \frac{0.017L}{2 \times 0.333} = 0,$$

from which $L = 120$ ft. For maximum length, $\mathbf{M}_2 = 1.0$,

$$\mathbf{M}_2 = \frac{V_2}{\sqrt{kp_2/\rho_2}}, \quad V_2 = \frac{G}{\rho_1}\left(\frac{p_1}{p_2}\right)^{1/k}, \quad \mathbf{M}_2 = \frac{(G/\rho_1)(p_1/p_2)^{1/k}}{\sqrt{kp_2/\rho_2}}.$$

From the relation $p_1/p_2 = (\rho_1/\rho_2)^k$ we have $\rho_2 = 0.00107$. Hence

$$M_2 = \frac{0.798}{0.00237} \times \frac{(3)^{1/1.4}}{\sqrt{1.4 \times [(4.9 \times 144)/0.00107]}} = 0.85.$$

Therefore the length is not the maximum possible.

13–5   UNIFORM FLOW IN PRISMATIC OPEN CHANNELS

An open channel is a conduit in which a dense fluid flows under gravity with a definite interface separating it from an overlying lighter fluid. When the dense fluid is a liquid and the overlying fluid is a gas, the interface is called a *free surface*. Due to the large density difference between liquid and gases, the effect of the latter in the gravity field can be neglected. Here we will restrict ourselves to turbulent free-surface flows. Although water is the most common liquid flowing in channels, the conclusions we reach will apply generally to Newtonian liquids. There are exceptions, however, including some *super* liquids, such as certain cryogenic fluids which exhibit peculiar surface tension and capillarity effects.

Natural open channels vary in size, shape, and roughness and present irregular nonuniform sections to the flow. Artificial channels also vary in size but have a narrower range of roughnesses. Moreover, artificial channels usually are built with regular geometric shapes. They are called *prismatic* if the channel section and bottom slope are constant. Rectangles, trapezoids, triangles, circles, parabolas, and combinations of these shapes are commonly used as prismatic channel sections.

Flow in a given prismatic channel may be uniform or nonuniform. Nonuniform, or *varied*, flow will be *rapidly varied* or *gradually varied* depending on the rate of change of the fluid depth and velocity in the flow direction. With uniform flow, wall friction is in balance with head loss and hence controls the depth-velocity relation for a given flow rate. In gradually varied flow, the depth changes very slowly so that boundary friction very nearly balances the head loss. For rapidly varied flow, momentum and inertia forces dominate in establishing the behavior. Nonuniform flow will be treated in Chapter 14.

As discussed in Section 13–3, flows in noncircular conduits depart from the two-dimensional characteristics of circular tubes by exhibiting three-dimensional velocity profiles with secondary motions. Open channels are, in general, noncircular. However, many open channels are wide and the velocity-friction relations can be examined on a two-dimensional basis. Usually, the relations among energy flux, momentum flux, flow depth, and friction are treated most readily by a one-dimensional analysis. In the following we will present the general one-dimensional relations for prismatic open channels and examine the resistance problem for steady uniform flow, including the two-dimensional characteristics of wide channels.

13–5.1 One-dimensional open-channel equations

The one-dimensional total head or total energy per unit weight for each element of fluid in an open channel is

$$H = h + p/\gamma + V^2/2g$$

measured in (ft-lb)/lb or merely in feet. The terms are defined in Table 13–4 and illustrated graphically in Fig. 13–18, where a general case of open-channel flow is represented. Assuming the flow to be uniform or gradually varying in the flow direction so that accelerations normal to the bottom may be neglected, and neglecting static pressure variations due to turbulence, we find that the

TABLE 13–4

NOTATION FOR FLOW IN OPEN CHANNELS

A = cross-sectional area of channel

b = surface width
  = bottom width for rectangular channel

C = Chezy coefficient

$\dfrac{p}{\gamma} + h$ = piezometric head

H = total head
  $= \dfrac{p}{\gamma} + h + \dfrac{V^2}{2g}$

h = elevation above datum

h_0 = elevation of channel bottom

h_f = head loss due to surface resistance

H_L = total head loss

H_0 = specific head
  $= y_0 + \dfrac{V^2}{2g}$

L = length along slope $(dL = dx)$

n = roughness factor in Manning formula

P = wetted wall perimeter

q = $y_0 V$ = cfs/ft

Q = AV = cfs

$R_h = \dfrac{A}{P}$ = hydraulic radius

$S = -\dfrac{d(y_0 + h_0)}{dx}$ = slope of free surface

$S_H = -\dfrac{dH}{dx}$ = slope of energy grade line

$S_0 = \sin \alpha_0$ = bottom slope
  $= -\dfrac{dh_0}{dx}$

V = average velocity corresponding to depth y_0

V_c = critical velocity corresponding to critical depth y_c

V_N = average velocity corresponding to normal depth y_N

x = distance in flow direction

y_0 = actual depth
y_c = critical depth
y_N = normal depth

for $\alpha_0 < 10°$, depth is taken as vertical distance which is satisfactory approximation

Conjugate depths = depths before and after a hydraulic jump

Alternate depths = subcritical and supercritical depths at the same specific head

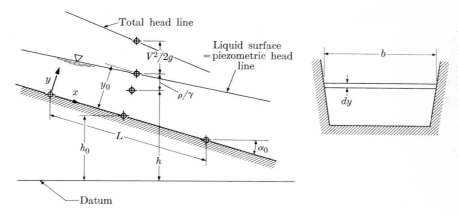

FIG. 13-18.   Notation for one-dimensional flow in prismatic open channels.

piezometric head $(p/\gamma + h)$ will be constant over a normal to the channel floor. Using h_0 as the bottom elevation, we define angle α_0 of the sloping bottom by

$$\sin \alpha_0 = -dh_0/dx = S_0 = \text{bottom slope,}$$

where x is the coordinate along the slope in the flow direction. Then, if y_0 is the depth, the pressure head on the channel floor $(p/\gamma)_0 = y_0 \cos \alpha_0$. The one-dimensional head equation becomes

$$H = h_0 + y_0 \cos \alpha_0 + V^2/2g. \qquad (13\text{-}61)$$

For small slopes (say $\alpha_0 < 10°$ or $S_0 < 0.018$), $\cos \alpha_0 \approx 1$. If we define the sum of the head terms measured above the channel floor as the *specific head* $H_0 = y_0 + V^2/2g$, Eq. (13-61) can be written for small slopes as

$$H = h_0 + y_0 + V^2/2g = h_0 + H_0. \qquad (13\text{-}62)$$

The difference in total head between two stations a distance L apart will be

$$H_1 - H_2 = (h_0 + H_0)_1 - (h_0 + H_0)_2 = H_L. \qquad (13\text{-}63)$$

Thus H_L is the head loss defined as

$$H_L = \int_0^L \frac{dH}{dx}\, dx, \qquad (13\text{-}64)$$

with

$$dH/dx = -S_H = \text{energy grade line slope.}$$

Differentiating Eq. (13-62) with respect to x, we have

$$\frac{dH}{dx} = \frac{dh_0}{dx} + \frac{dH_0}{dx}, \qquad \text{and} \qquad -S_H = -S_0 + \frac{dH_0}{dx}.$$

Rearranging gives us

$$\frac{dH_0}{dx} = \frac{dH_0}{dy_0}\frac{dy_0}{dx} = S_0 - S_H, \tag{13-65}$$

or

$$\frac{dy_0}{dx} = \frac{S_0 - S_H}{dH_0/dy_0}. \tag{13-66}$$

Equation (13-66) is the basic differential equation for one-dimensional open-channel flow. Its integrals provide the steady-flow depth-velocity-slope relations along a prismatic channel which are discussed in Chapter 14.

For steady uniform flow, both depth and velocity are constant. Hence the free surface, total head line, and channel bottom are parallel. Thus

$$S = S_H = S_0.$$

Then, from Eq. (13-65), we have

$$\frac{dH_0}{dx} = 0, \qquad H_0 = \text{const.}$$

The total head equation (13-63) reduces to

$$h_{0_1} - h_{0_2} = H_L = S_H L = S_0 L = h_f. \tag{13-67}$$

Free surfaces are subject to gravity waves, and the behavior at the interface depends on the ratio of the flow velocity V to the speed, or celerity, c of the wave. For elementary gravity waves in depths which are small compared to the crest-to-crest wavelength, the wave velocity is

$$c = \sqrt{gy_0}. \tag{13-68}$$

Then we have

$$V/\sqrt{gy_0} = \mathbf{F} = \text{Froude number for channels.} \tag{13-69}$$

When $\mathbf{F} = 1$, the velocity $V = c$ is called the critical velocity. For $\mathbf{F} < 1$, $V < c$, and the flow is called *subcritical*. For $\mathbf{F} > 1$, $V > c$, and the flow is *supercritical*.

Finally, computations will show that in a channel of uniform section, roughness, and slope, there will be a unique depth for uniform flow at each rate of discharge. This depth is determined by the balance between the loss of potential energy and the energy dissipation through boundary-friction resistance. It is called the *normal* depth. In treating nonuniform flow, it is useful to distinguish between the nonuniform condition at a discharge Q and the uniform, or normal state, which might exist in the same channel for the same Q. Where this distinction is necessary, we use the subscript N, for example, y_N, V_N, R_{h_N}, corresponding to $S_H = S_0$. In treating uniform flow, the distinction is not necessary and the subscript will be omitted.

13–5.2   Head-loss equations

The boundary shear stress for uniform open-channel flow and for gradually varied flow at the same flow depth and same velocity are very nearly identical. Consequently, it is the practice to use uniform-flow head-loss equations for both uniform and gradually varied nonuniform flows. Two head-loss relations are widely used: the *Darcy* equation expressed in the form for noncircular conduits, and the *Chezy-Manning* formulas.

Darcy equation. Using Eq. (13–34), we have

$$h_f = f \frac{L}{4R_h} \frac{V^2}{2g},$$

where

$$R_h = \text{hydraulic radius} = \frac{\text{flow cross section}}{\text{wetted perimeter}}$$

$$f = f(V, \rho, \mu, k, \text{channel size, channel shape}).$$

Here k is the wall roughness while R_h is a measure of channel size and shape. For channel sections having ratios of area to wetted perimeter close to that of a circumscribing circle or semicircle, we can evaluate the friction factor f from the pipe-friction diagram, Fig. 13–12, using Reynolds number $(4VR_h)/\nu$ and relative roughness $(k_s/4R_h)$. Note that a semicircle has the same hydraulic radius as a full circle. As channels become very wide or otherwise depart radically from the circle, or semicircle, the pipe-friction factors become less applicable. These pipe-friction factors are applied to gradually varied flow by using the total Reynolds number and local relative roughness to select the value of f.

Most open channels are physically large compared to pipes and other closed ducts. Consequently, the Reynolds number tends to be very large. At the same time, the roughness of open channels tends to be much greater than for pipes, and the relative roughness remains large even though the hydraulic radius increases. As a result, the turbulent flow is often in the fully rough regime, and f is a constant dependent only on the relative roughness.

Chezy and Manning formulas. Most friction-loss data for open channels have been evaluated using a formulation published by Chezy as early as 1769. It is commonly written

$$V = C\sqrt{R_h S}, \qquad\qquad (13\text{–}70)$$

where

$$V = \text{average velocity over the flow section,}$$
$$R_h = \text{hydraulic radius,}$$
$$S = S_0 = S_H = -dH/dx,$$
$$C = \text{flow resistance factor}$$
$$= C(V, \rho, \mu, k, \text{channel size, channel shape}).$$

TABLE 13–5

TYPICAL VALUES OF MANNING'S n

Surface	Range	Normal value
Planed wood	0.010–0.014	0.012
Unplaned wood	0.011–0.015	0.013
Finished concrete	0.011–0.013	0.012
Unfinished concrete	0.013–0.016	0.015
Vitrified sewer pipe	0.010–0.017	0.013
Glazed brickwood	0.011–0.015	0.013
Brick in cement mortar	0.012–0.017	0.015
Concrete pipe	0.012–0.016	0.015
Cast iron and wrought iron	0.012–0.017	0.015
Riveted and spiral steel pipe	0.013–0.020	0.017
Earth, straight and uniform	0.017–0.025	0.022
Corrugated metal flumes	0.023–0.030	0.025
Dry rubble	0.025–0.035	0.030
Rock cuts, smooth and uniform	0.025–0.035	0.033
Earth with weeds and stones	0.030–0.040	0.035

Rearranging the Darcy equation, we get

$$V = \sqrt{8g/f}\,\sqrt{R_h h_f/L}, \tag{13–70a}$$

so that the Chezy C is related to the Darcy f by

$$C = \sqrt{8g/f}, \tag{13–71}$$

since, for uniform flow, $S = h_f/L$.

For gradually varied flow, the Chezy C, like the Darcy f, is assumed to be a function of the local velocity and roughness conditions. Note that in the English system, the units of C are $\sqrt{ft}$/sec. For $f \approx 0.02$, C has the magnitude of approximately $114\ \sqrt{ft}$/sec. Moreover, C increases when f decreases.

Almost all experimental data for C have been for sizes, roughnesses, and flow velocities that produce the fully rough flow regime with the resistance factor independent of Reynolds number. In 1889 Manning presented a formula which expressed the flow velocity in terms of a roughness parameter as well as R_h and S. In the currently used form with English units, this formula is

$$V = (1.49/n)R_h^{2/3}S^{1/2}, \tag{13–72}$$

where n is a roughness coefficient known as Manning's n. Compared with Chezy's equation, we see that

$$C = (1.49/n)R_h^{1/6}. \tag{13–73}$$

This equation was derived from tests with channels of different roughnesses and cross sections at Reynolds numbers large enough to give fully rough turbulent-flow conditions. Under these conditions, n is a constant for each surface type. Thus the flow-resistance factor C is proportional to a relative roughness $(R_h^{1/6}/n)$. Typical values of n are given in Table 13-5.

For smooth surfaces, the Chezy C will not be a constant but will depend on the Reynolds number. In this case, the Manning expression for C, Eq. (13-73), can be used only if n is a variable quantity. Manning's n then ceases to be a roughness factor and becomes a function of the Reynolds number. Such a functional relationship has not been established, and, for this reason, n values for smooth surfaces like glass and lucite are omitted from Table 13-5. It is recommended that the Darcy equation and the Moody diagram be used for very smooth open channels.

13-5.3 Velocity and friction laws for two-dimensional channels

Let us consider first an open channel whose width is many times its depth. The flow is approximately two-dimensional, and the fully developed velocity profiles for steady uniform flow are logarithmic as found for pipes. The velocity-defect form is widely used *for two-dimensional channels* as follows:

$$\frac{U - \bar{u}}{\sqrt{g R_h S_0}} = -\frac{2.3}{\kappa} \log \frac{y}{y_0}, \tag{13-74}$$

where

$$\sqrt{g R_h S_0} = \sqrt{\tau_0/\rho} = u_*. \tag{13-75}$$

The latter is obtained by combining Eqs. (13-10), (13-67) and (13-70a). For uniform flow, the depth becomes $y_0 = y_N$. For two-dimensional flow, the hydraulic radius R_h becomes equal to the flow depth. This equation is applied to the outer region, $y/\delta = y/y_0 > 0.15$, and holds for both rough and smooth walls. The Kármán constant κ is frequently given the value 0.4, although there is evidence that the best value is lower. For example, Elata and Ippen [14] found $\kappa = 0.376$ for a gradually varied flow along a smooth horizontal channel. Vanoni [15] found values in the range $\kappa = 0.368$ to 0.397 for fixed sand-grain roughness when $0.003 < k_s/y_0 < 0.012$. The friction factor will be given closely by Eqs. (13-36a) and (13-36b) for two-dimensional channels, provided the depth y_N for uniform flow is used in place of $B/2$.

For channels which are not wide enough to give two-dimensional flow, similar universal velocity equations are not applicable. For head loss, it is the practice, as with noncircular closed ducts, to use pipe-friction factors for channel sections that do not depart radically from the circle. Darcy's equation is then used to solve the problem. Alternatively, the Chezy C can be used. For large Reynolds numbers with fully rough turbulent flow, the Manning roughness factor n is a constant for each surface type, and C can be calculated from Eq. (13-73).

13–5.4   Uniform-flow computations

The three basic problems for open channel flow are:

(a) CHANNEL SLOPE (HEAD LOSS)

    Given: Q, L, ν, size, shape, and roughness. *Find:* $S_0 = S_H$.

(b) FLOW RATE

    Given: S_0, L, ν, size, shape, and roughness. *Find:* V and Q.

(c) SIZE (R_h FOR A GIVEN SHAPE)

    Given: S_0, Q, L, ν, shape, and roughness. *Find:* R_h.

These problems are solved with steps analogous to those used for pipe flow. In all three problems, the fundamental step is the determination of the friction coefficient, either the Darcy f or the Chezy C. In the first problem, f or C may be determined directly from the given data and either Fig. 13–12 for f or Table 13–5 and Eq. (13–73) for C. Then a solution is obtained by direct computation. For the second problem, the given data will allow computation of the quantity

$$\mathbf{R}f^{1/2} = \frac{D^{3/2}}{\nu}\left[\frac{2gh_f}{L}\right]^{1/2} = \frac{8R_h^{3/2}}{\nu}\sqrt{2gS}. \qquad (13\text{–}76)$$

The value of f is determined directly by means of the $\mathbf{R}f^{1/2}$-scale of Fig. 13–12. Alternatively, C is obtained from Eq. (13–73) using R_h and n, and a direct solution of the Chezy equation gives V. The third problem can be solved directly, when the given data are used to compute

$$\mathbf{R}f^{1/5} = \frac{4}{\nu}\left[\frac{gh_f Q^3}{8\pi^3 L}\right]^{1/5} = \frac{4}{\nu}\left[\frac{gSQ^3}{8\pi^3}\right]^{1/5}. \qquad (13\text{–}77)$$

Thus, with the $\mathbf{R}f^{1/5}$-scale of Fig. 13–12, the friction factor is obtained and R_h may be computed from the Darcy equation. A direct solution based on the Chezy-Manning formulation is possible, but is inconvenient because the velocity V is related to the flow rate Q and the hydraulic radius R_h by a quadratic or higher-order equation. For cases where the Manning n and the Chezy-Manning equation are more appropriately used, a trial method of solution is more efficient.

Example 13–8: Calculation of Q for a rectangular channel

Water is flowing at a temperature of 60°F down an open channel constructed of smooth concrete. The channel has a rectangular section 10 ft wide and is on a slope $S_0 = 0.001$. The steady uniform-flow depth is 4 ft. Compute the discharge rate Q.

Given:

$$S_0 = 0.001 = S = h_f/L,$$
$$b = 10 \text{ ft} = \text{width}, \quad A = 40 \text{ ft}^2,$$
$$y_N = 4 \text{ ft} = \text{depth}, \quad R_h = 2.22 \text{ ft},$$
$$\nu_{60} = 1.22 \times 10^{-5} \text{ ft}^2/\text{sec} \quad (\text{Table 1–3}).$$

Method No. 1. Darcy with pipe friction f; direct solution

For smooth concrete, we use $k_s = 0.001$ ft (from Fig. 13–12). From Eq. (13–76),

$$\mathbf{R}f^{1/2} = \frac{8(2.22)^{3/2}}{1.22 \times 10^{-5}} [2g \times 0.001]^{1/2} = 5.53 \times 10^5,$$

$$k_s/4R_h = \frac{0.001}{4 \times 2.22} = 1.13 \times 10^{-4}.$$

From Fig. 13–12, $f = 0.0128$. Using Eq. (13–70a), we have

$$V = \left[\frac{8g}{0.0128} 2.22 \times 0.001 \right]^{1/2} = 6.7 \text{ ft/sec},$$

$$Q = AV = 40 \times 6.7 = 268 \text{ ft}^3/\text{sec}.$$

Note that fully rough flow at the given $k_s/4R_h$ is achieved at approximately $\mathbf{R} = 10^7$, when $f = 0.0125$. This is close enough to the 0.0128 value so that Reynolds-number effects can be ignored and the Chezy-Manning formula can be used as shown below.

Method No. 2. Chezy-Manning; direct solution

For smooth concrete, we use $n = 0.012$ (from Table 13–5). From Eq. (13–73),

$$C = \frac{1.49}{0.012} 2.22^{1/6} = 142.$$

Using the Chezy equation, we have

$$V = 142\sqrt{2.22 \times 0.001} = 6.7 \text{ ft/sec}.$$

(Alternatively, we may use Eq. (13–72) and compute V directly without identifying the value of C.)

Example 13–9: Calculation of flow depth. Trial procedure using Chezy-Manning formula

A smooth concrete rectangular channel is 10 ft wide on a slope $S_0 = 0.001$. Water at 60°F is flowing at the rate $Q = 250$ ft³/sec. Compute the flow depth $y_0 = y_N$.

Given:

$$S_0 = 0.001 = S = h_f/L, \quad b = 10 \text{ ft} = \text{width},$$
$$\nu_{60} = 1.22 \times 10^{-5} \text{ ft}^2/\text{sec} \quad \text{(Table 1–3)},$$
$$n = 0.012 \quad \text{(Table 13–5)}.$$

This problem is solved most efficiently with a trial solution. A depth is assumed, and the corresponding uniform-flow rate is computed by means of Eq. (13–72). Successive trials are repeated until the computed flow rate equals 250 ft³/sec.

To obtain an approximate depth, we assume two-dimensional flow and note that $R_h \to y_N$ as $b \to \infty$. Then, from Eq. (13–72), we get a trial y_N,

$$y_N = \left[\frac{Q/b}{(1.49/n)S^{1/2}} \right]^{3/5} = \left[\frac{250/10}{(1.49/0.012)(0.001)^{1/2}} \right]^{3/5} = 3.17 \text{ ft.}$$

TABLE 13–6

y_N	$A = 10y_N$	$R_h = \dfrac{10y_N}{10 + 2y_N}$	$R_h^{2/3}$	$Q = 3.93 A R_h^{2/3}$	Remarks
3.2	32	1.95	1.56	196	Low. Both A and R_h must be increased
4.2	42	2.28	1.73	286	High
3.6	36	2.09	1.64	231	Low
3.8	38	2.16	1.67	250 ←	*Solution* $y_N = 3.8$ ft.

We proceed as in the tabulations given in Table 13–6, using Eq. (13–72) and continuity, to calculate

$$Q = AV = (1.49 A/n) R_h^{2/3} S^{1/2}.$$

The same procedure can be used for any other channel section.

PROBLEMS

13–1. A pipeline conveys crude oil at 35°F (density $= 1.65$ slugs/ft^3; viscosity $= 3.3 \times 10^{-4}$ lb-sec/ft^2). The pipe has a 24-in. inside diameter. The flow rate is 31.4 cfs. Assume the walls are smooth and compute (a) the head loss per foot of length of the pipeline; (b) the wall shear stress, τ_0; (c) thickness of the laminar sublayer, δ'.

13–2. Determine the maximum allowable roughness of the pipe walls in Problem 13–1 when the energy required for pumping is not to increase over that required for smooth walls.

13–3. The walls of the pipe in Problem 13–1 are cast iron with an equivalent sand-grain roughness of $k_s = 0.0008$ ft. Compute the head loss for 100 miles of pipe and compare the power consumption with that of a smooth-walled pipe.

13–4. A water main is to be designed to carry 10 cfs from a standpipe through a distance of 2 miles to connect into another main. The pressure intensity in the receiving main is 90 psig and the elevation of the main is $+900$ ft. The water-level elevation in the standpipe is at $+1210$ ft. A steel pipe of equivalent roughness $k_s = 0.00015$ ft is to be used. Find the necessary diameter when the water temperature is 40°F.

13–5. What flow rate can be handled by a horizontal 1-in. copper fuel line if the pressure drop over a 50-ft length is 10 psi? The fuel is benzene at 68°.

13–6. What is the flow rate (cfs) through a 36-in. riveted steel pipe (roughness $k_s = 0.003$ ft) carrying sea water to the condensers of a steam plant when the pipe friction loss is 0.02 ft per ft of pipe? The sea water viscosity is approximately 10^{-5} ft^2/sec.

13–7. Assume that the velocity distribution for turbulent flow in a pipe of radius R is of the following form, where y is measured from the wall:

$$\frac{\bar{u}}{\sqrt{\tau_0/\rho}} = \left(\frac{y}{R}\right)^n.$$

Noting that the total shear is linearly distributed across the pipe section, find the form of the mixing-length relationship which corresponds to these conditions ($l = \phi[n, y, R]$). Compare the result with the assumption $l = \kappa y$.

13–8. Consider the relations for flow in pipes given by Eqs. (13–20) and (13–28) for smooth walls and for rough walls. (a) Integrate over the cross section to find the average velocity $V = Q/A$ and find the relation for V/u_* for both smooth and rough walls. (b) Using the results of (a) find expressions for $1/\sqrt{f}$ for both smooth and rough walls.

13–9. Crude oil is pumped downhill at the rate of 1.0 cfs through a 6-in. steel pipe. A stretch 1000 ft long is lying on a 1:10 slope. What will be the difference in pressures between the upper and lower ends of the 1000-ft stretch if the oil kinematic viscosity is 1.28×10^{-3} ft²/sec and the specific gravity $= 0.85$?

13–10. On a hot day the temperature rise of the oil in Problem 13–9 causes its viscosity to be reduced to 4/10 its original value. What will be the pressure difference for 1 cfs flow?

13–11. For the conditions of Problem 13–10, what will be the flow rate for the same pressure change as that required in Problem 13–9?

13–12. Gas is pumped through a smooth 10-in. pipe with a pressure head drop of 1/2 in. of water per 1000 ft. The dynamic viscosity of the gas is 2.1×10^{-7} lb-sec/ft², and its specific weight is 0.06 lb/ft³. Find the flow rate.

13–13. A conduit 4 ft in diameter is to carry water at the rate of 75 cfs with a minimum expenditure of energy. What is the order of magnitude of the permissible roughness k_s for a commercial pipe if $T = 60°F$?

13–14. What size galvanized pipe will be required to carry a flow of 2 cfs of water at 60°F without exceeding a head loss of 3.5 ft/100 ft of pipe?

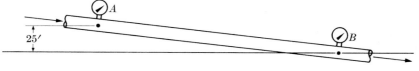

FIGURE 13–19

13–15. (a) When a new 8-in. cast-iron water main was first installed, pressure gauges were placed at points A and B, 1000 ft apart on a straight sloping length of pipe, with point A being upstream from B and located 25 ft higher than B (Fig. 13–19). When the gauge at A registered 45 psig and the flow rate through the new pipe was measured to be 2.5 cfs, what was the pressure reading in psig on the gauge at point B?

(b) Some years later, when the flow through this same pipe was measured to be 2.0 cfs, the gauges at point A and B read 42 and 44 psig, respectively. What was the absolute roughness, k_s, of the pipe at this time? State the percentage of increase in the absolute roughness from its original value for the new pipe. Assume that the temperature of the water is 60°F.

13–16. A 6-in. horizontal pipeline ($k_s = 0.0004$ ft) transports various petroleum products from a refinery to storage tanks. Gasoline (SG $= 0.68$, $\nu = 6 \times 10^{-6}$ ft²/sec) is known to be following kerosene (SG $= 0.81$, $\nu = 3 \times 10^{-5}$ ft²/sec) through the pipe with a fairly sharp interface between the two liquids somewhere between points A and B (Fig. 13–20). The steady flow rate is 1.60 cfs. At one instant of time pressure gauge readings are made at stations A and B which are located 3600 feet apart. Given that $p_A = 65$ psig and $p_B = 15$ psig at this instant, locate the position of the interface with respect to station A.

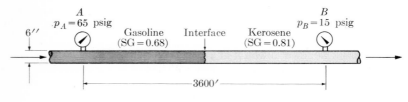

FIGURE 13–20

13–17. The change in pressure intensity in 100 ft of horizontal 2-in. pipe carrying a fluid of high viscosity is 30 psi. Determine (a) the total shear force on the pipe wall, and (b) the magnitude of the shear stress.

13–18. Linseed oil at 68°F flows through a smooth 3-in. pipe at the rate of 1.0 cfs. (a) Determine the laminar sublayer thickness δ'. (b) Find the velocity at $y = \delta'$.

13–19. Velocity measurements in a very rough pipe give the following results:

y/R	$\bar{u}$, ft/sec
0.3	10
0.6	11

Determine the roughness of this pipe in terms of equivalent sand grain roughness.

13–20. A 2-in. horizontal pipeline is carrying gasoline (SG $= 0.69$, dynamic viscosity $= 10^{-5}$ lb-sec/ft^2). The flow rate is 40 gal/min and the pressure difference between two points 100 ft apart is 1.3 psi. (a) Determine the absolute roughness of the pipe. (b) Determine the power required for pumping per 1000 ft of pipe. Compare this with the power required if the pipe were smooth.

13–21. Consider the flow of water in a 1000-ft section of a well-insulated pipeline. The rate of flow is 2.0 ft^3/sec in a cast iron pipe 4 in. in diameter. At section (1) the temperature of the water is 60°F. Calculate the temperature at section (2), 1000 ft downstream. Given that the pipe slopes upward at an angle of 10°, calculate the change in pressure between the two sections.

13–22. A comparison is to be made of the costs of pumping water through 20-in. and 24-in. pipe lines. (a) Find the ratio of costs if the pipe roughness is $k_s = 0.001$ ft and $Q = 20$ cfs at $T = 60°F$. (b) Will the cost ratio increase or decrease as the flow rate increases?

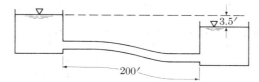

FIGURE 13–21

13–23. Water (60°F) is to be transported at the rate of 0.35 cfs between two large tanks as shown in Fig. 13–21. It is proposed to use cast-iron pipe having a length of 200 ft. The water surfaces in the tanks are maintained at a constant difference of 3.5 ft. Determine the minimum pipe diameter required to meet these specifications.

13–24. Ventilation in the London subway system is achieved by the piston effect of the trains in the tunnels. Both train and tunnel are of essentially circular section and the clearance is small. The tunnels are 10 ft in diameter and are constructed of concrete for which the equivalent sand-grain roughness may be assumed to be 0.05 ft. Assume that the induced flow of air in the tunnel is steady and that the station areas are large enough so that the pressure in the stations is constant and equal to atmospheric pressure at all times. The distance between stations is 5000 ft, and only one train at a time is in a tunnel between adjacent stations. (a) Given that the train is moving at a speed of 50 ft/sec, determine the change in pressure between the front and rear of the train due to pipe friction effects. (Neglect leakage of air along the train.) (b) Sketch the longitudinal pressure distribution when the train is halfway between stations. (c) What power is required to produce the flow of air in the tunnel?

13–25. Consider a steady, two-dimensional, turbulent flow between horizontal parallel plates 1 ft apart. The fluid kinematic viscosity $\nu = 1 \times 10^{-5} \text{ ft}^2/\text{sec}$, the density $\rho = 2 \text{ lb-sec}^2/\text{ft}^4$, and the shear stress at the wall $\tau_0 = 0.02 \text{ lb/ft}^2$. The turbulent velocity distribution is given by

$$\bar{u} = u_*[3.0 \ln (u_* y/\nu) + 5.5],$$

where $u_* = \sqrt{\tau_0/\rho}$ and y is measured from the wall. (a) Find the ratio between the turbulent shear stress and the molecular shear stress at the point $y = 0.25$ ft. (b) What is the magnitude of the velocity gradient at the wall? (c) Determine the mean velocity and the Reynolds number of the flow. (d) Determine the magnitude of the pressure gradient $\partial \bar{p}/\partial x$.

13–26. Air at 60°F and atmospheric pressure flows through a ventilating duct 2 ft in diameter under a pressure head drop of 0.2 in. of water per 1000 ft of pipe. The rate of flow is 1500 cfm. Determine the pipe roughness in terms of the equivalent sand roughness.

13–27. Compute the Mach and Reynolds numbers for air in a 3-in. pipe at a section where

$$V = 350 \text{ ft/sec}, \qquad p = 105 \text{ psia}, \qquad T = 60°F.$$

13–28. A 3-in. horizontal steel pipe carries ammonia vapor isothermally at 60°F. The flow rate is 0.25 lbm/sec. Find the head loss between two sections where the pressure is 65 and 60 psia, respectively.

13–29. A horizontal steel-pipe system carries oxygen isothermally at 60°F. At one cross section, $p = 20$ psia and the pipe diameter is 3 in. At a second cross section, $p =$ atmospheric and the pipe diameter is $1\frac{1}{2}$ in. For a rate of flow of 0.1 lbm/sec, find the heat transfer between the two sections.

13–30. Air at 60°F flows isothermally through a horizontal 8-in. steel pipe. The pressure drop along the pipe is 14 psi in 12,000 ft. Find the mass rate of flow if the initial pressure equals 70 psia.

13-31. Carbon dioxide at 60°F flows isothermally through a $1\frac{1}{2}$-in. galvanized iron pipe. The pressure drop along the pipe is 0.04 psi/ft. Find the mass flow rate if the initial pressure equals 60 psia.

13-32. Methane at 60°F enters a well-insulated horizontal pipe. Find the mass flow rate for a pressure drop of 25 psi/1000 ft of 12-in. diameter steel pipe if the initial pressure equals 100 psia.

13-33. Air flows with Mach number 0.6 into a duct having a hydraulic radius of 0.25 ft. The Mach number at the duct exit is 1.0. Assuming adiabatic flow, find the length of the duct if the average friction factor $\bar{f} = 0.02$.

13-34. Air enters a duct with a velocity of 150 ft/sec and temperature of 100°F. Determine the minimum duct diameter required for a mass flow rate of 1 lbm/sec over a length of 800 ft and at a constant temperature. The duct is smooth.

13-35. What uniform flow in cfs will occur in an earth-lined ($n = 0.020$) trapezoidal canal having a bottom width of 10 ft, sides sloping 1 vertical to 2 horizontal, and a slope of 0.0001 if the depth of flow is 6 ft?

13-36. An earth canal of trapezoidal cross section having side slopes 1.5 horizontal to 1.0 vertical is to carry a uniform flow of 234 cfs at a depth of 4 ft and a velocity of 4.5 ft/sec. Assuming that the earth lining corresponds to a roughness factor $n = .025$, determine the required bed slope.

13-37. A rectangular open channel 18 ft wide and 4 ft deep has a slope of 1 in 1000 and is lined with rubble masonry ($n = 0.017$). The amount of water discharged is to be increased as much as possible without changing the channel slope or the rectangular form of the section. The dimensions of the section may be changed but the amount of the lining (i.e., wetted perimeter) must not be more than for the 18-ft by 4-ft section. Compute: (a) discharge of original channel; (b) new dimensions of channel; (c) ratio of new discharge to old.

13-38. Given a flow of 20 ft³/sec in a rectangular channel whose depth is one-half its breadth. The channel surface is smooth cement. What are the necessary channel dimensions and slope for uniform flow if a mean velocity of 8 fps is to be obtained? (a) Solve using Manning's n. (b) Solve using roughness k_s and friction factor f.

13-39. A chemical process involves the steady uniform flow of crude oil in a rectangular open channel. The channel is 1 ft wide. The oil depth is 1 ft and the velocity is 4 ft/sec. The wall roughness k_s is 0.0013 ft. Find the channel slope. Crude oil properties:

$$\gamma = 53.7 \text{ lb/ft}^3, \qquad \mu = 1.86 \times 10^{-4} \text{ lb-sec/ft}^2, \qquad \nu = 1.1 \times 10^{-4} \text{ ft}^2/\text{sec}.$$

13-40. A liquid with a free surface is flowing steadily down an inclined smooth plate at a constant depth y_0 as shown in Fig. 13-22. The specific weight of the liquid is

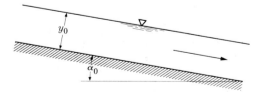

FIGURE 13-22

60 lb/ft^3, the flow is two-dimensional with a depth $y_0 = 1.5$ ft and the angle $\alpha_0 = 10°$. The kinematic viscosity is $\nu = 2.0 \times 10^{-5}$ ft^2/sec. (a) Apply the momentum equation in the direction of flow and determine the magnitude of the shear stress, τ_0, in lb/ft^2. (b) Determine the maximum permissible grain size of the roughness for the surface of the plate consistent with the assumption of a "smooth" surface. (c) Determine the quantity of liquid flowing in cubic feet per second per foot of width. (d) What is the velocity at the surface of the liquid?

13–41. A steel pipe ($k_s = 0.00015$ ft.) is 12 in. in diameter and is carrying water (60°F) in uniform motion with the pipe half filled. The flow rate is 1.60 cfs. (a) Determine the value of Manning's n for these conditions. (b) Determine the correct slope of the pipe line for uniform flow.

REFERENCES

1. BARBIN, A. R., and J. B. JONES, "Turbulent Flow in the Inlet Region of a Smooth Pipe. *Trans. ASME*, **85**, D, 29, (1963).

2. NIKURADSE, J., "Untersuchungen über die Geschwindigkeitsverteilung in turbulenten Strömungen," *VDI-Forschungsh.*, 281 (1926); and J. NIKURADSE, "Turbulente Strömung in *nichtkreisförmigen* Röhren," *Ing. Arch.*, **1**, 306 (1930).

3. LAUFER, J., "The Structure of Turbulence in Fully Developed Pipe Flow," *Nat. Advisory Comm. Aeron.*, Rept. 1174 (1954).

4. NIKURADSE, J., "Gesetzmässigkeiten der turbulenten Strömung in glatten Röhren," *VDI-Forschungsh.*, 356 (1932); and H. REICHARDT, "Vollständige Darstellung der turbulenten Geschwindigkeitsverteilung in glatten Leitungen," *ZAMM*, **31**, 208 (1951).

5. NIKURADSE, J., "Strömungsgesetze in rauhen Röhren," *VDI-Forschungsh.*, 361 (1933).

6. MOODY, L. F., "Friction Factors for Pipe Flow," *Trans. ASME*, **66**, 8 (1944).

7. CARLSON, L. W., and T. F. IRVINE, JR., "Fully Developed Pressure Drop in Triangular Shaped Ducts," *Trans. ASME*, Series C, **83**, 4 (1961). p. 441.

8. YEN, J. T., "Exact Solution of Laminar Heat Transfer in Wedge-Shaped Passages with Various Boundary Conditions," *WADC* TR 57–224, Wright-Patterson AFB, Ohio (1957), p. 6.

9. LAUFER, J., "Investigation of Turbulent Flow in a Two-dimensional Channel," *Nat. Advisory Comm. Aeron.*, Rept. 1053 (1951).

10. ROUSE, H., *Elementary Mechanics of Fluids*, John Wiley and Sons, Inc., New York, 1946, p. 214.

11. REICHARDT, H., "Über die Geschwindigkeitsverteilung in einer geradlinigen turbulenten Couetteströmung," *ZAMM Sonderhefte*, 526–529 (1956).

12. KEENAN, J. H., and E. P. NEUMAN, "Measurements of Friction in a Pipe for Subsonic and Supersonic Flow of Air," *Trans. ASME, J. Appl. Mech.*, **13**, 2 (1946).

13. SHAPIRO, A. H., *The Dynamics and Thermodynamics of Compressible Fluid Flow*, Vol. II, The Ronald Press Co., New York, 1954, Chapter 28.

14. ELATA, C., and A. T. IPPEN, "The Dynamics of Open Channel Flow with Suspensions of Neutrally Buoyant Particles," *Mass. Inst. Technol., Hydrodynamics Lab.*, Rept. TR 45 (1961).

15. VANONI, V. A., "Transportation of Suspended Sediment by Water," *Trans. ASCE.*, **3,** 67 (1946).

CHAPTER 14

Nonuniform Flow in Conduits

14-1   INTRODUCTION

A flow is characterized as nonuniform if the velocity vector at successive points along a streamline changes direction or magnitude. The majority of nonuniform flows are caused by one or more of the following:

(a) *Increase or decrease in fluid velocity and pressure.* It is frequently necessary to alter the size or shape of a flow conduit in order to change either velocity or pressure. In the design of a high-speed wind tunnel, for example, it is uneconomical from a power-loss standpoint to maintain the high test-section velocities throughout the circuit. Hence the flow is accelerated ahead of the test section and decelerated in a downstream section.

(b) *Flow metering devices.* A change in cross-sectional area produces a nonuniform flow and results in a pressure change which is a function of the volume or mass rate of flow. Hence, by measuring the pressure differential, the rate of flow may be determined. It is usually desirable to achieve the area changes with as little energy dissipation as possible.

(c) *Flow control devices.* Variation of flow rate may be accomplished by a device which is designed to produce energy dissipation, e.g., a valve accomplishes this by creating a separation zone downstream of an abrupt area change.

(d) *Change in flow direction.* A bend or elbow introduces nonuniformities into the flow and may cause a considerable amount of energy dissipation unless properly designed.

(e) *Flow around immersed objects.* An object immersed in a fluid generally produces a nonuniform flow in the surrounding fluid. Usually we are interested in the resultant force on the object due to the flow field. This subject is treated in detail in Chapter 15.

This chapter will discuss steady nonuniform flow of compressible and incompressible fluids in closed conduits as well as steady nonuniform free-surface flow. When frictional effects are considered, the flows will be assumed to be turbulent.

14-2   INCOMPRESSIBLE NONUNIFORM FLOW IN CLOSED CONDUITS. FLOW FEATURES AND LOSSES

It is helpful to consider nonuniform flows in terms of regions of accelerating and decelerating flows. Since we are concerned only with steady motions, these accelerations or decelerations are due to spatial variations in the velocity vector.

As discussed in Section 2–1, these are known as convective accelerations as opposed to the local accelerations which depend on the time rate of change of velocity at a point.

The distinction between accelerating and decelerating flow regions is of importance because of a basic concern with the type of pressure gradient which exists along the boundary. Discussion of the effects of boundary curvature in Sections 10–3 and 12–4 indicated that separation of the flow is possible if the pressure increases along the boundary in the direction of flow. Analysis and design of nonuniform conduits must be based on the knowledge of whether or not separation will occur. In Section 7–4.1 it was shown that gravity plays no part in an enclosed system, other than to alter the magnitude of the pressure by an amount which depends on the height above an arbitrary datum. Hence we may consider enclosed nonuniform flows in a horizontal plane. If the direction of flow, x, is inclined to the horizontal, it is the dynamic pressure gradient, $\partial p_d/\partial x$,* which determines whether or not separation will occur. From Eq. (7–20) it follows that

$$\frac{\partial p_d}{\partial x} = \frac{\partial}{\partial x}\,(p + \gamma h), \tag{14–1}$$

where h is the height above the datum.

Important concepts of nonuniform motion may be illustrated by considering the flow from left to right in the convergent conduit shown in Fig. 14–1. If frictional effects in the short transition section between the two regions of uniform flow are ignored, we may apply the one-dimensional Bernoulli equation (4–26) between sections (1) and (2), namely,

$$\frac{p_1}{\gamma} + \frac{V_1^2}{2g} = \frac{p_2}{\gamma} + \frac{V_2^2}{2g}. \tag{14–2}$$

From the continuity condition, it is obvious that the average velocity $V_2 > V_1$, and hence $p_2 < p_1$. Thus, by applying a one-dimensional analysis, we find that a favorable (decreasing) pressure gradient exists and conclude that there is no danger of separation. The fact that this conclusion is wrong points to one of the dangers of using a one-dimensional analysis to answer questions regarding details of the flow pattern between sections (1) and (2). The difficulty is not overcome by applying Eq. (14–2) from section (1) to some intermediate section in the transition. Since the streamlines within the transition are curved, there must be a pressure gradient normal to the streamlines (see Section 6–8), and the one-dimensional analysis provides information only on the average pressure at a section. If Fig. 14–1 represents a cross section through a two-dimensional transition, the streamline pattern may be found by assuming irrotational flow and by drawing the flow net. This is a graphical solution of the Laplace equation

---

* In order to simplify the notation from this point on through the remaining chapters, the temporal mean pressure will be denoted by p rather than $\bar{p}$, even though the flow may be turbulent.

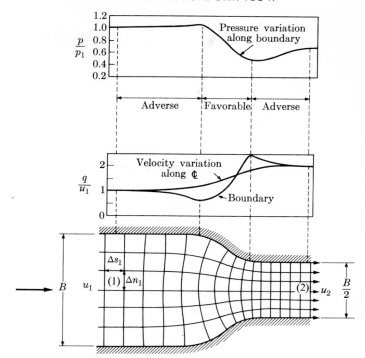

FIG. 14–1.   Velocity and pressure distribution in a convergent flow.

described in Sections 6–6 and 9–3.3. The ratio of the velocity at any point to the reference velocity u_1 is found by applying the continuity equation between any pair of adjacent streamlines,

$$u_1 \, \Delta n_1 \approx u \, \Delta n.$$

Hence

$$\frac{u}{u_1} \approx \frac{\Delta n_1}{\Delta n} \approx \frac{\Delta s_1}{\Delta s}. \tag{14–3}$$

The accuracy of the process increases as the streamlines and equipotential lines are drawn closer together. The velocity variations both along the centerline and on the boundary of the transition, computed from Eq. (14–3), are shown in Fig. 14–1. The pressure variation along the boundary is calculated from Eq. (6–61) and is also shown. Thus far we have ignored frictional effects. However, along the boundaries of the converging section, there are two zones of small adverse pressure gradient and a zone of favorable pressure gradient of much larger magnitude. Separation is therefore possible for a real fluid if the magnitude of either of the adverse pressure gradients is large enough to cause a velocity reversal at the wall. Transparent models of transition sections may be tested to determine whether separation will occur. Separation zones may be made visible by the injection of dye or smoke at the walls.

It should be emphasized that the flow net will not indicate the streamline pattern if separation does occur, since there is no mechanism which can cause separation in an irrotational flow. For example, if the direction of flow in the transition of Fig. 14–1 is reversed, the irrotational-motion flow net, the velocity distribution, and the pressure variations are unchanged. However, the designations "adverse" and "favorable" on the pressure-gradient plot must be interchanged. It can be seen that the magnitude of the adverse gradient is now much larger than in the converging flow. Separation would undoubtedly occur in the diverging flow, and the streamline pattern would change to that shown schematically in Fig. 14–2.

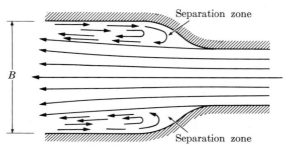

FIG. 14–2.   Separation in a divergent flow.

The primary difference between the convergent and divergent flows is in the magnitude of the energy dissipation associated with the separation in the divergent flow. Even though the conduit section becomes uniform immediately downstream of the transition, the separated flow requires a much larger distance to reestablish a normal pressure and velocity distribution within the conduit. Two factors are responsible for the large energy dissipation associated with the separation. First, the maintenance of the circulatory motion in the region of separated flow requires a continual supply of energy from the main stream. Second, the large velocity gradient at the surface of separation generates large-scale turbulence which is swept downstream and ultimately dissipated into heat. In other words, the production of turbulence by separation is much greater than that which would occur under normal wall shear in an unseparated flow. As the separated fluid gradually expands to fill the larger conduit, there will be a gradual reduction of mean velocity and a corresponding increase in the mean pressure. Quantitative results, both experimental and analytical for some basic nonuniform geometries, are given in the following sections.

14–2.1   Energy dissipation in nonuniform flow

In dealing with the nonuniform behavior of fluids which may be treated as incompressible, it is convenient to determine the energy dissipation in terms of the unit weight of fluid flowing. This includes both the heat transferred to the surroundings and the increase in internal (heat) energy of the fluid. The energy dissipation per unit weight is defined by the one-dimensional energy equation, Eq. (4–24a), and is commonly called a "head loss," although the loss is only in the mechanical-energy terms since the total energy of the system is conserved. In the absence of shaft work, Eq. (4–24a) may be written between an upstream section (1) and a downstream section (2) as

$$\left(\frac{p_1}{\gamma} + h_1 + \frac{V_1^2}{2g}\right) - \left(\frac{p_2}{\gamma} + h_2 + \frac{V_2^2}{2g}\right) = H_{L_{1-2}}. \tag{14–4}$$

It is important to note that the energy equation is used simply to *define* the quantity which we call head loss. Therefore, a theoretical analysis of head losses must be based on the continuity equation and the momentum equations of motion. The point is sometimes confusing, since in short transition regions without separation, we frequently neglect energy dissipation due to wall shear and employ the frictionless Bernoulli equation as an analytical equation relating velocity and pressure. We have shown in Eqs. (6–71) and (6–72) that for frictionless incompressible flow, the momentum and energy equations are similar.

In many cases, short nonuniform sections occur as part of a system of uniform-flow conduits, as shown in Fig. 14–3. If Eq. (14–4) is used to determine the head loss, we have to subtract the length l of the nonuniform flow region in calculating the pipe friction loss in the downstream conduit. To avoid this difficulty, the total head or energy grade line (see also Fig. 4–7) may be extended upstream until it meets the uniform-flow total-head gradient in the upstream conduit. The loss charged to the transition is the vertical difference in the two gradients and is assumed to occur in essentially zero length.

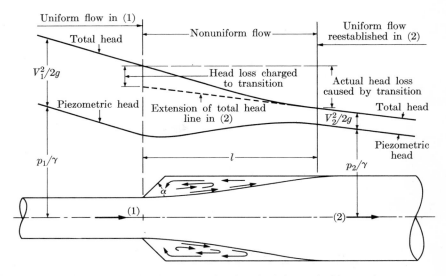

FIG. 14–3. Total and piezometric head gradients for an abrupt expansion.

With reference to the similitude discussion for enclosed flows in Section 7–4.1, we can reason that the dimensionless form of the energy dissipation per unit weight should be a function of the boundary geometry and a characteristic Reynolds number of the flow. The energy dissipation in nonuniform flow is primarily a result of turbulence generation in zones of separation, and we would expect a significant dependence on the Reynolds number only if the point of separation changed due to viscous effects.

The energy dissipation per unit weight, H_L, may be made dimensionless by dividing by a kinetic energy per unit weight, $V^2/2g$. As a characteristic parameter we choose the value of $V^2/2g$ which exists in the uniform-flow region. In a transition between two uniform-flow regions, the higher value of $V^2/2g$ is generally chosen. Therefore, a dimensionless head-loss coefficient K_L is defined as

$$\frac{H_L}{V^2/2g} = K_L = f(\text{geometry, Reynolds number}). \qquad (14\text{--}5)$$

In many transitions in which separation occurs, the point of flow separation is governed primarily by the boundary geometry. The value of K_L then depends only on the geometry unless the flow approaches laminar conditions.

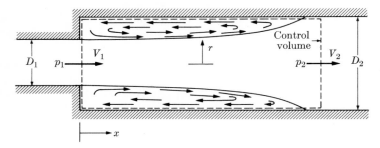

FIG. 14–4.   Control volume for an abrupt expansion.

14–2.2   Expanding flows

The simplest case of an expanding flow is the abrupt enlargement shown in Fig. 14–3. The angle of an abrupt enlargement has a secondary influence on the separated flow, so we can take a 90° angle as typical and consider a control volume of the type shown in Fig. 14–4. The control volume should extend far enough downstream to be in a region of parallel flow. We use the x-component of the one-dimensional momentum equation (4–34). The body force F_{b_x} is zero, and the summation of shear forces on the walls is neglected. Equation (4–34) becomes, with $K_m = 1$,

$$F_{p_x} = (p_1 - p_2)A_2 = \rho Q(V_2 - V_1). \qquad (14\text{--}6)$$

The continuity equation may be written, at either section,

$$Q = V_1 A_1 = V_2 A_2. \tag{14-7}$$

If we substitute these into Eq. (14–4), the head loss becomes

$$H_{L_{1-2}} = \frac{(V_1 - V_2)^2}{2g}, \tag{14-8}$$

or, in dimensionless form, using Eqs. (14–5) and (14–7), we have

$$\frac{H_{L_{1-2}}}{V_1^2/2g} = K_L = \left[1 - \left(\frac{D_1}{D_2}\right)^2\right]^2. \tag{14-9}$$

Experimental results confirm Eq. (14–9) within a few percentage points. This indicates that the boundary shear stresses in the separation zone are small in comparison with the loss due to the production and dissipation of turbulence by the expanding jet. If D_2 in Fig. 14–4 becomes very large in relation to D_1, we have a conduit discharging into a large tank or reservoir. As is expected, Eq. (14–9) shows that all of the kinetic energy entering the tank is dissipated, since $K_L = 1$ as $D_1/D_2 \to 0$.

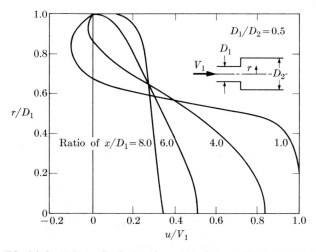

FIG. 14–5. Velocity distributions downstream from an abrupt expansion [1].

Experimental velocity profiles [1] for various sections downstream from the abrupt expansion of Fig. 14–4 (diameter ratio $D_1/D_2 = \frac{1}{2}$) are shown in Fig. 14–5. A profile of the separation zone and the spatial distribution of mean pressure below the expansion are also shown in Fig. 14–6. While the separation zone extends only about five diameters downstream, the pressure does not become uniformly distributed across the section until much farther downstream.

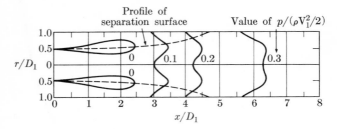

FIG. 14-6. Separation zone and pressure distribution downstream from an abrupt expansion [1].

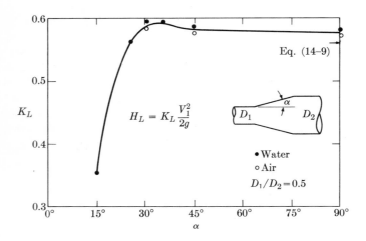

FIG. 14-7. Head loss in wide-angle diffusers [1].

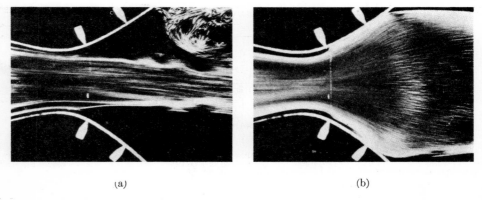

FIG. 14-8. Flow in a diverging channel: (a) separation under normal conditions; (b) prevention of separation by suction removal of fluid at the boundaries [4].

Measurements [1] of head loss in diffusers having expansion angles (α in Fig. 14–3) less than 90° are shown in Fig. 14–7. The design of a conical diffuser for minimum energy dissipation requires that the expansion angle be small enough to prevent separation and large enough to avoid unnecessary losses due to excessive length and frictional dissipation. The optimum angle, α, is of

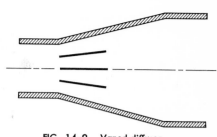

FIG. 14–9. Vaned diffuser.

the order of 3° to 4° and is dependent on the velocity distribution at the entrance to the diffuser [2]. The performance of wide-angle diffusers can be improved by the removal of fluid by suction along the diverging walls [3]. The continuous removal of the low-momentum fluid can prevent separation, since it is this fluid that first flows backward. Figure 14–8 [4] shows a photograph of a wide-angle diffuser both with and without suction. Another method of improving performance is to use interior vanes which effectively change the wide-angle expansion into a number of small-angle expansions, as shown in Fig. 14–9. Experimental investigations [5] have shown that the flow through a vaned diffuser is more stable when the length of the vanes is less than the total length of the diffuser.

14–2.3   Converging flows

The basic features of a convergent flow are shown in Fig. 14–10 [6]. The free stream contracts downstream from the wall over a distance of approximately eight times the step height. Below this point, the stream expands and reaches the lower boundary at $x/b \approx 17$. The primary energy dissipation occurs in the region in which the flow is expanding.

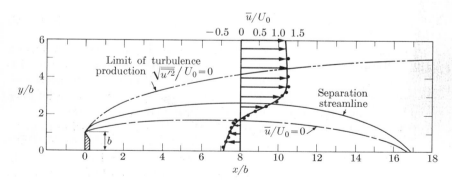

FIG. 14–10   Separation zone downstream from a wall [6].

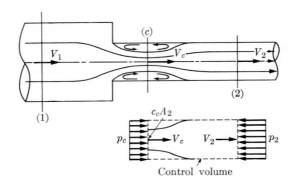

FIG. 14-11.   **Flow pattern and control volume for an abrupt contraction.**

By a similar line of reasoning we can deduce the flow pattern for the abrupt contraction shown in Fig. 14-11. If it is assumed that all of the energy dissipation occurs in the zone of expanding flow, Eq. (14-8) may be used between sections (c) and (2). Thus

$$H_L = \frac{(V_c - V_2)^2}{2g}. \tag{14-10}$$

The area of the contracted section is given by a contraction coefficient c_c. Thus

$$c_c A_2 V_C = A_2 V_2,$$

and Eq. 14-10 may be written

$$\frac{H_L}{V_2^2/2g} = K_L = \left(\frac{1}{c_c} - 1\right)^2. \tag{14-11}$$

Experiments give c_c, in terms of the area ratio A_2/A_1, as

$$c_c = 0.62 + 0.38(A_2/A_1)^3. \tag{14-12}$$

In the limit, when A_1 is very large compared with A_2, we have flow from a large tank into a square-edged conduit. Under these conditions, $K_L \approx 0.4$. A relatively small amount of rounding, as in a bell-mouthed entrance, effectively reduces K_L to 0.05 or less. The loss coefficient for a well-designed free-discharge nozzle at the end of a pipe is approximately $K_L = 0.03$.

A *valve* is a device which is designed to produce energy dissipation in order to control rate of flow. A common type is the gate valve which consists of a disk having approximately the same shape as the conduit section. The gate or disk can be raised or lowered so as to obstruct a certain portion of the cross-sectional area. The flow pattern is similar to that of Fig. 14-10, in which b might represent the obstruction caused by the gate. As the gate is closed further, both the separation zone and the energy dissipation become larger. The following values of K_L are appropriate for a gate valve in a circular pipe:

	Open	$\frac{3}{4}$ Open	$\frac{1}{2}$ Open	$\frac{1}{4}$ Open
$K_L = \dfrac{H_L}{V^2/2g} =$	0.2	1.15	5.6	24.0

If the flow in a conduit takes place either by gravity or under the action of a pump, there is a fixed amount of energy available for dissipation. For a given conduit, the maximum rate of flow will occur when all of the available energy is dissipated in pipe friction. If a valve is partially closed, the dissipation by the valve reduces the amount that can be dissipated by friction, and the velocity and discharge are thereby reduced.

14–2.4   Flow meters

Two types of pipe-line flow meters which are widely used are the Venturi and the orifice. All of the basic features of these meters are contained in the diagram and discussion of Fig. 14–11. The Venturi meter is used to measure flow rates when power saving by minimizing energy dissipation is an important factor. Therefore, the Venturi meter is shaped to eliminate all of the separation zones shown in Fig. 14–11. A typical shape of such a meter is shown in Fig. 14–12. An orifice meter is a plate with an opening (usually concentric) placed between flanges of a pipe, as shown in Fig. 14–13. The orifice meter is much cheaper, but it causes considerably more energy dissipation than the Venturi meter. If we neglect energy dissipation between the region of upstream parallel

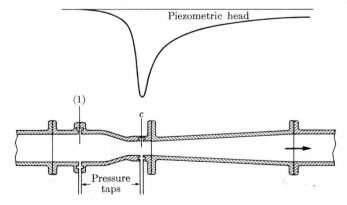

FIG. 14–12.   Venturi meter.

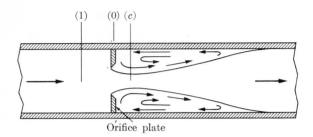

FIG. 14–13.　Orifice meter.

flow (1) and the section of maximum contraction (c), the frictionless energy equation (14–2) and the continuity condition give the flow rate Q in terms of the pressure differential and areas at the parallel flow sections 1 and c,

$$Q = A_c \sqrt{\frac{2(p_1 - p_c)/\rho}{1 - (A_c/A_1)^2}} \, . \tag{14–13}$$

In the Venturi, both A_1 and A_c are physical dimensions of the meter. However, to apply Eq. (14–13) to the orifice meter, A_c must be related to A_0, the diameter of the orifice-plate opening. A reasonable approximation for concentric orifices is obtained by using $A_c = c_c A_0$, where

$$c_c = 0.60 + 0.40(A_0/A_1)^2. \tag{14–14}$$

Experimental data on actual calibrations, location of pressure-measuring sections, and other types of flow meters are contained in an *ASME* publication [7], to which the reader is referred for more detailed information.

14–2.5　Change of flow direction

A bend in a uniform conduit generates secondary spiral motions, as illustrated in Fig. 10–1. The flow pattern within and downstream of a bend is extremely complex due to the possibility of separation in addition to the secondary motions. The reason for flow separation is again the existence of an adverse pressure gradient. As for the converging flow shown in Fig. 14–1, we can deduce the basic features of the boundary pressure gradients in a bend by means of the potential, or irrotational, motion approximation. Figure 14–14 shows the

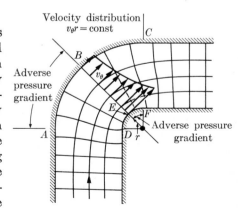

FIG. 14–14.　Flow in a 90° two-dimensional bend.

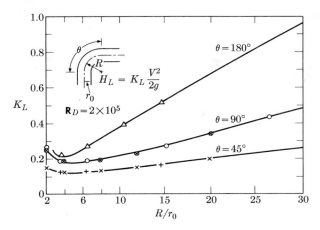

FIG. 14–15. Variation of head loss with relative radius of smooth pipe bends [8].

potential flow net for a two-dimensional 90° bend in a conduit. The velocity decreases on the outer boundary along AB and on the inner boundary along EF. This leads to zones of adverse pressure gradient and possible flow separation in these regions. The potential-flow velocity distribution along the axis of symmetry is that of an irrotational or free vortex given by Eq. (6–92), where $v_\theta r = $ const.

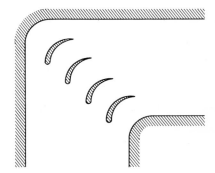

FIG. 14–16. Miter bend with turning vanes.

The energy dissipation in a conduit is increased by the introduction of a bend. The amount of the increase is called the head loss due to the bend, and its magnitude depends in a large measure on the extent to which secondary spiral flow and separation occur. Both the secondary flow intensity and the separation are related to the curvature of the bend. Experimental results [8] for the total head loss in smooth pipe bends having various relative radii are shown in Fig. 14–15. Minimum dissipation is achieved when the centerline radius of the bend is about four times the radius of the pipe. For a large-diameter conduit, a bend designed on this basis becomes quite large, and it is usually advantageous to design a vaned miter bend as shown in Fig. 14–16 [9].

14–2.6 Pipeline systems

The conduit friction equations in Chapter 13 were developed for steady flow in uniform sections. Fluids may be transported in a system of pipes or conduits made up of interconnected sections of uniform flow together with regions of nonuniform flow caused by transition sections, elbows, and valves. In the preceding we have shown how the energy dissipation for some of these non-uniform sections may be determined. The design of a pipeline system is then reduced to the problem of making an overall energy balance on a system-control volume such that the energy dissipated plus the energy remaining equals the initial energy. In a gravity-flow system, a certain amount of potential energy is available for dissipation or conversion. In nongravity systems, energy must be supplied by means of a pump. Energy removal from a pipeline system is not necessarily only a local dissipative process. A turbine, for example, re-moves energy from the flow and the connected generator converts it into elec-trical energy.

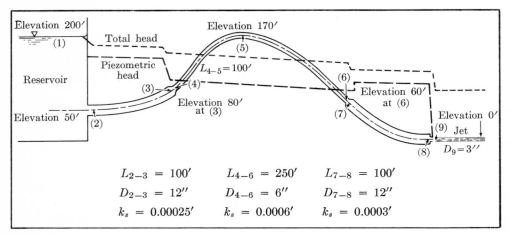

FIG. 14–17. Gravity-flow pipe system.

For a control volume enclosing the fluid in a given system we may make use of the energy balance given by Eq. (4–24a). Using subscripts 1 and n to refer to the inlet and outlet sections of the control volume, respectively, we can express this as

$$\frac{p_1}{\gamma} + h_1 + \frac{V_1^2}{2g} = \frac{p_n}{\gamma} + h_n + \frac{V_n^2}{2g} + \frac{\text{shaft work}}{\text{lb}} + \sum H_{L_{1-n}}. \qquad (14\text{–}15)$$

Because of general inaccuracies in the various energy-dissipation terms, it is usual to take the kinetic-energy flux correction, $K_e = 1$ for turbulent flow. The shaft-work term is negative for a pump and positive for a turbine.

The term $\sum H_{L_{1-n}}$ is the summation of the head losses in both uniform and nonuniform regions. Example 14-1 illustrates the method of calculating flow rate and pressure distribution in a system, shown in Fig. 14-17, consisting of pipes connected in series.

It should be remembered that H_L is the energy dissipation per pound of fluid flowing; consequently, if the conduit system has interior loops, we may include in $\sum H_{L_{1-n}}$ only the head loss in any one of the possible flow paths. This is illustrated in Fig. 14-18, in which we may choose to evaluate $\sum H_{L_{1-n}}$ along either the upper loop or the lower loop. At point (2) the flow will proportion itself between the two loops so that the head loss between (2) and (5) will be identical. The computation of the flow division between loops is shown in Example 14-2.

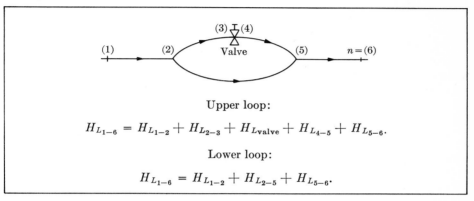

Upper loop:

$$H_{L_{1-6}} = H_{L_{1-2}} + H_{L_{2-3}} + H_{L\text{valve}} + H_{L_{4-5}} + H_{L_{5-6}}.$$

Lower loop:

$$H_{L_{1-6}} = H_{L_{1-2}} + H_{L_{2-5}} + H_{L_{5-6}}.$$

FIG. 14-18.   Pipe system with interior loop.

Calculations of pressure distributions and flow rates for complex pipe networks involving many interior branches and loops are generally performed on analog or digital computers. The references cited [10] may be consulted for specific details.

Example 14-1: Gravity-flow pipe system

Water (60°F) flows from a large reservoir through a series of pipes of different sizes as shown in Fig. 14-17. The pipeline system terminates with a nozzle and a jet discharging into the atmosphere at point (9). All of the transitions, except for the nozzle, are abrupt contractions or expansions. As a first approximation, it will be assumed that the pipe flows are fully developed over their entire length. The elevations of various points in the system, together with the length, diameter, and relative roughness of each pipe, are specified in Fig. 14-17. If we use the relative roughness as a guide, a value of f for each pipe may be assumed from Fig. 13-12. Thus

$$f_{2-3} = 0.015, \qquad f_{4-6} = 0.017, \qquad f_{7-8} = 0.016.$$

The expression for $\sum H_{L_{1-9}}$ may be written as follows:

$$\sum H_{L_{1-9}} = K_{L_{1-2}}\frac{V_2^2}{2g} + \left(f\frac{L}{D}\right)_{2-3}\frac{V_2^2}{2g} + K_{L_{3-4}}\frac{V_4^2}{2g} + \left(f\frac{L}{D}\right)_{4-5}\frac{V_4^2}{2g} + \left(f\frac{L}{D}\right)_{5-6}\frac{V_4^2}{2g}$$

$$+ K_{L_{6-7}}\frac{V_4^2}{2g} + \left(f\frac{L}{D}\right)_{7-8}\frac{V_7^2}{2g} + K_{L_{8-9}}\frac{V_9^2}{2g}.$$

From the continuity equation, we have

$$A_2 V_2 = A_4 V_4 = A_7 V_7 = A_9 V_9.$$

Hence all the velocity terms in the $\sum H_{L_{1-9}}$ equation may be expressed in terms of the jet velocity, V_9, and an area or diameter ratio. Therefore,

$$\sum H_{L_{1-9}} = \frac{V_9^2}{2g}\left[K_{L_{1-2}}\left(\frac{D_9}{D_2}\right)^4 + \left(f\frac{L}{D}\right)_{2-3}\left(\frac{D_9}{D_2}\right)^4 + K_{L_{3-4}}\left(\frac{D_9}{D_4}\right)^4\right.$$

$$+ \left(f\frac{L}{D}\right)_{4-5}\left(\frac{D_9}{D_4}\right)^4 + \left(f\frac{L}{D}\right)_{5-6}\left(\frac{D_9}{D_4}\right)^4$$

$$\left. + K_{L_{6-7}}\left(\frac{D_9}{D_4}\right)^4 + \left(f\frac{L}{D}\right)_{7-8}\left(\frac{D_9}{D_7}\right)^4 + K_{L_{8-9}}\right].$$

The energy equation (14–15) for the system-control volume between points (1) and (9) is

$$0 + 200 + 0 = 0 + \frac{V_9^2}{2g} + 0 + \sum H_{L_{1-9}}.$$

If the previous expression for $\sum H_{L_{1-9}}$ is substituted in the above, the velocity V_9 is the only unknown. The following values of K_L from Section 14–2 are used:

$K_{L_{1-2}} = 0.40,$ $K_{L_{3-4}} = 0.35,$ abrupt contraction (Eqs. 14–11, 14–12),
$K_{L_{6-7}} = 0.55,$ abrupt expansion (Eq. 14–9),
$K_{L_{8-9}} = 0.03,$ nozzle.

From the computations,

$$\frac{V_9^2}{2g} = 122.4 \text{ ft.} \qquad \therefore V_9 = 89 \text{ ft/sec and } Q = 4.35 \text{ cfs.}$$

Also,

$$V_2 = V_7 = 5.5 \text{ ft/sec,}$$

and

$$V_4 = 22.2 \text{ ft/sec.}$$

The corresponding pipe Reynolds numbers are

$$\mathbf{R}_2 = \mathbf{R}_7 = 4.5 \times 10^5, \qquad \mathbf{R}_4 = 9.1 \times 10^5.$$

The exact values of the friction factor may now be found from Fig. 13–12. Thus

$$f_{2-3} = 0.016, \qquad f_{4-6} = 0.017, \qquad f_{7-8} = 0.017.$$

These values are very close to the original assumptions for f. No significant change in the results is obtained by correcting the computations. A convenient form for arranging the computations is shown in Table 14–1. This arrangement makes it possible to determine the pressure heads at the numbered points in the system. Of critical importance is the magnitude of the pressure at the summit (point 5). In general, subatmospheric pressures are to be avoided, since dissolved gases may come out of solution and form a gas pocket. This will result in additional head loss and flow surging. Large pipes with relatively thin walls may collapse under subatmospheric pressure. For the pipe system in Fig. 14–17, the pressure head at the summit is 7.4 ft below atmospheric. The choice of a slightly larger pipe diameter between sections (4) and (6) would eliminate the subatmospheric pressure at the summit.

TABLE 14–1

	FLOW CALCULATION							GRADE LINES		
(1)	(2)	(3)	(4)	(5)	(6)	(7)	(8)	(9)	(10)	(11)
Point	Loss	K_L	D (in)	$(D_9/D)^4$	$(K_L)_9$	ΔH_L (ft)	H (ft)	$V^2/2g$ (ft)	h (ft)	p/γ (ft)
1							200.0	0	200	0
	Entrance	0.40	12	0.004	0.002	0.2				
2							199.8	0.5	50	149.3
	$f\,L/D$	1.50	12	0.004	0.006	0.7				
3							199.1	0.5	80	118.6
	Contraction	0.35	6	0.063	0.022	2.7				
4							196.4	7.7	80	108.7
	$f\,L/D$	3.40	6	0.063	0.214	26.1				
5							170.3	7.7	170	−7.4
	$f\,L/D$	3.40	6	0.063	0.321	39.2				
6							131.1	7.7	60	63.4
	Expansion	0.55	6	0.063	0.035	4.3				
7							126.8	0.5	60	66.3
	$f\,L/D$	0.80	12	0.004	0.006	0.7				
8							126.1	0.5	0	125.6
	Nozzle	0.03	3	1	0.030	3.7				
9							122.4	122.4	0	0
				Σ =	0.636	77.6				

$$Note: \quad 200 = (1 + 0.636)\frac{V_9^2}{2g}. \qquad \therefore \frac{V_9^2}{2g} = 122.4 \text{ ft.}$$

Example 14–2: Division of flow rate between parallel loops

In the pipe system shown in Fig. 14–18, the following pipe characteristics are known:

$$L_{2-3} = 2000 \text{ ft}, \qquad L_{4-5} = 2000 \text{ ft}, \qquad L_{2-5} = 3000 \text{ ft},$$
$$D_{2-3} = 8 \text{ in}, \qquad D_{4-5} = 8 \text{ in}, \qquad D_{2-5} = 12 \text{ in},$$
$$k_s = 0.0005 \text{ ft}, \qquad k_s = 0.0005 \text{ ft}, \qquad k_s = 0.001 \text{ ft}.$$

The gate valve is in the fully open position, and hence $K_{LV} = 0.2$. The total rate of flow of water (60°F) through the pipe system is 10 cfs. Determine the volume rate of flow through each loop and the change in the piezometric head between points 2 and 5.

One procedure which is applicable to any number of pipes in parallel is to assume a discharge through any one of the loops. For the assumed discharge, say in pipe 2–5, the head loss H_{L2-5} may be computed. The flow rate in the remaining pipe loop may then be determined, since the head loss between 2 and 5 is the same for both loops. If the combined discharges of the loops do not equal the total discharge which was specified originally, a second trial value of the discharge in pipe 2–5 may be found by proportion. Thus

$$Q''_{2-5} = Q'_{2-5}(\textstyle\sum Q / \sum Q'),$$

where $\sum Q$ is the specified total flow rate (10 cfs), Q'_{2-5} is the first assumption for the flow in 2–5, and $\sum Q'$ is the total flow rate calculated on the basis of the first assumption. The second assumption for the flow distribution is Q''_{2-5}, and the process is repeated until the desired accuracy is obtained.

Let us assume that $Q'_{2-5} = 6$ cfs. Then $V'_{2-5} = 7.65$ fps, $\mathbf{R}'_{2-5} = 7.0 \times 10^5$, $(k_s/D)_{2-5} = 10^{-3}$, and $f'_{2-5} = 0.020$ from Fig. 13–12. Therefore, the lower loop loss is merely the pipe friction

$$h'_{f_{2-5}} = \left(f \frac{L}{D} \frac{V^2}{2g} \right)'_{2-5} = 54.5 \text{ ft}.$$

Since the head loss is identical in the two loops,

$$H'_L = h'_{f_{2-5}} = h'_{f_{2-3-5}} + H_{LV},$$

where

$$H_{LV} = \text{valve head loss} = K_{LV} \frac{V^2_{2-3-5}}{2g}.$$

Hence

$$(V^2)'_{2-3-5} = \frac{2gH'_L}{[(fL/D)'_{2-3-5} + K_{LV}]},$$

and $K_{LV} = 0.2$. Assuming that $f'_{2-3-5} = 0.02$, we have

$$V'_{2-3-5} = 5.42 \text{ fps}, \qquad Q'_{2-3-5} = 1.90 \text{ cfs}, \qquad R_{2-3-5} = 3 \times 10^5,$$

and $(k/D)_{2-3-5} = 0.00075$. Hence $f'_{2-3-5} = 0.02$, and the above value of f' is correct.

The total discharge on the basis of the first assumption is

$$\Sigma Q' = 6.0 + 1.90 = 7.90 \text{ cfs.}$$

Using the second trial value of $Q''_{2-5} = 6.0(10/7.90) = 7.60$ cfs, and repeating the calculation of head loss in the pipe loop 2–5, we obtain

$$Q''_{2-5} = 7.60 \text{ cfs,} \qquad \text{then} \qquad V''_{2-5} = 9.70 \text{ fps.}$$

From Fig. 13–12 we have $\mathbf{R}''_{2-5} = 8.9 \times 10^5$ and $f''_{2-5} = 0.020$. Therefore

$$h''_{f_{2-5}} = 87.7 \text{ ft.}$$

If $Q''_{2-5} = 7.60$ cfs is the correct flow rate in loop 2–5, then the flow rate in loop 2–3–5 would be $10 - 7.60 = 2.40$ cfs. The computations may be checked by calculating the head loss in loop 2–3–5 for $Q''_{2-3-5} = 2.40$ cfs, and observing whether it agrees with the value of 87.7 ft determined for loop 2–5. Thus

$$Q''_{2-3-5} = 2.40 \text{ cfs,} \qquad \text{then} \qquad V''_{2-3-5} = 6.87 \text{ fps,}$$

$$\mathbf{R}''_{2-3-5} = 3.7 \times 10^5, \qquad \text{hence} \qquad f''_{2-3-5} = 0.019.$$

Therefore

$$H''_L = \left[\left(f \frac{L}{D} \right)_{2-3-5} + K_{L_V} \right] \left[\frac{V^2}{2g} \right]''_{2-3-5} = 83.5 \text{ ft.}$$

A slight change in the flow rate distribution such that

$$Q_{2-5} = 7.55 \text{ cfs,} \qquad Q_{2-3-5} = 2.45 \text{ cfs}$$

will bring the two head losses into agreement at 85.5 ft.

14–3 COMPRESSIBLE NONUNIFORM FLOW

Various aspects of compressible flow phenomena have been discussed in earlier chapters, namely, thermodynamic definitions in Chapter 1, one-dimensional continuity, energy and momentum equations in Chapters 4 and 6, dynamic similitude in Chapter 7, and frictional effects in uniform compressible flow in Chapter 13. In this section, we will draw together many of these topics in an introductory treatment of steady nonuniform, compressible flow. It is expected that a more detailed treatment would be undertaken in advanced subjects in the field of gas dynamics.

As in the previous section on incompressible flow, we shall be primarily concerned with the analysis of spatial variations of flow properties in enclosed systems. The basic difference between compressible and incompressible homogeneous fluid flow is in the introduction of the fluid density as a variable which is in turn related to pressure and temperature through an equation of state. Since all fluids, whether liquid or gas, are somewhat compressible, it is important to develop a quantitative measure of the conditions under which the approximation of incompressibility may be employed. For example, liquids in unsteady motion may undergo large pressure changes which require analysis as a compressible medium. On the other hand, low-speed gas flows may frequently be analyzed as an incompressible flow. This section will be primarily concerned with the steady nonuniform flow of gases. We shall follow the general practice of considering only dynamic pressure changes and neglecting body forces due to gravitational attraction.

14–3.1 Multidimensional analysis in gas dynamics

The mathematical analysis of gas flows in either two or three dimensions is usually restricted to the isentropic flow of a perfect gas. Under the restriction of constant entropy, the flow process must be both adiabatic (no external heat transfer) and reversible (no frictional dissipation). This is equivalent to an inviscid irrotational flow, if it is assumed that the motion was generated in a fluid initially at rest. The conditions for irrotationality, Eqs. (6–17), do not involve the density and apply to both compressible and incompressible flows. For a two-dimensional flow in the xy-plane, the irrotationality condition is

$$\frac{\partial v}{\partial x} = \frac{\partial u}{\partial y}. \tag{14–16}$$

The Navier-Stokes equations for a compressible fluid flow are given by Eqs. (6–24). By neglecting body forces and setting the viscosity equal to zero, we may write these for nonviscous steady flow in the xy-plane as

$$-\frac{\partial p}{\partial x} = \rho\left(u\,\frac{\partial u}{\partial x} + v\,\frac{\partial u}{\partial y}\right), \qquad -\frac{\partial p}{\partial y} = \rho\left(u\,\frac{\partial v}{\partial x} + v\,\frac{\partial v}{\partial y}\right). \tag{14–17}$$

If we substitute the condition for irrotational flow, Eq. (14–16), these equations become

$$\frac{\partial p}{\partial x} = -\frac{\rho}{2}\,\frac{\partial}{\partial x}\,(u^2 + v^2), \qquad \frac{\partial p}{\partial y} = -\frac{\rho}{2}\,\frac{\partial}{\partial y}\,(u^2 + v^2). \tag{14–18}$$

The compressible-flow continuity equation must also be satisfied, and Eq. (6–1a) may be written for a steady two-dimensional motion as

$$u\,\frac{\partial \rho}{\partial x} + v\,\frac{\partial \rho}{\partial y} + \rho\left(\frac{\partial u}{\partial x} + \frac{\partial v}{\partial y}\right) = 0. \tag{14–19}$$

It should be noted that the equation of motion relates pressure and velocity gradients while the continuity equation relates density and velocity gradients. The pressure and density gradients may be eliminated by means of the definition of the speed of sound, Eq. (1–15a),

$$c^2 = dp/d\rho.$$

This may be expressed in terms of partial derivatives as

$$c^2 = \frac{(\partial p/\partial x)\, dx + (\partial p/\partial y)\, dy}{(\partial \rho/\partial x)\, dx + (\partial \rho/\partial y)\, dy}.$$

Hence it follows that

$$\frac{\partial \rho}{\partial x} = \frac{1}{c^2} \frac{\partial p}{\partial x}, \qquad \frac{\partial \rho}{\partial y} = \frac{1}{c^2} \frac{\partial p}{\partial y}. \tag{14–20}$$

The velocity gradients may be written in terms of the velocity potential function ϕ defined by Eqs. (6–47a). The equations of motion (14–18) and continuity (14–19) may then be combined to give an equation of the form

$$\left[1 - \left(\frac{\partial \phi}{\partial x}\right)^2 \frac{1}{c^2}\right] \frac{\partial^2 \phi}{\partial x^2} + \left[1 - \left(\frac{\partial \phi}{\partial y}\right)^2 \frac{1}{c^2}\right] \frac{\partial^2 \phi}{\partial y^2} - \frac{2}{c^2} \frac{\partial \phi}{\partial x} \frac{\partial \phi}{\partial y} \frac{\partial^2 \phi}{\partial x \partial y} = 0. \tag{14–21}$$

When c becomes very large, Eq. (14–21) reduces to the Laplace equation, Eq. (6–49),

$$\frac{\partial^2 \phi}{\partial x^2} + \frac{\partial^2 \phi}{\partial y^2} = 0,$$

developed in a similar manner for incompressible flow. Thus the condition of irrotational incompressible fluid flow is equivalent to the assumption of an infinite speed of propagation of a small pressure disturbance in the fluid media.

The techniques for solving the linear Laplace equation for incompressible flow, such as the flow net, no longer apply to the compressible-flow nonlinear partial differential equation (14–21). Thus, even under the restrictions of isentropic perfect gas flow, the analysis becomes extremely complex. The available techniques for dealing with the nonlinear equation for multidimensional compressible flow may be grouped into two categories. Each of these is beyond the scope of this text and only a brief description will be included in this section. Detailed developments may be found in various treatises on gas dynamics [11].

The first method is based on a perturbation technique for linearizing Eq. (14–21). The method is appropriate to the analysis of flow about slender bodies in which departures from the free-stream velocity U_0 are small. The linearized form of Eq. (14–21) is given by

$$\left[1 - \left(\frac{\partial \phi}{\partial x}\right)_0^2 \frac{1}{c_0^2}\right] \frac{\partial^2 \phi}{\partial x^2} + \frac{\partial^2 \phi}{\partial y^2} = 0. \tag{14–22}$$

The subscript 0 refers to conditions in the undisturbed flow which is assumed to be uniform in the x-direction at some distance from the object causing the perturbation. Since $U_0 = -(\partial \phi/\partial x)_0$ and the Mach number (Eq. 7–38) is defined as the ratio of the local velocity to the local speed of sound, Eq. (14–22) may also be written as

$$(1 - \mathbf{M}_0^2) \frac{\partial^2 \phi}{\partial x^2} + \frac{\partial^2 \phi}{\partial y^2} = 0. \qquad (14\text{--}23)$$

Incompressible flow is therefore equivalent to a flow in which $\mathbf{M}_0^2 \ll 1$. A significant observation results from Eq. (14–23) when it is realized that the coefficient of the first term is positive when the flow is subsonic, $\mathbf{M}_0 < 1$, and negative when the flow is supersonic, $\mathbf{M}_0 > 1$. The subsonic form is of the elliptic or Laplace type, while the supersonic case takes the form of a hyperbolic differential equation generally known as the wave equation. The linearized approximation represented by Eq. (14–23) is not valid for Mach numbers close to unity (transonic flow) or for Mach numbers much larger than one (hypersonic flow). Solutions to the linearized equation of motion can be found for certain boundary conditions.

The second method of analysis, known as the *hodograph method*, is based on a substitution of variables which transforms the exact nonlinear equation into an exact linear equation. The independent variables in the new coordinate system are the velocity components. Although the method is exact, satisfaction of the boundary conditions in the hodograph plane is difficult, and approximate methods have been developed to overcome this difficulty.

Before leaving the subject of multidimensional compressible flow, it is of interest to examine the classical problem of the acoustic-wave pattern produced by a moving disturbance. We may consider a source periodically generating an acoustic wave traveling with the sonic speed c. For a line source, cylindrical

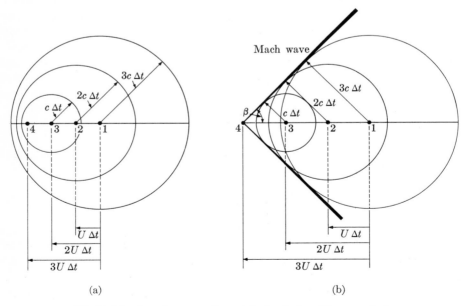

FIG. 14–19.   Acoustic wave patterns: (a) subsonic flow; (b) supersonic flow.

waves are produced, and for a point source, the generated waves are spherical. If the source is moving with a constant subsonic velocity, $U < c$, the resulting wave pattern at a certain instant is as shown in Fig. 14–19(a). Thus the pressure waves continually precede the source and eventually fill the entire fluid field. If the source is moving at a supersonic velocity, $U > c$, the wave pattern is as shown in Fig. 14–19(b), and pressure disturbances are felt only *inside* the cone or wedge formed by the envelope of the pressure waves. A simple geometric construction shows that the half-angle of the cone, β, is given by

$$\sin \beta = c/U = 1/\mathsf{M}. \qquad (14\text{--}24)$$

The angle β is called the Mach angle, and the line forming the envelope of the pressure disturbances is a Mach wave. The significant difference between the subsonic and supersonic cases is the existence of a *zone of silence* in the supersonic flow. In other words, fluid particles outside the Mach cone are unaware of the approach or passage of the disturbance. We shall see later that modifications to the above analysis are necessary if the disturbance is of finite magnitude; however, the basic concept of a zone of silence is still valid.

14–3.2   One-dimensional isentropic channel flow

As in the analysis of incompressible flows, we find that the one-dimensional methods provide much useful information on compressible fluids by avoiding the mathematical complexities of the multidimensional treatment.

For nonuniform flows in an enclosed system, the variation of the cross-sectional area in the direction of flow is known. If the area changes are gradual and streamline curvatures are small, mean values of the four basic flow properties, pressure, density, velocity, and temperature, may be assumed for each cross section. Since there are four unknowns, four equations are needed to solve one-dimensional isentropic-flow problems. These are the equation of state, the continuity equation, the energy equation, and the momentum equation. For convenience they are summarized below.

1. *Equation of state for a perfect gas* (Eq. 1–17):

$$p/\rho = RT. \qquad (14\text{--}25)$$

2. *Continuity equation* (Eq. 4–12):

$$\rho V A = \text{const},$$

or

$$d\rho/\rho + dV/V + dA/A = 0. \qquad (14\text{--}26)$$

3. *Energy equation:* In the absence of heat transfer and shaft work, the one-dimensional energy equation (Eq. 4–21) may be stated (neglecting gravitational effects) as

$$u + p/\rho + V^2/2 = \text{const}, \qquad (14\text{--}27)$$

where u is the internal energy per unit mass. Or, in differential form, the energy equation is

$$d(u + p/\rho) + V \, dV = 0. \tag{14–28}$$

Employing Eq. (1–11a), we obtain

$$c_p \, dT + V \, dV = 0. \tag{14–29}$$

4. *Momentum equation:* A convenient form of the momentum equation for one-dimensional flow is obtained by integrating the compressible Navier-Stokes equation for frictionless flow along a streamline as we did when deriving Eq. (6–68). This equation may also be interpreted for a stream tube coincident with the boundaries of the nonuniform flow by letting $q = V$, the mean velocity. Again, if we neglect gravitational effects, Eq. (6–68) may be written as

$$\int (dp/\rho) + V^2/2 = \text{const along stream tube}, \tag{14–30}$$

or, in differential form, as

$$dp/\rho + V \, dV = 0. \tag{14–31}$$

In Section 13–4.1 equations were presented which are similar to Eqs. (14–25) through (14–31), except that the latter are specialized for adiabatic and frictionless flow.

Eliminating the term $V \, dV$ which appears in both the energy equation (14–29) and the momentum equation (14–31), we have

$$c_p \, dT = dp/\rho. \tag{14–32}$$

The energy equation is valid for adiabatic flows with or without friction; however, the momentum equation holds only for frictionless flow. Therefore, Eq. (14–32) is valid for a frictionless adiabatic (i.e. isentropic) flow. The differential form of the perfect gas law is

$$d(p/\rho) = R \, dT, \tag{14–33}$$

and c_p may be expressed in terms of R by Eq. (1–19), namely

$$c_p = \frac{k}{k-1} R.$$

Then Eq. (14–32) may be written in the form

$$dp/p = k(d\rho/\rho), \tag{14–34}$$

or

$$p/\rho^k = \text{const.} \tag{14–35}$$

This is a derivation of the isentropic pressure-density relation given in Eq. (1–22a). We see that it incorporates both the energy and momentum equations, and Eq. (14–35) is frequently used in place of either the energy or momentum equation in the analysis of isentropic flow processes. For example, the momentum equation (14–30) may be integrated between two sections,

$$\frac{V_1^2 - V_2^2}{2} = \int_{p_1}^{p_2} \frac{dp}{\rho}. \tag{14-36}$$

By substituting the isentropic pressure-density relation, Eq. (14–35), we get

$$\frac{V_1^2 - V_2^2}{2} = \frac{k}{k-1} \frac{p_1}{\rho_1} \left[\left(\frac{p_2}{p_1} \right)^{(k-1)/k} - 1 \right], \tag{14-37}$$

or

$$\frac{V_1^2 - V_2^2}{2} = \frac{k}{k-1} \frac{p_2}{\rho_2} \left[1 - \left(\frac{p_1}{p_2} \right)^{(k-1)/k} \right]. \tag{14-37a}$$

The same equations may also be obtained by integrating the energy equation (14–29) along a stream tube using the isentropic pressure-density relation.

Stagnation conditions. As a quantitative measure of the influence of compressibility in steady motion, we may examine the measurement of velocity in a subsonic compressible flow by means of the stagnation tube shown in Fig. 6–12. Equation (14–37) may be applied along the streamline between the undisturbed free-stream velocity U_0 and the tip of the tube, where the velocity is zero. Thus

$$\frac{U_0^2}{2} = \frac{k}{k-1} \frac{p_0}{\rho_0} \left[\left(\frac{p_{stg}}{p_0} \right)^{(k-1)/k} - 1 \right]. \tag{14-38}$$

Since

$$c_0^2 = \frac{kp_0}{\rho_0} \quad \text{and} \quad \mathbf{M}_0 = \frac{U_0}{c_0},$$

the stagnation pressure is given by

$$\frac{p_{stg}}{p_0} = \left[1 + \mathbf{M}_0^2 \frac{k-1}{2} \right]^{k/(k-1)}. \tag{14-39}$$

Expanding Eq. (14–39) in a binomial series, we obtain

$$p_{stg} = p_0 + \frac{\rho_0 U_0^2}{2} \left[1 + \frac{\mathbf{M}_0^2}{4} + \frac{2-k}{24} \mathbf{M}_0^4 + \cdots \right]. \tag{14-40}$$

Equation (6–80), obtained for stagnation conditions in incompressible flow, may be expressed as

$$p_{stg} = p_0 + \rho(U_0^2/2). \tag{14-41}$$

Thus the compressibility effect on stagnation pressure is given by the bracketed quantity. For a free-stream Mach number of 0.2 in air, the difference is about one percent. Note that if both p_{stg} and p_0 are measured, the free-stream Mach number M_0 may be computed from Eq. (14–39). However, if the free-stream velocity is sought, a measurement of the free-stream temperature must also be made in order to determine the free-stream sonic speed c_0. Since

$$c_0 = \sqrt{k(p_0/\rho_0)} = \sqrt{kRT_0}, \qquad (14\text{–}42)$$

usually it is more convenient to measure the stagnation temperature than the free-stream temperature. In this case the two temperatures are related by integrating the energy equation (14–29) between a point in the free stream and the stagnation point. Thus

$$c_p(T_{stg} - T_0) = U_0^2/2. \qquad (14\text{–}43)$$

Again using $c_p = [k/(k-1)]R$ and (14–42), we may write this as

$$\frac{T_{stg}}{T_0} = 1 + \frac{k-1}{2}\,\mathbf{M}_0^2. \qquad (14\text{–}44)$$

Variable-area flow. We may investigate isentropic flows in variable-area channels by considering a converging duct connected to a large reservoir, as shown in Fig. 14–20. It will be assumed that the duct discharges the gas jet into atmospheric pressure, p_a, through a parallel wall throat of area A_t with a mean velocity V_t and pressure p_t. If the reservoir area is large compared with the throat area, the velocity in the reservoir will be negligibly small. In other words, stagnation conditions may be assumed within the tank. The effect of area changes can be shown by using the differential forms of the momentum and continuity equations. Using $dp = c^2 d\rho$ from Eq. (1–15a), we may rewrite Eq. (14–31) as

$$c^2\,\frac{d\rho}{\rho} + V\,dV = 0.$$

The term $d\rho/\rho$ may be eliminated in the continuity equation (14–26) with the result that

$$\frac{dV}{V}\,(\mathbf{M}^2 - 1) = \frac{dA}{A}. \qquad (14\text{–}45)$$

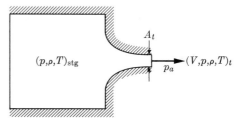

FIG. 14–20. Flow from a reservoir through a convergent duct.

For subsonic flow ($\mathbf{M} < 1$), dA must be negative when dV is positive. Therefore a decreasing area in the direction of flow is necessary for an increasing velocity just as with incompressible flow. In supersonic flow, the reverse is true. When the local Mach number exceeds unity, the decrease in density is so rapid for a given speed increase that the area must increase in the direction of flow. Equation (14–45) also indicates that at a throat (where $dA = 0$), two possibilities exist, either $dV/V = 0$ or $\mathbf{M} = 1$. Thus, while sonic velocity can occur at a parallel wall section, the existence of such a section does not guarantee that sonic velocity will occur. If $\mathbf{M} < 1$, the condition $dV/V = 0$ indicates that the velocity reaches a maximum where the channel area is a minimum. If $\mathbf{M} > 1$, the velocity is a minimum where the area is a minimum.

Returning to the flow geometry shown in Fig. 14–20, we may obtain the necessary condition for sonic velocity at the throat by applying Eq. (14–37a) between the stagnation region in the reservoir ($V_1 = 0$) and the throat where $V_2 = V_t$. If we remember that $c^2 = kp/\rho$, Eq. (14–37a) gives

$$-\frac{V_t^2}{2} = \frac{c_t^2}{k-1}\left[1 - \left(\frac{p_{\text{stg}}}{p_t}\right)^{(k-1)/k}\right].\tag{14–46}$$

When $V_t = c_t$,

$$\frac{p_t^*}{p_{\text{stg}}} = \left(\frac{2}{k+1}\right)^{k/(k-1)},\tag{14–47}$$

where the asterisk indicates sonic conditions.

For air, with $k = 1.40$, we have $p_t^*/p_{\text{stg}} = 0.528$.

For subsonic jets discharging into the atmosphere at the throat, the jet pressure p_t must equal the atmospheric pressure p_a. Hence, if we visualize a sequence of events in which the tank pressure p_{stg} is slowly increased above atmospheric pressure, the jet velocity V_t will increase from zero to values given by Eq. (14–37), so long as $p_a/p_{\text{stg}} \geq p_t^*/p_{\text{stg}}$. From Eq. (13–41) the mass rate of flow per unit area at the throat is $G_t = \rho_t V_t$. From Eqs. (14–35) and (14–37), with $p_2 = p_t = p_a$, we have

$$G_t = \sqrt{[2k/(k-1)]p_{\text{stg}}\rho_{\text{stg}}(p_a/p_{\text{stg}})^{2/k}[1 - (p_a/p_{\text{stg}})^{(k-1)/k}]}.\tag{14–48}$$

When the ratio of the atmospheric pressure to the stagnation pressure is equal to p_t^*/p_{stg}, the velocity at the throat is sonic. Since the channel is entirely convergent ahead of the throat, there is no way for the throat velocity to exceed the sonic speed even if the pressure differential between the atmosphere and the reservoir is increased.

When $p_a/p_{\text{stg}} < p_t^*/p_{\text{stg}}$, the exit pressure p_t^* in the sonic jet must be greater than the surrounding atmospheric pressure, since p_t^* is the fixed percentage of p_{stg} given by Eq. (14–47). The pressure adjustment from p_t^* to the ambient pressure takes place downstream of the nozzle through a series of compression and expansion waves. The maximum mass rate of flow per unit area is designated as G_t^*, which indicates sonic conditions at the throat. The value is given by Eq. (14–48) with the variable (p_a/p_{stg}) replaced by the fixed ratio p_t^*/p_{stg} of Eq. (14–47). Thus

$$G_t^* = \left(\frac{2}{k+1}\right)^{1/(k-1)}\sqrt{\left(\frac{2k}{k+1}\right)p_{\text{stg}}\rho_{\text{stg}}}.\tag{14–49}$$

Equation (14–49) shows that a convergent tube designed to produce sonic velocity at the throat may be used as a flow meter. For a given throat area, the mass-flow rate is a function only of the stagnation pressure and density (or temperature).

If we wish to produce a supersonic stream, the geometry of Fig. 14–20 must be modified by the addition of a *diverging* section downstream from the parallel wall throat, as shown in Fig. 14–21. The converging-diverging section may also be operated entirely in the subsonic range. In this case the pressure drops and the velocity increases in the converging section, while the pressure increases and the velocity decreases in the diverging section. The pressure and velocity changes are therefore similar to that of a Venturi meter in incompressible flow. This condition is illustrated by the pressure curve s-t_1-e_1 in Fig. 14–21. So long as the flow may be assumed to behave isentropically, the mass rate of flow per unit area is given by Eq. (14–48) with $p_a = p_{e_1}$. If the exit pressure is further reduced to p_{e_2}, sonic velocity is reached at the throat. Velocities remain subsonic at all other sections and the mass rate of flow per unit area is a maximum at the throat. The pressure curve is given by s-t^*-e_2, showing *compression* in the diverging section. Since the throat velocity is now sonic and the downstream area is increasing, it must also be possible for an isentropic *expansion* to take place at the same mass rate of flow. As shown in Fig. 14–21, there is one exit pressure (p_{e_4}) which produces this condition. The pressure curve is s-t^*-e_4,

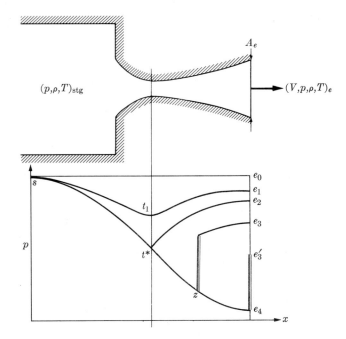

FIG. 14–21.   Reservoir flow through a converging-diverging duct.

and the velocities in the diverging portion are supersonic with the maximum Mach number occurring at the exit.

For a fixed stagnation pressure, isentropic flows are possible for all exit pressures between p_{e_0} and p_{e_2} and at p_{e_4}. If the exit pressure is changed to p_{e_3}, the resulting pressure curve is s-t^*-z-e_3. The flow is supersonic between the throat and z where a pressure jump occurs across a normal shock wave. After the shock wave the flow is subsonic. In the following section it is shown that a shock wave is always accompanied by energy dissipation; therefore, this type of flow is not isentropic. When the exit pressure is reduced to $p_{e_3'}$, the normal shock is at the exit section and the pressure curve is s-t^*-e_4-e_3'. The exit pressure must therefore be at least as low as $p_{e_3'}$ in order to have completely supersonic flow within the diverging section. Ambient pressures between $p_{e_3'}$ and p_{e_4}, or below p_{e_4}, are reached by either oblique shock or expansion waves in the supersonic jet downstream from the exit section. This jet is no longer in the realm of one-dimensional flows. Frictional effects in boundary-layer development and the possibility of flow separation due to adverse pressure gradients also complicate the internal flow patterns.

In Fig. 14–22, the pressure diagram of Fig. 14–21 is labeled to show the several regimes of subsonic and supersonic flow and of nozzle and diffuser action. Note that diffuser action (recompression) occurs everywhere on the right-hand side of the critical velocity point in the throat except on the line for isentropic expansion.

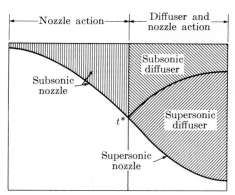

FIG. 14–22. Regimes of flow in a converging-diverging duct.

14–3.3 Shock waves

Normal shocks in gases. It has been shown that the continuous motion of a compressible fluid which satisfies the continuity, adiabatic energy, and inviscid momentum equation is isentropic. However, in real fluids it is observed that sudden, finite changes of pressure, density, temperature, and velocity may occur in a conduit. Such discontinuities, known as shock waves, cannot be explained by the isentropic-flow theory. We may employ a one-dimensional control volume enclosing a stationary shock normal to the flow as shown in Fig. 14–23. The uniform-flow properties ahead of the shock are designated by subscript 1, and the uniform properties downstream by subscript 2. Since the thickness of the shock wave is extremely small, we may reasonably assume that across the shock there is no gain or loss of heat to or from the surroundings. We therefore use the one-dimensional adiabatic energy equation (14–29) integrated between sections 1 and 2,

$$c_p(T_2 - T_1) + \frac{V_{n2}^2 - V_{n1}^2}{2} = \text{const.} \tag{14-50}$$

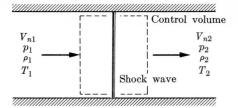

FIG. 14–23. One-dimensional control volume for normal shock wave.

Using Eq. (1–19) and the perfect gas law, we obtain

$$\frac{V_{n2}^2 - V_{n1}^2}{2} = \frac{k}{k-1}\left(\frac{p_1}{\rho_1} - \frac{p_2}{\rho_2}\right). \tag{14-51}$$

The continuity equation for the constant-area flow, Eq. (4–12) becomes

$$\rho_1 V_{n1} = \rho_2 V_{n2}. \tag{14-52}$$

We depart from the earlier isentropic-flow analysis by using the one-dimensional form of the momentum equation which allows energy dissipation. Thus, from Eq. (4–33), neglecting the gravitational body force, we have

$$F_{p_x} + F_{s_x} = (V_x \rho Q)_2 - (V_x \rho Q)_1. \tag{14-53}$$

Since the longitudinal extent of the control volume is extremely small, the frictional shear, F_{s_x}, may also be neglected. Therefore, since

$$A_1 = A_2,$$

$$F_{p_x} = (p_1 - p_2)A$$

and Eq. (14–53) becomes

$$p_1 - p_2 = \rho_2 V_{n2}^2 - \rho_1 V_{n1}^2. \tag{14-54}$$

It is important to note that Eq. (14–54) does not preclude internal energy dissipation even though the boundary shear stresses are not included. The momentum equation (14–54) may be divided by the continuity equation (14–52) and multiplied by $(V_{n2} + V_{n1})$ to give

$$\left(\frac{1}{\rho_2} + \frac{1}{\rho_1}\right)(p_1 - p_2) = V_{n2}^2 - V_{n1}^2. \tag{14-55}$$

Substituting Eq. (14–55) into the energy equation (14–51), we obtain

$$\left(\frac{1}{\rho_2} + \frac{1}{\rho_1}\right)\left(\frac{p_1 - p_2}{2}\right) = \frac{k}{k-1}\left(\frac{p_1}{\rho_1} - \frac{p_2}{\rho_2}\right). \tag{14-56}$$

Either the pressure or the density ratio may be factored out to yield the Rankine-Hugoniot relations for the normal shock,

$$\frac{p_2}{p_1} = \frac{\dfrac{k+1}{k-1}\dfrac{\rho_2}{\rho_1} - 1}{\dfrac{k+1}{k-1} - \dfrac{\rho_2}{\rho_1}}, \qquad (14\text{–}57)$$

or

$$\frac{\rho_2}{\rho_1} = \frac{1 + \dfrac{k+1}{k-1}\dfrac{p_2}{p_1}}{\dfrac{k+1}{k-1} + \dfrac{p_2}{p_1}} = \frac{V_{n1}}{V_{n2}}. \qquad (14\text{–}58)$$

The fact that the shock is not an isentropic process may be readily shown by comparing the isentropic pressure-density relation,

$$\frac{p_2}{p_1} = \left(\frac{\rho_2}{\rho_1}\right)^k,$$

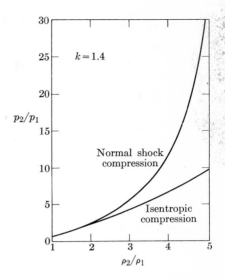

FIG. 14–24. Comparison of pressure-density relation for normal shocks and isentropic flow processes.

with Eq. (14–57) as shown in Fig. 14–24. Note that the pressure ratio across the shock becomes infinite when the density ratio

$$\frac{\rho_2}{\rho_1} = \frac{k+1}{k-1}.$$

When $k = 1.4$, the limiting density ratio is 6. However, the adiabatic assumption is violated in actual flows before this condition is reached.

The Rankine-Hugoniot relations may also be expressed in terms of the upstream Mach number, since $\mathbf{M}_{n1} = V_{n1}/c_1 = V_{n1}/\sqrt{kp_1/\rho_1}$. Hence

$$\frac{p_2}{p_1} = \frac{2k}{k+1}\mathbf{M}_{n1}^2 - \frac{k-1}{k+1}, \qquad (14\text{–}59)$$

and

$$\frac{\rho_2}{\rho_1} = \frac{(k+1)\mathbf{M}_{n1}^2}{(k-1)\mathbf{M}_{n1}^2 + 2}. \qquad (14\text{–}60)$$

If the upstream flow is sonic, $\mathbf{M}_{n1} = 1$ and $p_2/p_1 = \rho_2/\rho_1 = 1$.

Therefore, for all supersonic upstream flows, both the pressure and density show an increase across the shock. The shock-wave equations are valid only for initially supersonic flows. This may be shown by evaluating the entropy change across the shock. The difference in entropy between any two states in a perfect gas is given by

$$S_2 - S_1 = c_v \ln \frac{p_2}{p_1}\left(\frac{\rho_1}{\rho_2}\right)^k.$$

The entropy change across the shock is then

$$\frac{S_2 - S_1}{c_v} = \ln\left[\frac{2k\mathbf{M}_{n1}^2 - (k-1)}{k+1}\right]\left[\frac{(k-1)\mathbf{M}_{n1}^2 + 2}{(k+1)\mathbf{M}_{n1}^2}\right]. \qquad (14\text{–}61)$$

For an initial Mach number close to unity, this can be expressed as the series

$$\frac{S_2 - S_1}{c_v} = \frac{2k(k-1)}{(k+1)^2}\frac{(\mathbf{M}_{n1}^2 - 1)^3}{3} + \cdots. \qquad (14\text{–}61a)$$

Thus $S_2 - S_1$ is positive (entropy increase) for $\mathbf{M}_{n1} > 1$. If the initial Mach number were less than unity, Eq. (14–61a) would indicate a decrease in entropy. This violates the second law of thermodynamics, and it is concluded that abrupt expansion shocks leading to pressure and density decreases are not possible.

The relation between the Mach numbers on either side of the shock wave may be found from the equations previously developed. Thus

$$\mathbf{M}_{n2}^2 = \frac{1 + [(k-1)/2]\mathbf{M}_{n1}^2}{k\mathbf{M}_{n1}^2 - (k-1)/2}. \qquad (14\text{–}62)$$

Equation (14–62) shows that when $\mathbf{M}_{n1}$ is greater than unity, $\mathbf{M}_{n2}$ is less than unity. Thus the normal shock always represents a transition between supersonic and subsonic motion.

In this discussion, the shock wave was assumed to be stationary, with the gas approaching the shock from left to right with a velocity V_{n1} as shown in Fig. 14–23. Another useful frame of reference is obtained by superimposing on this diagram a uniform velocity V_{n1} from right to left. Thus the gas in region 1 has zero velocity, and the shock advances into the undisturbed gas with a velocity V_{n1}. The gas behind the shock has a velocity $(V_{n1} - V_{n2})$ in the same direction as the moving shock. From Eq. (14–59) it can be seen that for a shock of finite strength, $p_2/p_1 > 1$, the Mach number $\mathbf{M}_{n1}$, and hence the velocity of shock propagation V_{n1}, is greater than the speed of sound in the undisturbed fluid.

Normal shocks in liquids. Since liquids are compressible, although to a much smaller degree than gases, shock waves may also occur in a liquid. Shock waves may result from an underwater explosion or they may occur in pipe lines due to the failure of a pump or the sudden closure of a valve. In the latter cases, the phenomenon, known as "water hammer," is the one-dimensional equivalent of the normal compression shock in a gas. In an infinite body of liquid or in a perfectly rigid conduit, the speed of propagation of a small pressure disturbance, c, is given by Eq. (1–15b) in terms of the bulk modulus of elasticity E_v of the liquid (see Tables 1–2, 1–3, and 1–5):

$$c = \sqrt{E_v/\rho}.$$

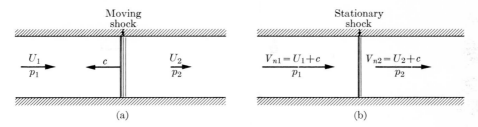

FIG. 14–25. Unsteady and steady pressure-wave propagation: (a) moving pressure wave; (b) stationary pressure wave.

For a large range of pressure, E_v and ρ change very little in liquids, and the speed of propagation of the pressure wave is essentially constant. This is in contrast to the gas shock in which the speed of propagation increases with an increase in the pressure change across the shock.

Since both density and temperature may be considered to be constants in the liquid shock, the number of flow variables is reduced from four to two. Therefore, only the continuity and momentum equations are needed in the analysis. We consider the one-dimensional problem of a pressure wave in a liquid propagating to the left with a velocity c in a rigid conduit. The fluid is moving to the right with velocities U_1 and U_2 upstream and downstream of the shock, as shown in Fig. 14–25(a). This unsteady flow can be reduced to an equivalent steady flow by changing the frame of reference. If a uniform velocity c from left to right is superimposed on Fig. 14–25(a), the steady flow of Fig. 14–25(b) is obtained. The result is exactly the same as in Fig. 14–23 which was used in the analysis of the gas shock. Therefore we may employ the continuity and momentum equations (14–52) and (14–54) which apply equally to gases and liquids. Combining and writing $V_{n1} = c + U_1$ and $V_{n2} = c + U_2$, we have

$$\frac{p_1 - p_2}{\rho(c + U_1)} = U_2 - U_1. \tag{14-63}$$

In liquids, the sonic speed c is of the order of 4000 to 5000 ft/sec. Therefore, for all practical purposes $c + U_1 \approx c$, and

$$p_2 - p_1 = \rho c(U_1 - U_2). \tag{14-63a}$$

Thus a decrease in liquid velocity causes an increase in pressure by an amount proportional to ρc, a property of the liquid. For water, the product

$$\rho c = \sqrt{\rho E_v} = \sqrt{(1.94)(320{,}000)(144)} = 9150 \text{ lb-sec/ft}^3,$$

and it is apparent that large pressure increases may be created by moderate changes in velocity.

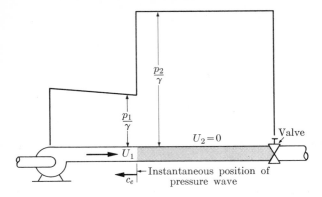

FIG. 14–26. Pressure rise due to rapid valve closure.

Figure 14–26 shows the pressure rise caused by the rapid closure of a valve in a pipe line. In this case $U_2 = 0$, and the pressure change given by Eq. (14–63a) is a maximum. Due to the elasticity of the conduit, the velocity of wave propagation in the liquid is reduced to c_e. The reduced speed of the pressure wave is a function of both the conduit and fluid properties. The kinetic energy per unit volume of fluid in the flow ahead of the pressure wave must equal the sum of the work per unit volume done in compressing the liquid and in expanding the conduit walls. The ratio of the reduced speed to that in the infinite medium is

$$\frac{c_e}{c} = \frac{1}{\sqrt{1 + E_v D/Et}},\tag{14–64}$$

where D, E, and t refer to the conduit diameter, modulus of elasticity, and wall thickness, respectively. For a steel conduit filled with water, $c = 4700$ ft/sec, $E = 30 \times 10^6$ psi, and, for $D/t = 100$, the value of c_e is 3340 ft/sec. Equation

FIG. 14–27. Oblique shocks in air (courtesy National Physics Laboratory, Teddington, Great Britain).

(14–63a) may be used to compute the pressure change for an elastic conduit with a rapid valve closure:

$$p_2 - p_1 = \rho c_e U_1.$$

More detailed information may be found in the reference cited [12].

Oblique shocks in gases. The normal shock wave is a limiting case of the oblique shock which may be formed by the abrupt deflection of a boundary through an angle θ in a supersonic gas flow. In the oblique shock shown in Figs. 14–27 and 14–28, the normal shock relations may be used for the components of the velocity normal to the shock V_{n1} and V_{n2}, whereas the actual flow directions are V_1 and V_2.

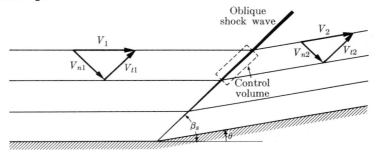

FIG. 14–28. Oblique shock geometry.

From the momentum equation (14–54) applied to the control volume normal to the shock in Fig. 14–28, we have

$$p_1 - p_2 = \rho_2 V_{n2}^2 - \rho_1 V_{n1}^2,$$

and parallel to the shock, we have

$$0 = V_{t2}\rho_2 V_{n2} - V_{t1}\rho_1 V_{n1}. \tag{14–65}$$

By comparing Eqs. (14–52) and (14–65), it is apparent that $V_{t1} = V_{t2}$. It follows from the geometry of Fig. 14–28 that

$$V_{t1} = V_{n1}/\tan \beta_s = V_{t2} = V_{n2}/\tan (\beta_s - \theta), \tag{14–66}$$

whence, from the continuity Eq. (14–52), we have

$$\frac{\rho_2}{\rho_1} = \frac{\tan \beta_s}{\tan (\beta_s - \theta)}. \tag{14–67}$$

The Mach number of the flow upstream of the oblique shock is related to the Mach number normal to the shock,

$$\mathbf{M}_{n1} = \mathbf{M}_1 \sin \beta_s.$$

Therefore, Eqs. (14–60) and (14–67) may be combined to form

$$\tan \theta = \frac{-\cot \beta_s (1 - \mathbf{M}_1^2 \sin^2 \beta_s)}{(1 - \mathbf{M}_1^2 \sin^2 \beta_s) + [(k+1)/2]\mathbf{M}_1^2}. \tag{14–68}$$

Equation (14–68) gives the angle of the oblique shock for a given wall angle θ and initial Mach number $\mathbf{M}_1$. The plot of Eq. (14–68) in Fig. 14–29 shows that for a given initial Mach number and wall angle, there are two possible shock-wave angles. For the smaller value of β_s the flow downstream from the oblique shock generally remains supersonic. For the larger shock angle, the flow below the shock is subsonic. The latter condition is difficult to achieve since in a subsonic flow, p_2 and ρ_2 are no longer uniquely determined by the initial flow.

The same plot also shows that for a given initial Mach number, there is a maximum wall deflection angle θ for which an oblique shock may be obtained. For $\theta > \theta_{\max}$, a normal shock wave ($\beta_s = 90°$) will occur. When θ equals zero, the flow may be interpreted as an infinitesimal disturbance at point 0 in a supersonic flow. From Eq. (14–68), with $\theta = 0$, we have

$$\mathbf{M}_1^2 \sin^2 \beta_s = 1, \qquad \text{or} \qquad \sin \beta_s = \frac{1}{\mathbf{M}_1},$$

which is the same as Eq. (14–24). In this limiting case, the oblique shock reduces to a Mach wave, as shown in Fig. 14–19(b) and $\beta_s = \beta$.

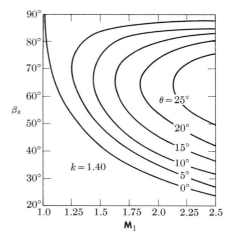

FIG. 14–29. Oblique shock relationships for air.

14–4 NONUNIFORM FREE-SURFACE FLOW

We conclude the discussion of steady nonuniform flows with some basic principles of spatial variations in free-surface motion. While we regain the simplification of incompressibility by dealing with free-surface liquid flows, we lose a considerable analytical advantage in terms of boundary conditions. The essential feature of free-surface flow is the knowledge that while the pressure is constant on a free surface, its location is initially unknown.

In the analysis of uniform free-surface flow in Chapter 13, it was found that uniform flow can occur only when the accelerating forces due to gravity are in balance with the decelerating forces due to boundary shear. In this case, the free surface is a plane parallel to the channel bottom, and its location is described by a constant equal to the depth of flow. In the analysis of a nonuniform free-surface flow it is helpful to decide whether the flow is to be considered as a rapidly or gradually varied flow.

In a rapidly varied flow, the streamline curvatures are such that the vertical accelerations are not negligible in comparison with the gravitational acceleration. In gradually varied flow, vertical accelerations may be neglected and the vertical pressure distribution may be assumed to be hydrostatic. This assumption permits a one-dimensional analysis of the fluid motion.

14–4.1 Rapidly varied flow

Analysis of *two-dimensional* rapidly varying free-surface flows may be made by assuming irrotational motion in some cases. If the flow originates from a fluid at rest and boundary-layer development on fixed boundaries is small in relation to the total mass of fluid, the assumption is reasonable for real fluids. As in the case of incompressible fluids in enclosed systems, the equation to be solved is the two-dimensional Laplace equation (6–53). An example of a simple, two-dimensional free-surface flow is the discharge from a sharp-edged slot shown in Fig. 14–30. A direct solution of the Laplace equation is not feasible since the configuration of the boundary streamlines is unknown. Since the flow is irrotational, the Bernoulli equation (6–61) holds throughout the flow,

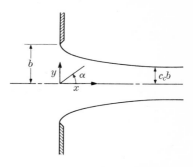

$$\frac{q^2}{2} + gh + \frac{p}{\rho} = \text{const.}$$

FIG. 14–30. Two-dimensional flow from sharp-edged slot.

Along the free surfaces p is constant and equal to the ambient pressure. If we set this reference pressure equal to zero, the general free-surface boundary condition follows from the Bernoulli equation,

$$\left[\frac{q^2}{2} + gh\right]_{\text{fs}} = \text{const.} \tag{14–69}$$

If the efflux velocity is large and the gravitational force acts in the y-direction, the downward deflection of the jet will be small close to the slot. When we neglect gravitational forces, the free-surface boundary condition reduces to

$$[q]_{\text{fs}} = \text{const.} \tag{14–70}$$

These two cases are treated in the following two subsections.

Free streamlines. No gravitational deflection. In Fig. 14–30 the streamlines which originate at the slot converge and become parallel at a section known as the *vena contracta*. If the slot opening is $2b$, the dimension of the free jet at the *vena contracta* is $c_c 2b$, where c_c is a contraction coefficient. The Laplace equation could be solved by the graphical construction of a flow net, as in confined flows, by assuming a shape for the free surface. The solution could be tested by observing whether the constant-velocity boundary condition of Eq. (14–70) is satisfied. If not, the assumed shape would have to be modified and the process repeated. This method is not practical, and Helmholtz showed that this problem could be solved by a conformal transformation into a hodograph plane [13]. The parametric equations for the jet boundaries are

$$\frac{y}{b/2} = 1 - \frac{2}{2 + \pi} (1 - \sin \alpha), \tag{14–71}$$

$$\frac{x}{b/2} = \frac{2}{2 + \pi} \left(\cos \alpha - \tfrac{1}{2} \ln \frac{1 + \cos \alpha}{1 - \cos \alpha} \right). \tag{14–72}$$

As $\alpha \to 0$, we find the thickness of the jet, $y \to c_c b/2$, and, from Eq. (14–71), $c_c = \pi/(2 + \pi) = 0.61$. Essentially the same degree of contraction is found for circular free jets where A_0 is the area of the orifice. Thus

$$A_{\text{jet}} = c_c A_0. \tag{14–73}$$

A generalization of the Helmholtz theory has been made by Von Mises [14] to show the effect of an aperture with inclined walls of finite height as indicated in Fig. 14–31. These values of the contraction coefficient can be used for symmetrical jets or for half jets along a horizontal boundary coincident with the centerline of the jet. In the latter case, the height a may also be the distance to a free surface upstream of an inclined or vertical wall. In this event, the structure, known as a sluice or tainter gate, is frequently used as a control of

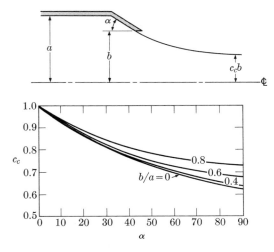

FIG. 14–31. Free streamline flow from an aperture.

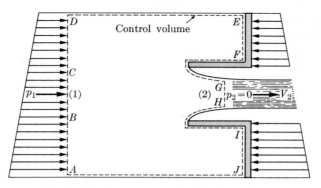

FIG. 14-32.   Momentum analysis of Borda mouthpiece.

discharge in open channels. The streamlines are parallel upstream of the control structure and again become parallel at the *vena contracta*. The one-dimensional continuity and momentum equations therefore may be applied between these sections to determine the flow rate and the horizontal force on the structure.

In one particular case the contraction coefficient for a free streamline jet without gravitational deflection may be computed from the one-dimensional continuity and momentum equations. This is the so-called Borda mouthpiece or short reentrant tube shown in Fig. 14-32. While the device is not of great importance, the analysis provides a useful exercise in the correct application of the one-dimensional equations. We apply the momentum equation (4-33) in the x-direction considering only pressure forces acting on the control volume. If the area of the approach section $A-D$ is large compared with the tube area, we may neglect the velocity at section $A-D$ and the longitudinal pressure gradient between sections $A-D$ and $E-J$. Thus the hydrostatic vertical pressure distributions on $A-D$ and $E-J$ are identical except for the unbalanced portion at F–I. Equation (4-33) becomes

$$p_1 A_t = V_2 \rho Q, \tag{14-74}$$

where A_t is the cross-sectional area of the reentrant tube. From the continuity equation, we have

$$Q = A_j V_2, \tag{14-75}$$

where A_j is the area of the free jet at section (2). The frictionless Bernoulli equation (6-69) relates velocities and pressures along the centerline of the jet,

$$\frac{p_1}{\rho} = \frac{V_2^2}{2}. \tag{14-76}$$

Therefore, $A_j = A_t/2$, and

$$c_c = \frac{A_j}{A_t} = 0.5. \tag{14-77}$$

This contraction coefficient is the minimum value which can be obtained for a free jet. It should be evident that the one-dimensional analysis may be applied in this special case only because the pressure distribution on all of the vertical boundaries may be assumed to be hydrostatic. As the reentrant tube is reduced to zero length, we obtain the flow pattern of Fig. 14–30 and the pressure distribution on the vertical boundary is no longer hydrostatic.

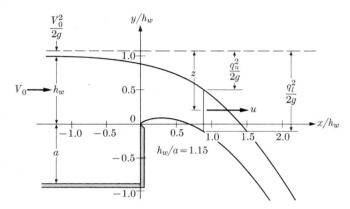

FIG. 14–33. Sharp-crested weir.

Free streamlines with gravitational deflection. In the previous section it was assumed that the free streamlines were not deflected by gravitational forces. Thus, in the flow shown in Fig. 14–30, we may apply Eq. (14–76) to determine the velocity V_2 in the *vena contracta* in terms of the pressure p_1 in the approach flow on the axis of the jet. Any number of flow rates are possible for the same free streamline geometry. If gravitational forces are included in the free streamline boundary condition, Eq. (14–69) must be used instead of Eq. (14–70). In this case it is not possible to alter the flow rate without simultaneously changing the free streamline geometry. This may be shown by considering a horizontal free surface behind the vertical aperture of Fig. 14–30. As the elevation of this surface approaches the half-height of the opening b, the upper free surface will detach and form a continuous streamline as shown in Fig. 14–33. The structure, known as a sharp-crested weir, is used to measure flow rates in open channels. The form of the relation between the discharge and the surface elevation h_w in the uniform-flow region upstream of the crest may be predicted. In the free falling jet,

$$u \sim \sqrt{2gz},$$

from Eq. (14–69). The discharge is then

$$Q \sim W \int u \, dz \sim W\sqrt{2g} \int \sqrt{z} \, dz \sim \tfrac{2}{3}\sqrt{2g} \, W z^{3/2},$$

where W is the length of the weir crest.

Since z and the limits of integration depend on the magnitude of h_w,

$$Q = C_d \tfrac{2}{3}\sqrt{2g}\, W h_w{}^{3/2}, \quad (14\text{–}78)$$

where C_d is a dimensionless discharge coefficient which, in potential flow, depends upon the geometric parameter h_w/a. The free streamline geometry and the discharge coefficient may be computed [15] by a numerical solution of an integral equation derived by conformal mapping. The computed discharge coefficients are compared with measured values [16] in Fig. 14–34. The diagram covers the complete range from a weir of infinite height, $h_w/a = 0$, to a sill of zero height, $a/h_w = 0$.

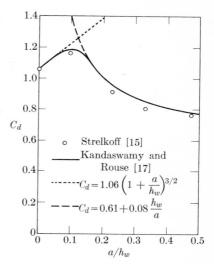

FIG. 14–34. Discharge coefficients for weirs and sills [16].

In the figure: o Strelkoff [15] — Kandaswamy and Rouse [17] — $C_d = 1.06\left(1 + \dfrac{a}{h_w}\right)^{3/2}$ — $C_d = 0.61 + 0.08\dfrac{h_w}{a}$

The spillway section of a dam is shaped to conform to the lower free streamline of the weir flow for a certain design head. The discharge coefficient for the design flow is not greatly modified by the lower solid boundary; however, a boundary layer develops and increases in thickness until it equals the depth of flow. The resistance of the fixed boundary opposes the acceleration of the free falling jet under gravity. If the spillway is long enough, the flow will become uniform at a constant depth.

14–4.2 Varied flow in short channel transitions

In this and the following sections, nonuniform free-surface flow is analyzed in a one-dimensional manner. It is assumed that the streamline curvatures are small and that the accelerations normal to the direction of motion are negligible. If the flow is in the x-direction, we conclude from the equation of motion in the y-direction that the vertical pressure distribution is hydrostatic. In the analysis of relatively short transitions, energy dissipation by boundary friction is frequently omitted as a first approximation. In certain other cases boundary shear may be ignored while energy dissipation due to turbulence production must be considered. The aim of this section is to give examples of these types of flow. A more detailed treatment may be found in the references cited [18].

Constant-energy flow in rectangular channels. Relationships for a frictionless, one-dimensional, nonuniform free-surface flow may be obtained by considering flow over a rise in the channel shown in Fig. 14–35. Upstream of the rise the flow is uniform and parallel to the channel bottom. Horizontal distances are given by the x-coordinate, and it is assumed that the distances measured along the actual bottom differ negligibly from the horizontal distance. The vertical

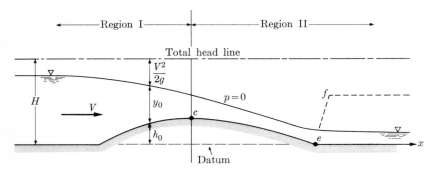

FIG. 14-35. One-dimensional frictionless free-surface flow.

component of velocity is ignored, and the horizontal velocity V is therefore uniform over the depth of flow, y_0. The total head, velocity, depth, and elevation of the channel bottom are related by the frictionless Bernoulli equation. Thus

$$H = V^2/2g + y_0 + h_0 = \text{const.} \qquad (14\text{-}79)$$

The equation of continuity is

$$q = Vy_0 = \text{const}, \qquad (14\text{-}80)$$

where q is the discharge per unit width. If Eqs. (14-79) and (14-80) are differentiated with respect to x, we have

$$\frac{V}{g}\frac{dV}{dx} + \frac{dy_0}{dx} + \frac{dh_0}{dx} = 0, \qquad (14\text{-}81)$$

$$y_0\frac{dV}{dx} + V\frac{dy_0}{dx} = 0. \qquad (14\text{-}82)$$

Eliminating dV/dx between the two equations, we obtain an equation for the slope of the free surface,

$$\frac{dy_0}{dx} = -\frac{dh_0/dx}{1 - V^2/gy_0} = -\frac{dh_0/dx}{1 - \mathbf{F}^2}, \qquad (14\text{-}83)$$

where $\mathbf{F}$ is the Froude number, $\mathbf{F} = V/\sqrt{gy_0}$.

In region I of Fig. 14-35, dy_0/dx is negative and dh_0/dx is positive. According to Eq. (14-83), the Froude number must therefore be less than unity throughout region I. At the crest of the rise (point c), $dh_0/dx = 0$, but $dy_0/dx \neq 0$, and therefore $\mathbf{F}_c = 1$, since dy_0/dx may remain finite when $dh_0/dx = 0$ only when it takes the form $0/0$. At the crest where $\mathbf{F}_c = 1$,

$$V_c = \sqrt{gy_c}, \qquad (14\text{-}84)$$

and the flow is said to be "critical." The flow in region I is subcritical, while in region II it is supercritical ($F > 1$) since both dy_0/dx and dh_0/dx are negative in region II. This free-surface flow is analogous to the compressible flow in the converging-diverging channel shown in Fig. 14–21. The frictionless free-surface flow in Fig. 14–35 corresponds to the isentropic compressible flow undergoing a supersonic expansion to an exit pressure p_{e_4}. As noted in Section 13–5.1, the velocity $\sqrt{gy_0}$ is the speed of propagation of surface waves caused by a small disturbance in a shallow liquid. Thus the Froude number, like the Mach number, may be interpreted as the ratio of the local fluid velocity to the local speed of propagation of a pressure disturbance.

From the continuity equation (14–80) we may write $q = V_c y_c$, and since $V_c = \sqrt{gy_c}$, it follows that

$$y_c = \sqrt[3]{q^2/g}. \qquad (14\text{–}85)$$

Thus for one-dimensional flow the critical depth is a function only of the discharge per unit width. Since the total head is constant along the channel, H in Eq. (14–79) may be evaluated at the critical-depth section and equated to the value of H at any other section. Thus

$$H = V_c^2/2g + y_c + (h_0)_c = V^2/2g + y_0 + h_0. \qquad (14\text{–}86)$$

The profile of the free surface throughout the transition may then be determined from

$$H - h_0 = y_c^3/2y_0^2 + y_0 = q^2/2gy_0^2 + y_0, \qquad (14\text{–}87)$$

which is obtained by substituting Eqs. (14–80), (14–84), and (14–85) into (14–86). The equation is cubic in y_0 and has three real roots, one of which is negative. The two positive roots will be seen to correspond to the subcritical and supercritical states of flow. The computation of surface profiles is facilitated by a graphical plot of Eq. (14–87) as shown in Fig. 14–36, in which the discharge per unit width, q, is the independent variable. The quantity $H - h_0$ is the specific head H_0 defined in Eq. (13–62) and represents the head referred to the channel bottom rather than to the datum plane. The advantage of the specific-head plot is that it is independent of the geometry of the transition for all rectangular channels. Equation (14–87) is also plotted in Fig. 14–36 in the dimensionless form,

$$\frac{H_0}{y_c} = \frac{H - h_0}{y_c} = \frac{1}{2}\left(\frac{y_c}{y_0}\right)^2 + \frac{y_0}{y_c}. \qquad (14\text{–}88)$$

Several interesting conclusions may be deduced directly from the specific-head diagram. If the discharge per unit width, q, and the total head H are fixed, and the initial flow is subcritical as in Fig. 14–35, an increase in bottom elevation will result in a decrease in depth. The diagram clearly shows that h_0 is limited since $H - h_0$ is a minimum when the flow becomes critical. If the bottom were raised by an additional amount, the discharge would be reduced assuming that the total head remained constant. If the initial flow was supercritical, an increase in the bottom elevation would result in an increase of depth. The

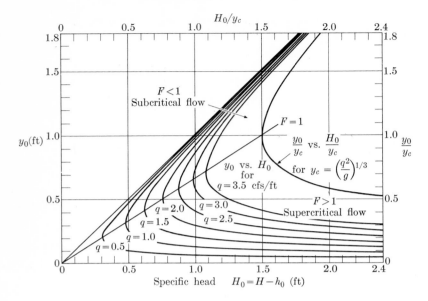

FIG. 14–36. Specific head diagram for rectangular channels.

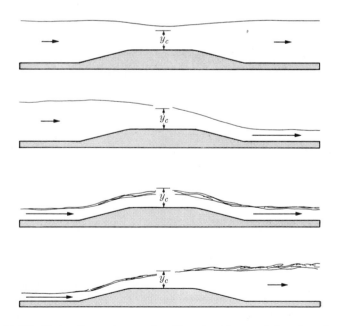

FIG. 14–37. Various flow regimes produced by a local increase in boundary elevation.

various flow regimes which may be produced by a local increase in boundary elevation are shown in Fig. 14–37. Further analogies with the compressible channel flow should now be evident. If the total head is fixed and the depth at section e (Fig. 14–35) downstream from the transition is gradually reduced, the discharge will increase from zero to a maximum when critical depth is reached at section c. A further decrease in the downstream depth will not increase the discharge. In addition, if the downstream depth is higher than the supercritical value corresponding to H, a free-surface shock wave will occur along the line e-f in Fig. 14–35. In this case, energy dissipation may no longer be neglected.

The method of analysis developed above may also be applied to the case where a change in width occurs in a rectangular channel. Under these conditions, $h_0 = 0$ and the depth change must occur along a vertical line, corresponding to $H_0 = $ const in Fig. 14–36, as the discharge per unit width changes. A study of the diagram will show that critical depth may be reached by narrowing the channel. A simultaneous change in width and bottom elevation may be treated by combining the two solutions. If a change in width occurs in a channel in which the flow is initially supercritical, the one-dimensional methods outlined above may not be used. For example, in a convergent channel an oblique free-surface shock will be generated in the converging section [19].

The methods discussed in this section are valid for rectangular channels; however, there are no fundamental difficulties in extending them to other cross-sectional shapes [18].

14–4.3 Gradually varied flow

The condition of uniform, free-surface flow with frictional resistance at the fixed boundaries was analyzed in Chapter 13. For a given discharge (Q), channel slope (S_0), and boundary roughness, there is only one depth at which uniform flow can occur. This depth is uniquely determined by Eq. (13–72) in terms of the Manning roughness parameter n; for a wide channel in which $R_h = y_N$ and $q = V_N y_N$,

$$q = \frac{1.49}{n} \, y_N^{5/3} S_0^{1/2}.$$

Therefore, the uniform flow depth y_N is

$$y_N = \left(\frac{nq}{1.49 S_0^{1/2}} \right)^{3/5}, \qquad (14\text{–}89)$$

and the free surface is parallel to the channel bottom.

Due to entrance and exit conditions and to changes of slope or structures within the channel, the uniform flow depths are approached asymptotically in an actual channel. The so-called normal depth y_N is therefore a useful characteristic length parameter in the analysis of steady nonuniform free-surface flow.

Since the depth changes are assumed to be gradual, they occur over long distances and frictional resistance may not be neglected as in the previous section which dealt with local flow transitions. If we refer to the diagrammatic sketch of a nonuniform free-surface flow in Fig. 13–18, the slope of the free surface is given by Eq. (13–66) as

$$\frac{dy_0}{dx} = \frac{S_0 - S_H}{dH_0/dy_0}.$$

(14–90)

Since the specific head for rectangular channels is

$$H_0 = q^2/2gy_0^2 + y_0,$$

(14–91)

the derivative with respect to the depth is

$$dH_0/dy_0 = 1 - q^2/gy_0^3 = 1 - \mathbf{F}^2.$$

(14–92)

Therefore, the slope of the free surface is

$$\frac{dy_0}{dx} = \frac{S_0 - S_H}{1 - \mathbf{F}^2}.$$

(14–93)

Equation (14–93) reduces to Eq. (14–83) when the flow is frictionless and $S_H = 0$. The additional assumption which is necessary to determine the free-surface profile from Eq. (14–93) concerns the slope of the total head line. It is assumed that the rate of head loss at any section in a nonuniform flow is the same as for a uniform flow having the same velocity and depth. This is equivalent to writing Eq. (14–89) as

$$y_0 = \left(\frac{nq}{1.49S_H^{1/2}}\right)^{3/5}.$$

(14–94)

Therefore, Eq. (14–93) becomes

$$\frac{dy_0}{dx} = \frac{S_0(1 - S_H/S_0)}{1 - \mathbf{F}^2} = S_0\left[\frac{1 - (y_N/y_0)^{10/3}}{1 - \mathbf{F}^2}\right].$$

(14–95)

Since $\mathbf{F}^2 = q^2/gy_0^3 = (y_c/y_0)^3$, we have

$$\frac{dy_0}{dx} = S_0\left[\frac{1 - (y_N/y_0)^{10/3}}{1 - (y_c/y_0)^3}\right].$$

(14–96)

As indicated by Eq. (14–96), the slope of the free surface becomes parallel to the bottom as y_0 approaches the normal depth y_N asymptotically. As y_0 approaches the critical depth y_c, the equation indicates that dy_0/dx becomes infinitely large. However, free-surface slopes at critical flow are not infinite, as can be seen from Fig. 14–35. The discrepancy is explained by the fact that the assumption of hydrostatic pressure distribution or negligible streamline curvature is not valid near the critical depth. Equation (14–96) is nevertheless very useful in establishing the general shape of gradually varied free-surface flows with boundary resistance. If the discharge q, roughness n, and bottom slope S_0

Mild slope: $V_0 < V_c$ Steep slope: $V_0 > V_c$

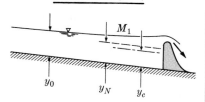 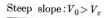 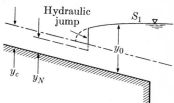

M_1, backwater curve S_1, rising transition

$y_0 > y_N > y_c$ $y_0 > y_c > y_N$

$\dfrac{dy_0}{dx}(+)$ $\dfrac{dy_0}{dx}(+)$

Increasing depth   Increasing depth

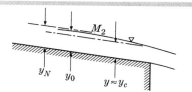 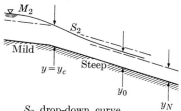

M_2, drop-down curve S_2, drop-down curve

$y_N > y_0 > y_c$ $y_c > y_0 > y_N$

$\dfrac{dy_0}{dx}(-)$ $\dfrac{dy_0}{dx}(-)$

Decreasing depth   Decreasing depth

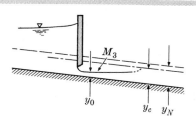 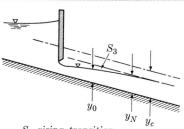

M_3, rising transition S_3, rising transition

$y_N > y_c > y_0$ $y_c > y_N > y_0$

$\dfrac{dy_0}{dx}(+)$ $\dfrac{dy_0}{dx}(+)$

Increasing depth   Increasing depth

FIG. 14–38. Surface profiles for open channels.

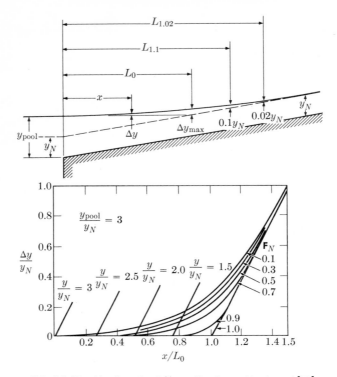

FIG. 14–39. Nondimensional M_1-profiles for a wide channel [21].

are known, the normal depth y_N and the critical depth y_c may be determined from Eqs. (14–85) and (14–89). Therefore the normal depth can be greater or less than the critical depth for the same discharge depending upon the slope and roughness. We will use this fact to define the following slope classifications:

Critical slope: The slope at which $V_N = V_c$; $y_N = y_c$ for a given discharge in a channel of given shape and roughness ($V_N =$ velocity at uniform flow).

Mild slope: A slope less than the critical slope so that $V_N < V_c$, $y_N > y_c$.

Steep slope: A slope greater than the critical slope so that $V_N > V_c$, $y_N < y_c$.

The actual local depth y_0 may be greater or less than the normal depth or critical depth depending upon the kind of upstream or downstream controls governing the flow. Considering the several possibilities for channels sloping in the direction of flow, we find that Eq. (14–96) indicates the types of surface profiles shown in Fig. 14–38.

For three-dimensional channel shapes, only the exponents in Eq. (14–96) change. Thus, for any prismatic shape,

$$\frac{dy_0}{dx} = S_0 \left[\frac{1 - (y_N/y_0)^P}{1 - (y_c/y_0)^Q} \right]. \tag{14–97}$$

The surface curve is found by integrating Eq. (14–97),

$$x = \frac{y_N}{S_0} \left[t - \int_0^t \frac{dt}{1 - t^P} + \left(\frac{y_c}{y_N} \right)^Q \int_0^t \frac{t^{P-Q}}{1 - t^P} dt \right] + \text{const}, \quad (14\text{–}97a)$$

where $t = y_0/y_N$. The above integrals have been tabulated by Chow [20] for a wide range of channel shapes. A family of surface curves [21] of the M_1-type is shown for various uniform-flow Froude numbers in Fig. 14–39. The starting point for such calculations should be at a control section where the depth is known. The surface profile for nonuniform flow in any prismatic channel may be calculated in an approximate manner by writing Eq. (13–65) in finite-difference form and by making step-by-step calculations. Thus, in finite difference form, we have

$$\frac{\Delta H_0}{\Delta L} = S_0 - \bar{S}_H \quad \text{and} \quad \Delta L = \frac{\Delta H_0}{S_0 - \bar{S}_H}. \quad (14\text{–}98)$$

For rectangular channels, the computations are carried out by starting from a section where the depth is known. By assuming a small depth change, the change in specific head between the two sections may be computed from Eq. (14–91). The average slope of the energy gradient, $\bar{S}_H$, may be obtained by averaging the values of S_H obtained from Eq. (14–94) for both the initial and the assumed depth. The distance between the initial and assumed depth is then ΔL. Summing ΔL gives the surface profile of depth versus distance.

14–4.4   Free-surface shock waves

The compressible-flow shock wave discussed earlier in this chapter has a close analogy with the free-surface shock or "hydraulic jump." The free-surface shock is a standing wave in which the depth increases and the flow changes from supercritical to subcritical. In the normal hydraulic jump, the wave front is at right angles to the direction of the oncoming flow. The longitudinal thickness of the free-surface shock is much larger than in the compressible-flow case, being of the order of five times the depth change. Nevertheless the flow conditions both upstream and downstream of the shock may reasonably be assumed uniform, and one-dimensional methods of analysis may be employed between the two regions. Boundary shears may be neglected, but not internal energy dissipation due to the production of turbulence at the shock front. Following an analysis similar to that for the gas shock, we may use the one-dimensional momentum equation (14–53) and the continuity equation for the control volume shown in Fig. 14–40, the difference being that the fluid density is constant and the forces due to pressures are evaluated from the assumed hydrostatic pressure distributions in the uniform-flow regions. Therefore, from Eq. (14–53), neglecting boundary shear forces, in a channel of unit width, we find

$$\gamma y_1^2/2 - \gamma y_2^2/2 = V_{n2}\rho q - V_{n1}\rho q, \quad (14\text{–}99)$$

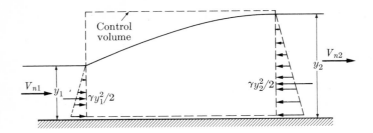

FIG. 14–40.   Normal free-surface shock or hydraulic jump.

and from the continuity equation, we have

$$q = y_1 V_{n1} = y_2 V_{n2}. \tag{14–100}$$

Combining these equations and introducing the Froude number $\mathbf{F}_{n1} = V_{n1}/\sqrt{gy_1}$, we have

$$(y_2/y_1)^2 + y_2/y_1 - 2\mathbf{F}_{n1}^2 = 0. \tag{14–101}$$

This equation has both positive and negative roots; however, only the positive one is of physical significance. Thus the depth ratio across the shock is

$$y_2/y_1 = \tfrac{1}{2}(\sqrt{1 + 8\mathbf{F}_{n1}^2} - 1). \tag{14–102}$$

As in the gas-shock counterpart, $\mathbf{F}_{n1}$ must be greater than unity since the jump can occur only from supercritical to subcritical flow. The free-surface surge or moving hydraulic jump may be obtained by changing the frame of reference as outlined in the discussion of the normal gas shock.

     The oblique gas shock also has its analog in the free-surface oblique hydraulic jump. Shocks of this type occur, as shown in Fig. 14–41, if a lateral boundary is deflected into a flow which is initially supercritical. Referring to the oblique shock geometry of Fig. 14–28 and Eq. (14–66), we may use the continuity Eq. (14–100) to obtain the depth change across the free-surface oblique shock,

$$y_2/y_1 = \tan \beta_s/\tan (\beta_s - \theta). \tag{14–103}$$

Combining Eqs. (14–102) and (14–103), and noting that $\mathbf{F}_{n1} = \mathbf{F}_1 \sin \beta_s$, we have

$$\tan \theta = \frac{\tan \beta_s \left(\sqrt{1 + 8\mathbf{F}_1^2 \sin^2 \beta_s} - 3\right)}{(2 \tan^2 \beta_s - 1) + \sqrt{1 + 8\mathbf{F}_1^2 \sin^2 \beta_s}} \tag{14–104}$$

as the counterpart of Eq. (14–68). This relation between the wall angle θ, the shock angle β_s, and the Froude number has been verified experimentally [22], and a plot of the equation is given in Fig. 14–42.

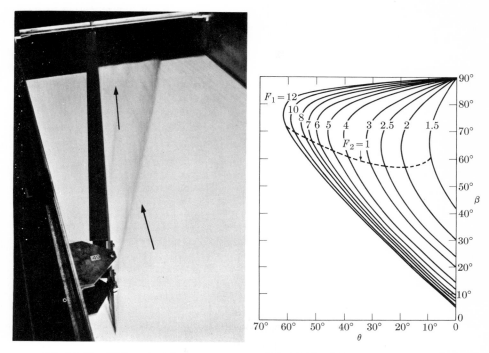

FIG. 14–41. Oblique shock in water.   FIG. 14–42. Oblique shock relations for water.

The formal analogy between two-dimensional supersonic gas flow and super-critical free-surface flow has been used for the experimental investigation of gas dynamic problems [23].

PROBLEMS

14–1. The flow net shown in Fig. 9–9 may be used to represent the two-dimensional, irrotational flow of an incompressible fluid under a barrier. In the downstream parallel-flow region water is flowing at a rate of 75 cfs/ft of width, the vertical height of the opening under the barrier is 3 ft, and the pressure is 15 psia on the horizontal boundary of the barrier. By scaling appropriate dimensions from the figure, compute the absolute pressure intensity in the upstream direction along the inner boundary of the barrier.

14–2. Assume that the flow in the converging section shown in Fig. 14–1 takes place in a horizontal plane. A small pressure tap has been drilled in the boundary in the downstream zone of parallel, uniform flow. (a) Determine the location (on the boundary) of the upstream pressure tap which will result in the maximum pressure differential between the two taps. (b) This device is to be used as a flow meter. Determine the equation for the velocity in the downstream uniform-flow section in terms of the differential pressure measured at the two taps and the density of the fluid flowing.

14-3. Is the flow-meter equation developed in Problem 14-2(b) independent of the direction of flow through the device? Explain your answer.

14-4. Figure 14-6 shows the variation of pressure intensity in the radial and axial directions following an abrupt expansion. Give a qualitative explanation for the radial variation of pressure at $x/D_1 = 3$. (Refer to Example 11-1.)

14-5. A conical diffuser section is to be designed to connect a pipe 3 in. in diameter to a 6-in. pipe carrying water (68°F) at a rate of 1 cfs. The rate of energy dissipation caused by the diffuser is not to exceed $\frac{1}{4}$ hp. What is the maximum expansion angle which can be used for the diffuser?

14-6. What determines the location of the separation streamline in Fig. 14-10?

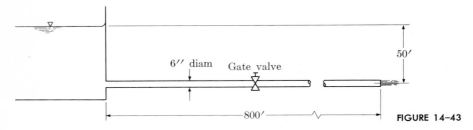

FIGURE 14-43

14-7. Water (68°F) is flowing from a large reservoir through 800 ft of smooth pipe 6 in. in diameter. The line discharges into the atmosphere at a point 50 ft below the water surface in the reservoir, as shown in Fig. 14-43. Compute the flow rate in the pipe for the following conditions: (a) gate valve fully open; (b) gate valve one-half open; (c) gate valve one-quarter open.

14-8. The head loss caused by an 8-in. valve (when in the fully open position) in a water-supply pipeline is to be determined. A valve obtained from the manufacturer is installed in an 8-in. line in a laboratory. The test line is straight and horizontal and has a total length of 50 ft. The valve is installed at the midpoint of the test line. Describe and illustrate, by means of a sketch, the test program which you would recommend. List all measurements which should be made including the location and number of pressure observation points along the pipeline. Indicate the dimensionless form in which the data should be presented.

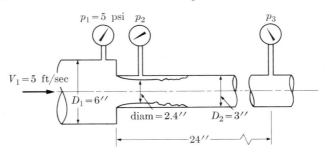

FIGURE 14-44

14-9. Derive the general relation for the Venturi and orifice flow meters given by Eq. (14–13).

14-10. A 6-in. horizontal water line suddenly contracts to a 3-in. diameter, as shown in Fig. 14–44. A pressure gauge 6 in. upstream from the contraction reads 5 psi when the mean velocity in the 6-in. pipe is 5 ft/sec. What will pressure gauges read (a) just downstream from the contraction (opposite the *vena contracta*), and (b) 24 in. down-stream when the diameter of the *vena contracta* is 2.4 in.? Neglect pipe friction.

14-11. The Venturi meter shown in Fig. 14–45 is inclined at an angle of 30° to the horizontal. The meter has an inlet diameter of $D_1 = 10$ in. and a throat diameter of $D_2 = 6$ in. A mercury-water differential manometer is connected to the pressure taps, and the deflection of the mercury is 1.00 ft. The pipeline is conveying water at 68°F. Determine the flow rate.

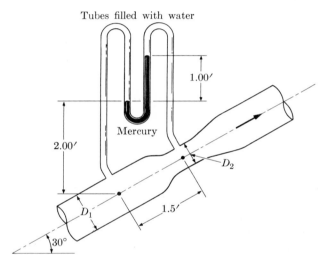

FIGURE 14–45

14-12. An orifice meter is to be installed in a horizontal pipeline 6 in. in diameter carrying oil (SG = 0.80). (a) Determine the diameter of the opening in the orifice plate, D_0, given that a pressure differential of 4 psi is to be obtained for a flow rate of 2.0 cfs. (b) Estimate the rate of energy dissipation caused by the orifice. The pressure taps are located at Sections (1) and (c) as shown in Fig. 14–13.

14-13. Plot the dimensionless pressure distribution $(p - p_0)/(\rho V_0^2/2)$ along the outer boundary of the two-dimensional bend shown in Fig. 14–14. (Here p_0 and V_0 are the pressure and velocity in the uniform approach flow.) Assume that the bend is in a horizontal plane.

14-14. A reservoir whose water (60°F) surface is at an elevation of 100 ft is connected to a second whose surface is at an elevation of 40 ft by pipes of the following sizes and lengths: The first 1000 ft of pipe is 12 in. in diameter. This branches into two pipes, each 6 in. in diameter and 2000 ft long, which later unite and discharge into a single pipe 12 in. in diameter and 2000 ft long. Assuming $k_s = 0.0004$ ft for all pipes, what will be the rate of flow? Include entrance and discharge losses but neglect losses at junction points.

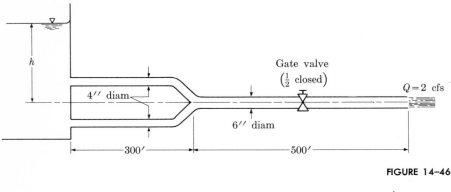

FIGURE 14–46

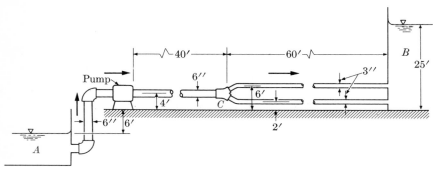

FIGURE 14–47

14–15. The pipe system shown in Fig. 14–46 is to convey water (60°F) at the rate of 2.0 cfs. The flow leaves the reservoir through two parallel pipes of equal diameter (4 in.) and length (300 ft). After merging into a pipe 6 in. in diameter (500 ft long) containing a gate valve (one-half closed) at its midpoint, the flow is discharged into the atmosphere. All pipes are galvanized iron. Determine the elevation of water in the reservoir (h) required to produce this flow.

14–16. The problem is to pump 0.75 cfs from reservoir A to reservoir B with the piping arranged as shown in Fig. 14–47. Before passing through the pump the flow goes through two 90° elbows ($R/D = 1.0$) and 12 ft of 6-in. pipe. The water (60°F) is discharged from the pump through 40 ft of 6-in. pipe and branches into two parallel 3-in. pipes 60 ft from reservoir B. The surfaces of both reservoirs are maintained at a constant elevation as shown. All pipes are galvanized iron ($k_s = 0.0006$ ft). Assuming a pump efficiency of 80%, find the motor horsepower required for the pump. Losses in fitting C may be neglected.

14–17. Two oil storage tanks A and B are connected to a lower tank C through the pipelines shown in Fig. 14–48. The oil has a kinematic viscosity of 5.8×10^{-5} ft²/sec and a density of 1.5 slugs/ft³. Assume that the surface elevations of all three tanks are steady and all pipes are hydraulically smooth. Determine the rate of flow into tank C.

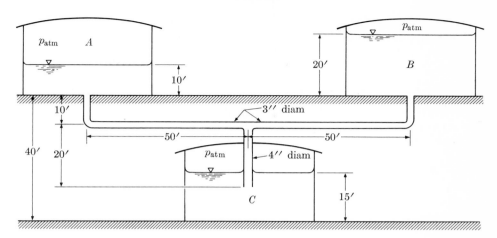

FIGURE 14–48

14–18. For a steady, inviscid, two-dimensional motion of a compressible fluid, the linearized equation of motion (Eq. 14–23) can be reduced to the Laplace equation

$$(\partial^2\phi/\partial x^2) + (\partial^2\phi/\partial\eta^2) = 0,$$

if $\eta = f(M_0, y)$ and $M_0 < 1$. Determine the relationship between η, M_0, and y. (*Note:* This is known as the Prandtl-Glauert transformation. By means of this transformation the compressible and incompressible characteristics of slender airfoils are comparable if the airfoil thickness is changed in proportion to the relation derived above.)

14–19. A supersonic jet plane is moving through the air at an elevation of 10,000 ft in standard atmosphere. A ground observer hears engine noise 5.11 sec after the plane has passed directly overhead. Estimate the Mach number at which the plane is traveling.

14–20. An aircraft is moving through the standard atmosphere at an elevation of 20,000 ft. A stagnation tube mounted on the craft indicates $p_{stg} = 11.0$ psia. Determine the speed of the aircraft.

14–21. Air is moving in a duct with a Mach number of 0.85 at a section where the static pressure $p = 30$ psia and $T = 60°F$. What will be the temperature rise at the tip of a pressure-measuring probe inserted into the flow?

14–22. Find the mass-flow rate of helium which discharges from a large reservoir at 1200 psia and 68°F into the atmosphere through a converging nozzle with a throat $\frac{1}{8}$ in. in diameter. What will be the temperature of the discharging jet?

14–23. Air in a reservoir at 900°F and 160 psia is released through a convergent-divergent nozzle to a pressure of 40 psia. The mass-flow rate is 15 lbm/sec. Find the throat and exit areas for the nozzle.

14–24. The nozzle required by the conditions of Problem 14–23 is used with the same air source but with an exit pressure of 65 psia. Compute the maximum velocity in the nozzle.

14–25. Inflow velocity of nitrogen to an isentropic nozzle is 900 ft/sec at 10 psia and 400°R. The gas is compressed to 12 psia at the nozzle exit. Find the exit velocity and temperature. What will be the ratio of exit-to-inlet cross-sectional areas?

14–26. The inflow velocity of nitrogen to an isentropic nozzle is 2000 ft/sec at 10 psia and 400°R. The gas is compressed to 12 psia at the nozzle exit. Find the exit velocity and temperature. Find the ratio of exit-to-inlet cross-sectional areas.

14–27. Oxygen is stored in steel bottles at 2000 psia and 60°F. Assume that the gas is released to the atmosphere isentropically by means of a nozzle. What would be the maximum possible mass-flow rate per unit area at the nozzle exit?

14–28. Air enters a diffuser at 80°F, 15 psia, and 1000 ft/sec velocity. The velocity at the diffuser exit is 300 ft/sec. Find the exit static pressure and the ratios of inlet-to-exit cross-sectional areas and Mach numbers.

14–29. Air at $T = 100$°F flows at Mach number 3 across a stationary normal shock. Find the ratios of pressure, temperature, and velocity across the shock.

14–30. Air moving at Mach number 2 in a duct flows across a normal shock. The pressure ahead of the shock is 15 psia. Find the pressure after the shock and the change of stagnation pressure across the shock.

14–31. The inflow of oxygen to a diverging passage is 2500 ft/sec at 14.7 psia, and 60°F. The ratio of exit-to-inlet area is 1.74. The exit pressure is 18 psia. Compute the maximum velocity and lowest pressure in the passage.

14–32. The inflow velocity of hydrogen to a diverging passage is at 2500 ft/sec, 14.7 psia, and 60°F. The ratio of exit-to-inlet areas is 1.74. The exit pressure is 18 psia. Compute the maximum velocity and lowest pressure in the passage. Explain any differences between the results of Problems 14–31 and 14–32.

14–33. Derive Eq. (14–64) for the velocity of a pressure wave in an elastic conduit filled with a liquid using only momentum and continuity between compressed liquid and expanded pipe.

14–34. The pipeline shown in Fig. 14–26 is constructed of steel with a wall thickness of 0.10 in. and an inside diameter of 10 in. Determine the maximum pressure increase due to a sudden valve closure, given that the pipeline is carrying water (60°F) at a velocity of 5 fps before closure.

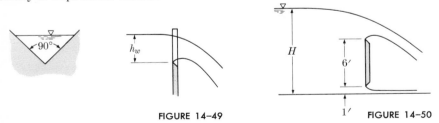

FIGURE 14–49   FIGURE 14–50

14–35. Determine the functional form of the relation between discharge and head h_w for a 90° V-notch weir (Fig. 14–49).

14–36. Determine the geometric shape (looking upstream) of a weir which has a linear relation between head h_w and discharge.

14–37. A combined sharp-crested weir and sluice are located in a channel 10 ft wide, as shown in Fig. 14–50. If the total discharge of the weir and sluice is to be 200 cfs, find the depth of water H upstream of the structure.

14-38. According to the elementary theory of inviscid free-surface flow, the depth along the horizontal portion of the structure shown in Fig. 14-51 should be constant and equal to the critical depth. Measurements indicate that the depth is critical at an intermediate section (1) and that the downstream depth decreases until at the overfall (2) it is approximately 0.7 of the critical depth y_c. The reason for the difference is the assumption that the pressure distribution is hydrostatic at all sections. Since the flow leaves the structure as a free overfall, the pressure at both surface and bottom at (2) must be atmospheric. Use the one-dimensional form of the steady-flow momentum equation for a control volume between (1) and (2) and obtain an expression for the depth ratio y_t/y_c. Assume that the pressure is atmospheric over the entire depth at (2).

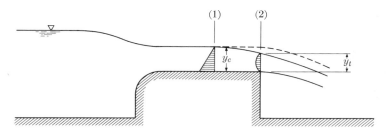

FIGURE 14-51

14-39. Determine the discharge per unit width in terms of y_c for Problem 14-38 by assuming that the structure is a sharp-crested weir of zero height having a head $h_w = y_c$ (see Figs. 14-33 and 14-34). Compare the answer with Eq. 14-85.

14-40. A rectangular open channel of constant width has its floor raised 0.15 ft at a given section. When the depth of the approaching flow is 0.5 ft, what rate of flow per unit width is indicated by: (a) A 0.025-ft drop in the surface elevation over the raised bottom? (b) A 0.27-ft increase in surface elevation over the raised bottom?

14-41. Water flows at a depth of 5 ft in a rectangular channel 10 ft wide. At a downstream section the width is reduced to 8 ft, the channel bed is raised 1 ft, and the water surface drops 0.5 ft. Neglect energy dissipation. (a) What is the flow rate? (b) What should the bed elevation be at the contracted section in order that the water surface have the same elevation upstream and downstream from the contraction?

14-42. A rectangular channel has a slope of 1 in 5000, a width of 25 feet, and a roughness $n = 0.012$. (a) Calculate the normal depth for a total discharge of 60 cfs. (b) Keeping the bottom gradient constant, determine the width to which the channel must be reduced to produce critical flow at this discharge.

14-43. The nonuniform-flow Eq. (14-96) is not valid for a horizontal channel since $S_0 = 0$ and $y_N = \infty$. Show that for this case ($S_0 = 0$), Eq. (14-90) may be integrated to obtain the following equation for the depth profile in a two-dimensional flow:

$$x = \left(\frac{1.49}{n}\right)^2 \frac{y_c^{4/3}}{g}\left[\frac{3}{4}\left(\frac{y_0}{y_c}\right)^{4/3} - \frac{3}{13}\left(\frac{y_0}{y_c}\right)^{13/3}\right] + \text{const.}$$

14–44. Water flowing under a sluice gate enters a concrete channel ($n = 0.015$) having a slope $S_0 = 0.001$. The sluice gate is operating under a high head and the lip of the gate is 1 ft above the bottom of the channel. The discharge rate is 20 cfs/ft of channel width. Calculate the distance to the point where the depth of water in the channel downstream of the gate is 1 ft.

14–45. Solve Problem 14–44 when the channel downstream of the sluice gate is horizontal. (a) Use the exact method developed in Problem 14–43. (b) Compare with the result obtained by the finite-difference method, using two steps.

14–46. Water is flowing uniformly in a rectangular channel 40 ft wide at a depth $y_N = 3$ ft. The slope of the channel bottom is $1/2000$ and $n = 0.020$. (a) Calculate the discharge. (b) A weir is installed across the channel which raises the depth at that section to 5 ft. Estimate the depth of flow at a section 500 ft upstream from the weir.

14–47. Water is discharged at the rate of 50 cfs/ft of width into a wide, horizontal concrete ($n = 0.015$) channel from under a sluice gate. The depth at the *vena contracta* is 1.50 ft. A change of flow conditions downstream causes a normal hydraulic jump to form with a depth just below the jump, $y_2 = 9$ ft. At what distance below the sluice gate will the jump occur? (Use the results of Problem 14–43.)

14–48. Use the one-dimensional form of the energy equation (4–24a) and derive an expression for the energy dissipation per pound of fluid across a normal hydraulic jump. Write the equation in the dimensionless form H_L/y_1 as a function of $\mathbf{F}_{n1}$.

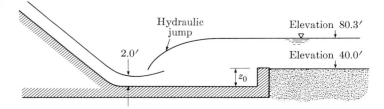

FIGURE 14–52

14–49. The spillway of a dam discharges 300 cfs/ft of width onto a horizontal apron (Fig. 14–52). The depth of the high-velocity flow at the toe of the spillway is 2.0 ft. Determine the necessary depth (z_0) of the horizontal channel below the original river bed to ensure that a hydraulic jump will form near the spillway toe.

14–50. Free-surface flow having a depth $y_1 = 0.10$ ft and an initial Froude number $\mathbf{F}_1 = 2.0$ approaches a wall with a deflection angle $\theta = 10°$ (see Fig. 14–28). Determine the angle β_s of the oblique shock wave and the depth of flow below the shock.

14–51. A supersonic stream of air of Mach number $\mathbf{M}_1 = 2.0$ approaches a wedge having an angle $\theta = 10°$. Determine the angle β_s of the oblique shock wave and the density ratio across the shock. Compare this result with the shock-wave angle and depth ratio for water in Problem 14–50.

REFERENCES

1. CHATURVEDI, M. C., "Flow Characteristics of Axisymmetric Expansions," *Proc. Am. Soc. Civil Engrs.*, **89**, HY-3 (May, 1963).

2. ROBERTSON, J. M., and H. R. FRASER, "Separation Prediction for Conical Diffusers," *Trans. ASME, Series D: J. Basic Eng.*, **82**, 1 (1960).

3. HOLZHAUSER, C. A., and L. P. HALL, "Exploratory Investigation of the Use of Area Suction to Eliminate Air Flow Separation in Diffusers Having Large Expansion Angles," *Nat. Advisory Comm. Aeron.*, Tech. note 3793 (Oct., 1956).

4. PRANDTL, L., and O. G. TIETJENS, *Applied Hydro- and Aeromechanics*, McGraw-Hill Book Co., New York, 1934.

5. KLINE, S. J., "On the Nature of Stall," *Trans. ASME, Series D: J. Basic Eng.*, **81**, 3 (1959).

6. ROUSE, H., "Energy Transformation in Zones of Separation," *Proc. Ninth Int. Assoc. for Hyd. Res.*, Dubrovnik (1961).

7. *Fluid Meters: Their Theory and Application*, American Society of Mechanical Engineers, New York, 5th ed., 1959.

8. ITO, H., "Pressure Losses in Smooth Pipe Bends," *Trans. ASME, Series D: J. Basic Eng.*, **82**, 1 (1960).

9. TURCHETTI, A. J., and J. M. ROBERTSON, "Design of Vaned-Turns for a Large Water Tunnel," *ASME*, Paper 48-SA-15 (1948).

10. McPHERSON, M. B., and J. V. RADZIUL, "Water Distribution Design and the McIlroy Network Analyzer," *Proc. Am. Soc. Civil Engrs.*, **84**, HY2 (April, 1958); Q. B. GRAVES and D. BRANSCOME, "Digital Computers for Pipeline Network Analysis," *Proc. Am. Soc. Civil Engrs.*, **84**, SA2 (April, 1958); R. S. GOOCH, "Electronic Computers in the Water Supply Field," *J. AWWA*, **52**, 311 (March, 1960).

11. SHAPIRO, A. H., *The Dynamics and Thermodynamics of Compressible Fluid Flow*, Vols. I, II, The Ronald Press Co., New York, 1953; H. W. LIEPMANN and A. E. PUCKETT, *Introduction to Aerodynamics of a Compressible Fluid*, John Wiley and Sons, Inc., New York, 1947; H. W. EMMONS (ed.), *Fundamentals of Gas Dynamics*, Vol. III of *High Speed Aircraft and Jet Propulsion*, Princeton University Press, 1958.

12. BERGERON, L., *Water Hammer in Hydraulics and Wave Surges in Electricity*, John Wiley and Sons, Inc., New York, 1961.

13. SOMMERFELD, A., *Mechanics of Deformable Bodies*, Vol. II, Academic Press, Inc., New York, 1964, p. 226.

14. VON MISES, R., "Berechnung von Ausfluss und Überfallzahlen," *Z. Ver. Dtsch. Ing. Band*, **61**, 447 (1917).

15. STRELKOFF, T. S., "Solution of Highly Curvilinear Gravity Flows," *Proc. Am. Soc. Civil Engrs.*, **90**, EM3 (June, 1964).

16. ROUSE, H., discussion of [15] appearing in *Proc. Am. Soc. Civil Engrs.*, **90**, EM5 (October, 1964).

17. KANDASWAMY, P. K., and H. ROUSE, "Characteristics of Flow over Terminal Weirs and Sills," *Proc. Am. Soc. Civil Engrs.*, **83,** HY4 (August 1957).

18. CHOW, V. T., *Open-Channel Hydraulics*, McGraw-Hill Book Co., New York, 1959; A. T. IPPEN, "Channel Transitions and Controls," in *Engineering Hydraulics* (H. Rouse, ed.), John Wiley and Sons, Inc., New York, 1950, pp. 39 ff.

19. IPPEN, A. T., R. T. KNAPP, J. H. DAWSON, H. ROUSE, B. V. BHOOTA, and E. Y. HSU, "High-Velocity Flow in Open Channels: A Symposium," *Trans. Am. Soc. Civil Engrs.*, **116** (1951).

20. CHOW, *op. cit.*, Chapter VIII.

21. VALLENTINE, H. R., "Characteristics of the Backwater Curve," *Proc. Am. Soc. Civil Engrs.*, **90,** HY4 (July, 1964).

22. IPPEN, A. T., and D. R. F. HARLEMAN, "Verification of Theory for Oblique Standing Waves," *Trans. Am. Soc. Civil Engrs.*, **121,** 678 (1956).

23. IPPEN, A. T., and D. R. F. HARLEMAN, "Certain Quantitative Results of the Hydraulic Analogy to Supersonic Flow," *Proc. Midwestern Conf. Fluid Mech.*, Second Conference, p. 219 (1952).

CHAPTER 15

Immersed Bodies.
Drag and Lift

15–1   INTRODUCTION

The motion of a body immersed in a fluid is related to the drag and lift
components of the resultant dynamic force exerted on the body by the fluid.
Drag, the resistance to motion, is the component of the resultant force in the
direction of the relative flow ahead of the body, and lift is the component
normal to this direction. Various aspects of fluid resistance on immersed objects
have already been treated in preceding chapters, where attention has been
primarily directed at problems which can be treated analytically.   The objective of this chapter is to develop an understanding of fluid resistance for
immersed bodies, particularly for the many important cases not susceptible
to an analytical solution.   The reader should also become familiar with some
of the vast amount of experimental information on aerodynamic and hydrodynamic drag and lift forces for symmetrical and nonsymmetrical bodies.
Interfacial and compressibility effects on fluid resistance are discussed, as well
as the problems of unsteady forces and flows.   On the basis of the theoretical
background which has been developed, we are in a position to explain the observed phenomena of fluid resistance.

The dimensionless force coefficients C_D and C_L, discussed in Section 8–3,
are useful parameters for expressing the steady-state dynamic force components
on an immersed body.   For convenience, they are summarized below:

$$\text{Total drag} = D = C_D\rho \, \frac{V_0^2}{2} \, A, \qquad (15\text{–}1)$$

$$\text{Total lift} = L = C_L\rho \, \frac{V_0^2}{2} \, A. \qquad (15\text{–}2)$$

The density ρ is that of the fluid in which the body is immersed, although
the body itself may be composed of another fluid. The velocity V_0 is the speed
of the undisturbed relative flow ahead of the body.   For a stationary object
in a steady-flow field, V_0 is the speed of the approach flow far enough upstream
so as to be unaffected by the presence of the object.   For a steadily moving
body in an otherwise stationary fluid, the equivalent flow field may be obtained
by superposition of a uniform velocity V_0 equal and opposite to the velocity
of the moving body. (It should be noted that the principle of equivalent flow

369

fields must be applied with caution. Under certain circumstances, differences in the level of turbulence in the free-stream approach flow may change the resistance of an immersed body by a large amount.) The area A is the area of the projection of the body on a plane normal to V_0 (except for flat plates, struts, and airfoils, where A is the largest projected area). From dimensional reasoning we see that the total drag and lift coefficients C_D and C_L represent a force divided by an area and by the dynamic pressure $\rho V_0^2/2$.

The concept of separation of the total drag into frictional and pressure components was introduced in Section 8–3. These principles should be reviewed, as they will be applied to some of the quantitative discussions to follow. Frictional drag refers only to drag due to the component of a frictional shear stress τ_0 on a boundary. On a flat plate, the total drag force is due to frictional drag, and the dimensionless drag coefficient C_f is defined by Eq. (8–25). On struts and elongated bodies of revolution, the same equation may be used to define that portion of the total drag due to frictional effects. In all cases, the area term A_f is the surface area of the body in question. The dimensionless drag coefficient C_D in Eq. (15–1) is defined in terms of the total drag force, and includes the components of both the frictional shear stress and the normal pressure distribution on the boundary. The latter contribution is usually called *pressure drag* or *form drag*. To avoid confusion regarding the area term in Eq. (15–1), the following convention will be adopted. The coefficient C_D without additional subscript refers to the projected area of the body on a plane normal to V_0. The coefficient C_{D_c} will designate a drag coefficient based on the maximum projected area or span times chord in the case of an airfoil.

From the concepts of dynamical similitude, we should expect the force coefficients to depend on the geometry and upon dimensionless parameters describing the dynamical behavior of the fluid motion. In general,

$$C_D = C_D(\text{geometry, } \mathbf{R}, \mathbf{F}, \mathbf{M}), \tag{15–3}$$

and

$$C_L = C_L(\text{geometry, } \mathbf{R}, \mathbf{F}, \mathbf{M}). \tag{15–4}$$

For various special cases, we can follow the reasoning of Chapter 7 to predict certain simplifications of the general functional dependence upon the several flow parameters. Consider the drag coefficient for the following cases:

Incompressible fluids in enclosed systems. For immersed bodies in homogeneous incompressible fluids of large extent or confined within fixed boundaries, C_D depends only on inertia and viscous forces* in addition to the geometry. Hence the resistance is a function of geometry and Reynolds number,

$$C_D = C_D(\text{geometry, } \mathbf{R}). \tag{15–5}$$

---

* The hydrostatic buoyancy of objects immersed in a fluid is considered separately from the dynamic lift and drag forces due to the motion of the fluid relative to the object.

Note that the geometry must include not only the geometric nature of the body (including surface roughness) but also the geometric relation of the body to any boundaries of the fluid system. Thus the drag on a sphere falling in a tube may differ from that of the same sphere falling in the same fluid of infinite extent.

Incompressible fluids in systems having an interface. The drag of objects moving on or near a liquid-liquid or liquid-gas interface is affected by viscous forces and by gravitational forces due to the propagation of energy by surface or interfacial wave motion. Thus the Froude number must be considered, and

$$C_D = C_D (\text{geometry, } \mathbf{R}, \mathbf{F}). \tag{15–6}$$

The gravitational effect decreases as the depth of submergence relative to the interface increases. Ultimately, the resistance coefficient reverts to the form of Eq. (15–5) for fluids of large extent.

Compressible fluids. The consideration of compressibility effects in fluid resistance is of importance in gas dynamics where the relative speed may approach or exceed the sonic velocity. In this case, energy is propagated away from the object by elastic waves. In the high subsonic and transonic regimes, the resistance depends upon both Reynolds and Mach numbers. At supersonic speeds, it is usually permissible to ignore the viscous forces and to assume that the resistance is a function of geometry and the Mach number. Thus

$$C_D = C_D (\text{geometry, } \mathbf{M}). \tag{15–7}$$

Note that if flows of different gases are to be considered, and if heat transfer is important, Eq. (15–3) is no longer general. In these cases, dimensionless parameters related to the thermodynamic properties of the gases must appear in the functional relationship.

All of the foregoing relationships for the drag coefficient apply equally to the lift coefficient.

15–2   HYDRODYNAMIC FORCES IN STEADY IRROTATIONAL MOTION

Consider what might be called the zeroth-order drag problem, namely, the resistance of a body of arbitrary shape in a steady uniform, incompressible, inviscid fluid. If the motion of a body in such a fluid is initiated from rest, the resulting fluid motions will remain irrotational in accordance with the discussion in Section 6–6.

As an example of irrotational motion about an immersed object, consider the two-dimensional flow in the x-direction past a stationary cylinder whose axis is normal to the direction of flow. The equation for the flow field is obtained by potential theory [1], and streamlines are given by constant values of the

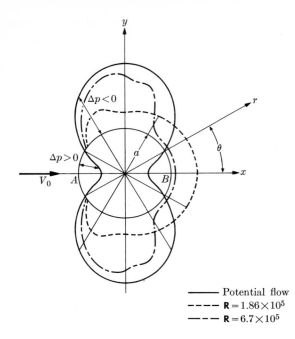

———— Potential flow
– – – – $R = 1.86 \times 10^5$
— – — $R = 6.7 \times 10^5$

FIG. 15–1.   Pressure distributions around a cylinder.

stream function ψ defined as

$$\psi = -V_0(r - a^2/r) \sin \theta, \tag{15–8}$$

where

$$a = \text{radius of cylinder,}$$
$$r, \theta = \text{cylindrical coordinates defined in Fig. 15–1,}$$
$$\text{and } r \geq a \text{ for the external streamlines.}$$

By Eqs. (6–8) and (6–10) the velocity field is

$$u = -\partial\psi/\partial y, \qquad v = \partial\psi/\partial x,$$

or

$$v_r = -\partial\psi/r\partial\theta, \qquad v_\theta = \partial\psi/\partial r. \tag{15–9}$$

It can be seen that Eq. (15–8) satisfies two sets of conditions:

$$(1) \quad r = a, \qquad \psi = 0 \text{ for all values of } \theta,$$
$$\theta = 0, \pi, \qquad \psi = 0 \text{ for all values of } r.$$

Thus the x-axis and the cylinder boundary form a continuous streamline.

$$(2) \quad r \to \infty, \qquad \psi \to -V_0 r \sin \theta = V_0 y.$$

Thus the streamlines at large distances from the cylinder are parallel to the x-axis and (since $u = -\partial\psi/\partial y$) correspond to undisturbed uniform flow in the $+x$-direction.

On the surface of the cylinder, the radial velocity is zero and the resultant velocity is identically the tangential velocity. Hence, letting $v_\theta = v_a$ at $r = a$ and using $v_\theta = \partial\psi/\partial r$, we have

$$v_a = -2V_0 \sin\theta. \tag{15-10}$$

The pressure on the cylinder surface is found from the Bernoulli equation (6–62) rewritten in terms of the dynamic pressure component.* For ambient conditions of $p_1 = 0$ and $q_1 = V_0$, we have

$$V_0^2/2g = v_a^2/2g + p_{d_a}/\gamma. \tag{15-11}$$

Hence, dropping the subscript d, we get

$$p_a = \rho(V_0^2/2)(1 - 4\sin^2\theta). \tag{15-12}$$

The potential-flow pressure distribution is symmetrical about both axes, as shown in Fig. 15–1. Since there are no tangential (viscous) surface stresses, any horizontal force on the cylinder must be due to the normal pressure distribution. It can be easily shown that the integration of the x-component of the pressure force yields a drag force of zero. Thus

$$D = -\int_0^{2\pi} p_a \cos\theta \, a \, d\theta = -\rho(aV_0^2/2)\int_0^{2\pi} (1 - 4\sin^2\theta)\cos\theta \, d\theta = 0. \tag{15-13}$$

In a similar manner, it can be shown that the lift force is also zero. In general, the hydrodynamic force on any immersed object in an infinite steady irrotational stream without circulation reduces to zero or a pure couple. The introduction of circulation, in the form of an irrotational vortex, for example, leads to the development of a lift force; however, the net drag force always remains zero. Even though the lift force is finite, the net work done is zero, since the lift force is at right angles to the velocity of the undisturbed flow.

If the cylinder were moving through the fluid field, fluid near the cylinder would be set in motion. The stream function for this case would be

$$\psi = -V_0 \frac{a^2}{r} \sin\theta. \tag{15-14}$$

It can be seen that Eq. (15–8) is obtained from this by superposing a velocity V_0 on the system, bringing the coordinates to rest relative to the cylinder. The relative fluid velocity is still given by Eq. (15–10). Then Eqs. (15–12) and (15–13), with the subsequent conclusions, apply to steady motion of bodies through otherwise stationary fluids.

---

* Using $p = (p_d + p_s)$ as in Eq. (7–20).

15–3 HYDRODYNAMIC FORCES IN ACCELERATED MOTION

When a body accelerates through a fluid, a force must be applied to accelerate the mass of the body and an added force must be applied to accelerate the mass of fluid set in motion by the body. The added force is a reaction between the body and the fluid and constitutes an additional resistance to the body motion over and above any real fluid drag that may be associated with a given velocity. We can demonstrate this by considering the kinetic energies involved and their rate of change during acceleration.

The total kinetic energy of the body and the fluid set in motion by it is

$$\text{KE} = \tfrac{1}{2}MV_0^2 + \tfrac{1}{2}\int_{\mho} \rho q^2 \, d\mho, \qquad (15\text{–}15)$$

where

$M = $ mass of body,

$q = \sqrt{u^2 + v^2 + w^2} = $ local fluid velocity,

$\mho = $ volume encompassing all the fluid in motion.

It is convenient to replace the integral over the fluid volume by an added kinetic energy defined as the product of a coefficient and the kinetic energy of fluid equal to the volume displaced by the body. Thus the effective kinetic energy is

$$\text{KE} = \tfrac{1}{2}(M + kM')V_0^2, \qquad (15\text{–}16)$$

where

$M' = $ mass of fluid equal to that displaced by the body,

$k = $ added mass coefficient.

The quantity kM' is known as the *added mass* and $(M + kM')$ as the *virtual mass*. With k known, the situation can be treated by an effective increase in the mass of the body. The added-mass coefficient depends on the body shape and the nature of the motion and may be calculated for irrotational motions. For a cylinder moving "broadside," $k = 1.0$, for a sphere, $k = 0.50$, and for an ellipsoid moving "end-on" with the major axis twice the minor axis, $k = 0.20$. Experimental observations for bodies oscillating harmonically in real fluids are in close agreement with the values computed for potential motions [2].

Now, for an accelerating body, the force F_a to accelerate the body plus fluid when multiplied by the forward velocity V_0 represents the power expended. This power must equal the time rate of increase of the kinetic energy of the system. Hence, taking k as a constant, we have

$$F_a V_0 = \frac{d(\text{KE})}{dt} = (M + kM')V_0 \frac{dV_0}{dt} \qquad (15\text{–}17)$$

or

$$F_a = (M + kM') \frac{dV_0}{dt}. \qquad (15\text{–}18)$$

Thus the presence of the fluid can be treated as an effective increase in the mass of the body for the purpose of determining the total accelerating force. The force due to the added mass, being the reaction between the body and the fluid, equals the integral over the body surface of the pressure-force component in the direction of motion. This is the extra force over that present if the body were accelerating in a vacuum. It is the resultant "external" force acting *on the fluid* to change the kinetic energy of the fluid. It vanishes for a steady velocity V_0.

A different but analogous result is obtained when the object is stationary and the fluid is accelerated past it. With a stationary body, it is obvious that a change in the mass of the body, assuming its size and shape remain constant, can have no effect upon the force exerted on the body by the fluid. However, as before, the kinetic energy of the fluid changes as a consequence of the resultant of all the external forces acting on the fluid. In this case, these include forces acting on the outer boundaries of the fluid-body system as well as the reacting force of the body on the fluid. We can see the existence of these by the following qualitative arguments.

Consider the linear acceleration of a mass of fluid in the x-direction. If we ignore the presence of the body for the moment, it should be recognized that if a fluid is accelerating in a given direction, there must be a pressure gradient in the same direction. Thus, from Eq. (6–24), the pressure gradient due only to acceleration is obtained by eliminating gravity and viscous effects. Thus

$$\frac{\partial p}{\partial x} = -\rho \frac{dV_0}{dt} = -\rho a_x, \tag{15–19}$$

where V_0 is the instantaneous local velocity.

Now consider a large volume υ enclosing a body of volume υ_b and through which the fluid is moving. The fluid volume is υ_f such that

$$\upsilon = \upsilon_f + \upsilon_b. \tag{15–20}$$

The net x-direction pressure force acting on the fluid is

$$\int_\upsilon \frac{\partial p}{\partial x}\, d\upsilon = \int_{\upsilon_f} \frac{\partial p}{\partial x}\, d\upsilon - \int_{\upsilon_b} \frac{\partial p}{\partial x}\, d\upsilon. \tag{15–21}$$

This acts on the outer boundaries of the fluid-body system minus the reacting force on the body. The effect on the object is a horizontal "buoyant force" which is analogous to the vertical buoyant force on an object in a hydrostatic pressure gradient. If the body has no influence on the streamlines and kinetic energy of the fluid, $\partial p/\partial x$ would be constant throughout the fluid, and the force of the fluid on the body would equal

$$F_a = -\upsilon_b(\partial p/\partial x). \tag{15–22}$$

To account for the effect of the body in modifying the velocity field and kinetic energy of the fluid, $\partial p/\partial x$ can be considered constant, and a correction can be made by increasing $\mathcal{V}_b$ by a quantity equal to the volume of an extra mass. Thus the acceleration force on the body becomes

$$F_a = -(\mathcal{V}_b + k\mathcal{V}_b)\frac{\partial p}{\partial x}$$

$$= (\mathcal{V}_b + k\mathcal{V}_b)\rho\frac{dV_0}{dt} \qquad (15\text{–}23)$$

$$= M'(1 + k)\frac{dV_0}{dt}.$$

The quantity kM' is an added mass analogous to that of the moving body. However, now the force interaction between body and fluid is the *total F*. It is this F which is the surface integral of the pressure-force component in the direction of motion. Note that Eq. (15–23) is identical to Eq. (15–18) for the total accelerating force of a moving body if in the latter the mass of the body equals the mass of the displaced fluid, that is, $M = M'$.

An important application of accelerating flows around fixed objects occurs in the determination of wave forces on offshore structures exposed to oscillatory wave motion. For wave motion, both fluid velocities and accelerations are harmonic functions of time. Of course, hydrodynamic drag forces due to fluid viscosity will be present in addition to the acceleration forces considered above.

15–4 DRAG OF SYMMETRICAL BODIES

In this section we consider the steady-state drag characteristics of bodies having an axis of symmetry parallel to the direction of relative motion between the body and an undisturbed fluid. By steady state we mean the time-averaged drag for a constant time-averaged relative velocity. For symmetric bodies, the streamlines around the body either have symmetry or the time average of asymmetries is zero. The distribution of hydrodynamic stresses over the body is therefore symmetric with respect to the direction of motion and the steady-state lateral force (lift or yaw) is zero. When asymmetries arise, the instantaneous forces are unsteady or periodic as we will discuss in a later section. From a geometric standpoint, the simplest case of a symmetrical body is a very thin smooth flat plate parallel to the direction of the oncoming flow. It will be helpful in understanding the resistance problem to investigate the variation of the total drag, beginning with the flat plate, and then considering bodies of increasing thickness. In addition, it is convenient to consider separately two- and three-dimensional bodies in a flow field of large extent so that the drag coefficients will depend only on body geometry and a characteristic Reynolds number.

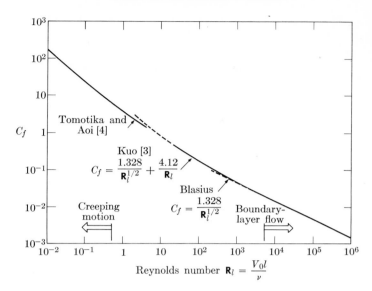

FIG. 15–2.   Flat-plate skin-friction drag coefficient for a wide range of Reynolds numbers.

15–4.1   Two-dimensional symmetrical bodies

Low Reynolds-number drag of flat plates. Two major aspects of flat-plate resistance have already been treated: the laminar boundary-layer drag in Chapter 10, and the turbulent boundary-layer drag in Chapter 12. For incompressible laminar flow, the Blasius flat-plate solution leads to Eq. (10–18) for the surface-resistance coefficient (for one side of an immersed plate), namely

$$C_f = \frac{1.328}{\mathbf{R}_l^{1/2}}. \tag{15–24}$$

This relation holds well for $\mathbf{R}_l > 10^4$ so long as the boundary layer remains laminar. Below $\mathbf{R}_l = 10^4$, the measured drag is larger than that calculated from the Blasius equation as shown in Fig. 10–6. This is due to a breakdown of the Prandtl boundary-layer approximation based on the assumption that the thickness of the boundary layer is very small, that is, $\delta \ll x$. A solution based on a perturbation technique [3] gives the equation (for one side of the plate)

$$C_f = \frac{1.328}{\mathbf{R}_l^{1/2}} + \frac{4.12}{\mathbf{R}_l} \tag{15–25}$$

which is valid to $\mathbf{R}_l = 10$ at the lower end and joins the Blasius solution in the range of $\mathbf{R}_l = 10^4$ (compare with the empirical equation in Fig. 10–6). When we approach the problem from the very low Reynolds-number range of

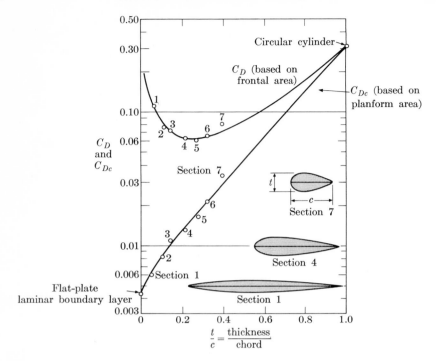

FIG. 15–3. Effect of thickness on drag coefficients of symmetric airfoils at constant Reynolds number $R_l = 4 \times 10^5$ [6].

creeping motions similar to the Stokes treatment of the sphere, drag coefficients valid up to $R_l = 1$ have been obtained [4]. A plot of the flat-plate drag coefficients in the range $10^{-2} < R_l > 10^4$ is shown in Fig. 15–2. The theoretical gap is small, and the mathematical difficulties in the range of Reynolds numbers from 1 to 10 are enormous, since all terms in the Navier-Stokes equation must be considered. A physical feeling for this Reynolds-number range is obtained by considering that a plate 1 ft long moving with a velocity of 0.1 ft/sec in glycerin has an R_l of approximately 10. The same plate in water would have an R_l of the order of 10^4.

Effect of thickness. The effect of thickness on the resistance of an immersed body in a fluid of large extent is shown by considering a family of symmetrical struts or airfoils described by the parameter t/c, where t is the thickness normal to the flow direction and c is the chord length parallel to the flow. The ratio t/c varies from zero (flat plate) to unity (circular cylinder). An example of such a family is the Joukowski profile for which the intermediate shapes are obtained mathematically by a conformal transformation [5] (or mapping) of a unit circle. This family of airfoils has the property that for irrotational flow, the velocity field and pressure distribution around the cylinder may also be

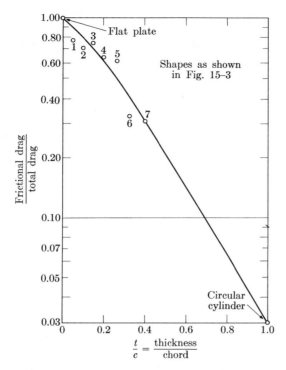

FIG. 15–4. Ratio of frictional to total drag versus thickness ratio of symmetric airfoils at constant Reynolds number $R_l = 4 \times 10^5$.

transformed into the flow and pressure fields around any of the airfoil sections. Thus measured pressure distributions on airfoils for real fluids may be compared with the irrotational pressure distributions computed for an ideal fluid.

To compare drag properties of immersed bodies, we will use the *total* drag which is the pressure and frictional drag summed over the entire body surface. Based on maximum projected area (planform: span times chord for a two-dimensional case), Eq. (15–1) becomes

$$C_{D_c} = \frac{\text{total drag}}{\frac{1}{2}\rho V_0^2 bc}, \tag{15–26}$$

where b = span, and c = chord.

Note that for a flat plate the coefficient defined by Eq. (15–26) is exactly twice the value of C_f in Eq. (15–24) or (15–25). A plot of C_{D_c} versus t/c is shown as the lower curve in Fig. 15–3 [6] for the Joukowski family of airfoils at the constant Reynolds number $R_l = 4 \times 10^5$. The drag coefficient increases from the limiting value for the flat plate to that of the circular cylinder. The increase is almost exponential for these airfoils at this Reynolds number.

Using these airfoils as fairings to struts (say around a circular cylinder of diameter $d = t$), we obtain the optimum chord length for a particular thickness by expressing the drag coefficient in terms of the projected area normal to V_0 [frontal area A in Eq. (8–27)]. Then we have

$$C_D = \frac{c}{t}\,(C_{D_c}), \tag{15–27}$$

and the corresponding values are shown in the upper curve in Fig. 15–3. For this particular airfoil family and Reynolds number, the minimum drag for a given thickness occurs when the chord is approximately four times the diameter of the cylinder enclosed in the fairing.

By direct measurement of the normal pressure distribution, Eq. (8–24) has been used to calculate the pressure drag for the family of symmetrical airfoil sections shown in Fig. 15–3. The frictional drag is therefore the difference between the measured total drag and the measured pressure drag. The ratio of frictional drag to total drag is shown in Fig. 15–4. In the long slender sections, the frictional drag accounts for 70 to 80% of the total; however, for the circular cylinder, it is only about 3% of the total. In the latter case, the boundary layer separates ahead of the maximum-thickness point of the cylinder leaving the entire aft portion in a low-pressure wake with consequent high form drag. The surface drag is determined almost entirely by the boundary layer up to the point of separation. Ideal (nonviscous) fluid theory predicts a symmetric pressure distribution and zero drag. The differences between ideal and actual fluids are illustrated in Fig. 15–1 and are discussed in the following paragraphs.

Separation and wake-formation effects. A composite plot of the steady-state drag of cylindrical bodies is given in Fig. 15–5. The diagram includes the various flow regimes encountered between very low and very high Reynolds numbers. These regimes are a consequence of the modification of the irrotational (potential)

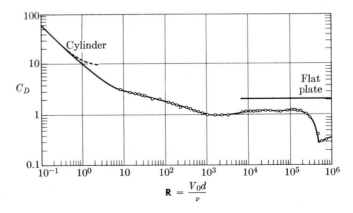

FIG. 15–5.   Drag coefficients for circular cylinders and flat plates normal to flow (two-dimensional).

flow pressure distribution in the case of real fluid flow. The modifications are viscous effects of considerable importance in the understanding of immersed body resistance.

In the very low Reynolds-number range ($\mathbf{R} < 1$), a behavior similar to the creeping motion past a sphere is to be expected. Although an analytical solution for the cylinder is available, its range of applicability is of little practical importance. As the Reynolds number becomes larger than about five, a separation of the laminar boundary layer takes place. The mechanism of the separation is related to the adverse pressure gradient and the curvature of the boundary as discussed in Section 10–3. The potential-flow pressure distribution in Fig. 15–1 shows the strong adverse pressure gradient near $\theta = 90°$. For $5 < \mathbf{R} < 50$, $\mathbf{R} = V_0 \, 2a/\nu$, a separated region encloses a stable eddy attached to the rear of the cylinder and is followed by a wavy layer of vorticity. The downstream modification of the flow field due to the presence of the object is known as a wake. In the above range, the wake is entirely laminar.

FIG. 15–6.   Vortices in the wake of a cylinder, $\mathbf{R} = 60$ [7].

At higher Reynolds numbers ($60 < \mathbf{R} < 5000$), the waves in the wake increase in amplitude and roll up into discrete vortices, as shown in Fig. 15–6 [7]. The eddy behind the cylinder is no longer stably attached but is alternately formed and detached from one side to the other. This phenomenon is known as a Kármán vortex street [8] and is characterized by a periodicity of vortex formation. Kármán has shown theoretically that the pattern of vortex formation is stable if the ratio of the separation h to the longitudinal spacing l (see Fig. 15–7) is $h/l = 0.28$. Measurements confirm this relationship in the early part of the wake, and at a larger distance the spacing h tends to increase. Of particular interest are the measurements of the frequency of vortex shedding which are expressed in terms of a dimensionless parameter known as the Strouhal number,

$$\mathbf{S} = nd/V_0, \tag{15–28}$$

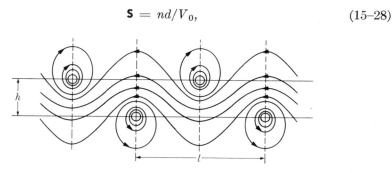

FIG. 15–7.   Kármán vortex street.

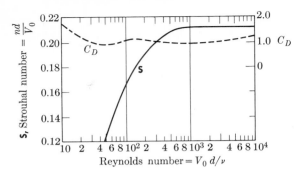

FIG. 15–8. Strouhal number and drag coefficient versus Reynolds number for circular cylinders [9].

where
$$n = \text{vortex shedding frequency,}$$
$$d = \text{cylinder diameter.}$$

As the Reynolds number increases, the frequency of shedding increases until, at $\mathbf{R} = 10^3$, $\mathbf{S} = 0.21$. The measurements are shown in Fig. 15–8 [9] together with the values for the steady-state drag coefficient in the same range of Reynolds numbers. Throughout the range of periodic vortex formation, the drag coefficient is essentially a constant $C_D \approx 1.0$. The lower limit of $\mathbf{R} = 50$ marks the end of creeping motions. The upper limit of $\mathbf{R} = 5000$ corresponds to the termination of laminar periodic conditions in the wake. The unsymmetrical vortex formation behind the cylinder gives rise to a lateral thrust or lift with a frequency n. If the cylinder is not rigidly supported, an oscillatory motion normal to the free-stream velocity will develop, especially if the frequency of vortex formation is close to the natural frequency of vibration of the object.

At Reynolds numbers beyond 5000, the wake may be considered to be entirely turbulent, although the boundary layer on the forward portion of the cylinder remains laminar. The separation point of the laminar boundary layer is very close to the point at which the adverse pressure gradient begins at $\theta = 90°$, or slightly ahead of it according to some measurements. The actual pressure distribution at $\mathbf{R} = 1.86 \times 10^5$ is shown in Fig. 15–1. In the Reynolds-number range $5 \times 10^3 < \mathbf{R} < 2 \times 10^5$, the flow pattern described above is maintained, and the drag coefficient is essentially constant at $C_D \approx 1.2$. The sharp drop in the drag coefficient in the vicinity of $\mathbf{R} = 5 \times 10^5$ occurs when the attached boundary layer on the forward half of the cylinder becomes turbulent. The separation of the turbulent boundary layer is delayed, owing to the increased kinetic energy and momentum near the boundary and the resulting ability to maintain attachment in a region of increasing pressure. A plot of the pressure distribution for $\mathbf{R} = 6.7 \times 10^5$ in Fig. 15–1 shows, in comparison with the laminar boundary-layer case, a closer agreement with the irrotational flow distribution. Therefore the total drag is reduced, since the negative pressure in the wake is smaller. The separation point for the turbulent boundary layer lies between a θ of 50° to 60°.

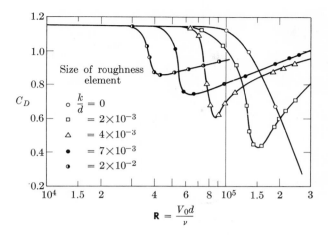

FIG. 15–9. Drag on circular cylinders of varying roughness [10].

The critical Reynolds number at which the boundary-layer transition takes place is strongly influenced by two factors; the free-stream turbulence level in the approach flow and the roughness of the cylinder. An increase in roughness or free-stream turbulence has the effect of decreasing the critical Reynolds number. Figure 15–9 [10] shows the effect of roughness on C_D in this region. Even a localized roughness such as wire or a sandpaper strip on the forward portion of the cylinder will have a considerable effect on the transition.

As a further example of separation and wake formation, the flow around a two-dimensional flat plate normal to the direction of the approach flow will be considered. In this case, the frictional drag is exactly zero, since the shear stresses can have no components in the direction of flow. The flow pattern in the vicinity of the plate and the pressure distribution are shown in Fig. 15–10. It is evident that there can be no shift of the separation point, and hence the pressure distribution is stable, and the drag coefficient should be independent of the Reynolds number outside of the range of creeping motions. Experiments indicate that $C_D = 2.0$ as shown in Fig. 15–5.

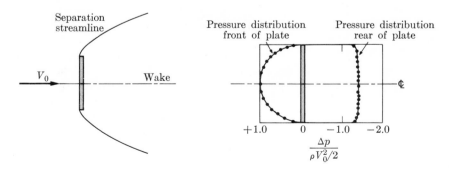

FIG. 15–10. Flow pattern and pressure distribution for flat plate normal to free stream.

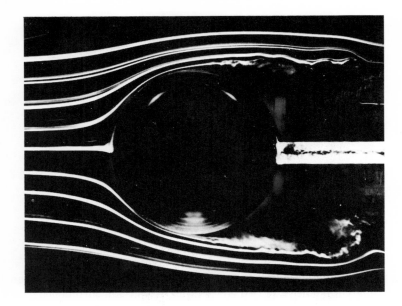

(a)

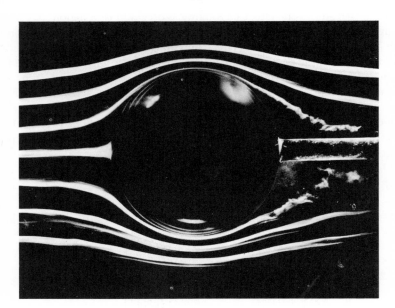

(b)

FIG. 15–11.   Smoke photographs of flow past a sphere: (a) separation of laminar boundary layer on a sphere, $R = 2.8 \times 10^5$; (b) separation of turbulent boundary layer on a sphere, $R = 3.9 \times 10^5$ [11].

15–4.2 Three-dimensional bodies

A summary of the drag coefficients for a sphere over a wide range of Reynolds numbers has been given in Fig. 9–5. The shape of the curve is very similar to that of the cylinder, and the three fundamental regimes of flow are evident: (1) creeping motion, (2) turbulent wake and laminar boundary layer (Fig. 15–11a) [11], and (3) turbulent wake and turbulent boundary layer (Fig. 15–11b). The critical Reynolds number for transition is again highly influenced by surface roughness and free-stream turbulence. In fact, smooth spheres may be used to compare the level of free-stream turbulence in various wind and water tunnels. Figure 15–12 [12, 13, 14] shows a correlation between the ratio of the rms-velocity fluctuation of the free stream (Eq. 11–4) to V_0 and the critical Reynolds number.

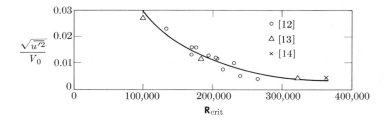

FIG. 15–12. Free-stream turbulence level for boundary-layer transition on a smooth sphere [12, 13, 14].

The effect of finite span or aspect ratio is to reduce the drag coefficient in comparison with the equivalent two-dimensional or infinite span body. Table 15–1 shows this effect for circular cylinders and flat plates normal to the stream. Table 15–2 compares selected two- and three-dimensional cases.

TABLE 15–1

EFFECT OF FINITE SPAN ON DRAG COEFFICIENT

Circular cylinder, $R = 88,000$		Flat plate normal to stream, $R = 68,000$ to $170,000$	
$\dfrac{\text{Length}}{\text{Diameter}}$	$\dfrac{C_D}{C_D \text{ for infinite span}}$	$\dfrac{\text{Length}}{\text{Breadth}}$	$\dfrac{C_D}{C_D \text{ for infinite span}}$
∞	1	∞	1
40	0.82	17.8	0.70
20	0.76	12.0	0.64
10	0.68	10.0	0.64
5	0.62	8.0	0.63
2.96	0.62	4.0	0.59
1.98	0.57	2.0	0.57
1.00	0.53	1.0	0.55

TABLE 15–2

Drag Coefficients

Object	c/d	L/d	$\mathbf{R} = V_0 d/\nu$	C_D
Circular cylinder,		1	10^5	0.63
normal to flow		5	10^5	0.74
		20	10^5	0.90
		∞	10^5	1.20
		5	$> 5 \times 10^5$	0.35
		∞	$> 5 \times 10^5$	0.33
Elliptic cylinder,	2	∞	4×10^4	0.60
normal to flow	2	∞	10^5	0.46
	4	∞	2.5×10^4 up to 10^5	0.32
	8	∞	2.5×10^4	0.29
	8	∞	2×10^5	0.20
Square cylinder, $\square$		∞	3.5×10^4	2.0
normal to flow $\diamondsuit$		∞	10^4 up to 10^5	1.6
Circular disk,		0	$> 10^3$	1.12
normal to flow				
Circular cylinder,		0	$> 10^3$	1.12
parallel to flow		1		0.91
		4		0.87
		7		0.99
Rectangular flat plate,		1	$> 10^3$	1.10
normal to flow		5		1.20
		20		1.50
		∞		2.00

L = length

d = maximum width of object measured normal to flow direction (equals minor axis of ellipsoid)

c = length of major axis of ellipsoid

The end effect reduces the drag force because of a reduction of the pressure near the edges of the object. For example, for a square plate normal to the flow, the pressure distribution shown in Fig. 15–10 occurs in both the horizontal and vertical directions. The drag coefficient for a circular disk normal to the flow is shown in Fig. 15–13 [15]. Note that at very low Reynolds numbers, the laminar-flow drag increases to very large values. For example, at $\mathbf{R} = 1, C_D \approx 30$, which is three times the drag of a cylinder at the same Reynolds number (see Fig. 15–5).

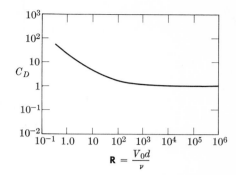

FIG. 15–13.   Drag coefficient C_D versus Reynolds number for a disk normal to the undisturbed flow [15].

$$R = \frac{V_0 d}{\nu}$$

Many experimental data are available on drag coefficients for both two- and three-dimensional bodies, including airship shapes, submarine hulls, ellipsoids, etc. A general summary of this information has been given by Hoerner [16].

15–5   LIFT AND DRAG OF NONSYMMETRICAL BODIES

Any nonsymmetric object will experience a lateral steady-state force as well as a steady-state drag when subjected to a relative flow. Consequently, any such object is potentially a lifting device. However, only certain shapes will produce a large lift-drag ratio and hence be efficient enough to be practical. These are airfoils or hydrofoils. In this section we will consider the steady-state lift and drag properties of airfoils.

15–5.1   Circulation and lift

It is known that an object, such as a ball, which is both translating and rotating through a fluid experiences a lift force. This real-fluid effect can be modeled mathematically by superposition of irrotational motions of ideal fluids. The simpler two-dimensional problem involving a cylinder is obtained by combining the stream function for uniform flow past a cylinder of radius a, Eq. (15–8), with that for an irrotational vortex rotating in the clockwise direction with circulation $-\Gamma$ [Eq. (6–97) with a negative sign]. Therefore, the combination of the uniform flow and the *bound* vortex is given by the sum of the stream functions,

$$\psi = -V_0 \left(r - \frac{a^2}{r} \right) \sin \theta - \frac{\Gamma}{2\pi} \ln r. \tag{15–29}$$

The streamlines are shown in Fig. 15–14. The velocity on the surface of the cylinder is given by

$$v_\theta = \left. \frac{\partial \psi}{\partial r} \right|_{r=a} = v_a = -2V_0 \sin \theta - \frac{\Gamma}{2\pi a}.$$

With the aid of the Bernoulli equation, as used in the development of Eq. (15–12), the pressure distribution around the cylinder is determined as

$$p_a = \rho \frac{V_0^2}{2} \left[1 - \left(2 \sin \theta + \frac{\Gamma}{2\pi a V_0} \right)^2 \right]. \tag{15–30}$$

The total drag force is zero since there is no frictional drag with an ideal fluid, and the pressure drag is

$$D_p = - \int_0^{2\pi} p_a \cos \theta \, a \, d\theta \equiv 0.$$

However, the lift force is finite and equals

$$L = - \int_0^{2\pi} p_a \sin \theta \, a \, d\theta = \rho V_0 \Gamma. \tag{15–31}$$

The lift produced is directly proportional to the velocity of translation and to the magnitude of the circulation Γ. This simple result is known as the *Kutta-Joukowski theorem* and holds not solely for circles but *applies to cylinders of any shape*, including nonsymmetric bodies.

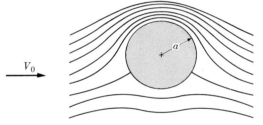

V_0

FIG. 15–14. Streamlines for uniform flow around a circular cylinder with circulation.

With real fluids, circulatory flow can be induced by rotating the cylinder. The resulting boundary layer will cause a rotary motion in the fluid which when superposed on a translating cylinder creates a lift proportional to the circulation and the translation velocity. This is called the Magnus effect, and its magnitude will depend on the Reynolds number as well as the translational and rotational velocities of the cylinder. With real fluids, the drag is not zero but is finite with both frictional- and pressure-drag components.

15–5.2 Two-dimensional airfoils

Let us consider a steady-flow velocity V_0 past a two-dimensional airfoil at an angle of attack which will produce lift efficiently. If the motion is begun from rest, it will be without circulation at the start. The initial flow pattern is similar to that of the zero-circulation potential flow shown in Fig. 15–15(a). Without circulation, the forward and rear points of zero velocity (stagnation points) occur at A and B. However, this requires an infinite velocity at the sharp trailing

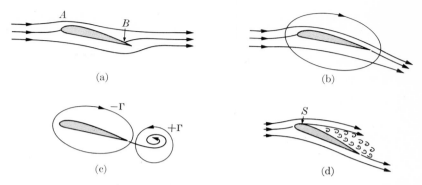

FIG. 15–15. Development of circulation about an airfoil: (a) potential flow; (b) real fluid flow; (c) real fluid flow showing starting vortex; (d) airfoil in stalled condition.

edge where the lower surface fluid would be forced to reverse direction and flow toward B. Instead, the real-fluid stream separates at the trailing edge causing a counterclockwise vortex to develop. The potential-flow stagnation point B is not realized, and the streamlines hug the upper surface more closely. The vortex grows during the acceleration to V_0, detaches, and sweeps aft. The steady-state streamline field, beyond a thin boundary layer, will be as shown in Fig. 15–15(b). The vortex, called a *starting vortex*, will have a circulation Γ dependent on the angle of attack and V_0. At the instant of detachment, the circulation about a contour enclosing the airfoil, but not the starting vortex, will be $-\Gamma$ as indicated in Fig. 15–15(c). The net circulation is zero, unchanged from the at-rest condition. Although this flow configuration is a real-fluid effect, it is described approximately by a mathematical model of superposed potential flows. The circulation $-\Gamma$ at the airfoil is represented by a bound-potential vortex as already described for the circular cylinder. Also the lift is proportional to the circulation and translation velocity as given by the Kutta-Joukowski theorem (Eq. 15–31) for pure potential motions. As for circular cylinders, the ideal-fluid potential-flow drag is zero, but the real-fluid drag is finite.

In the section on two-dimensional symmetric bodies, it was indicated that symmetric struts and surrounding streamlines could be mapped from flow around a circular cylinder using conformal transformation procedures. Similarly, asymmetric airfoils can be mapped from circles and corresponding streamlines from flow about circles. The real-fluid velocity field will be modeled approximately if the flow past the circle includes a circulation Γ of strength (magnitude) adjusted to make the velocity finite and tangential at the trailing edge. The potential-flow lift for the ideal fluid will be given by Eq. (15–31). The details of these methods [17] are largely of historical interest since they are limited to the incompressible subsonic regime and to airfoils of infinite span. However, the relation between the behavior of airfoils in real fluids in comparison with flows constructed by superposition of potential motions is of fundamental importance in understanding the development of lift.

If an airfoil is loaded too heavily, for example by increasing the angle of attack, the boundary layer on the upper surface will separate causing a large turbulent wake with loss of lift and increased drag.   This stalled condition, shown in Fig. 15–15(d), cannot be modeled easily with potential flows because location of the separation point S depends on the boundary layer.

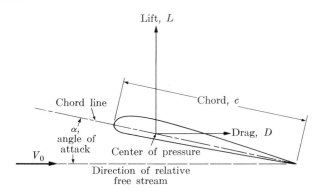

FIG. 15–16.   Airfoil geometry.

The force characteristics of an airfoil usually are determined by wind-tunnel tests.   The geometric definitions for an airfoil are given in Fig. 15–16.   The angle of attack is the angle between the chord line and the direction of the relative free stream.   Experimental data for a two-dimensional subsonic airfoil section are given in Fig. 15–17 [18] in which C_{D_c}, C_L, the ratio of lift to drag and the location of the center of pressure, are given as a function of the angle of attack.   The optimum lift-to-drag ratio for this airfoil occurs at an angle of attack of about $1\frac{1}{2}°$, and the lifting force increases linearly with angle of attack up to about 12°. Beyond this point, the drag increases sharply and the lift falls, indicating a stalled condition due to a separation of the flow from the upper surface of the airfoil.   At low angles of attack, the force coefficients are relatively insensitive to changes in the Reynolds number, $V_0 c/\nu$.   However, the angle of stall, the maximum lift, and the minimum-drag coefficient are functions of the Reynolds number.   The order of magnitude of this effect is given by [19]

$$\frac{(C_{L\text{max}})_1}{(C_{L\text{max}})_2} \approx \frac{(C_{D_c\text{min}})_2}{(C_{D_c\text{min}})_1} \approx \left[\frac{\mathbf{R}_1}{\mathbf{R}_2}\right]^{0.12}. \tag{15–32}$$

15–5.3   Three-dimensional effects

While the two-dimensional characteristics are useful in inter-comparisons of airfoil or hydrofoil performances, the three-dimensional nature of most lifting surfaces must be considered.   The aspect ratio (AR) of a wing is the ratio of the

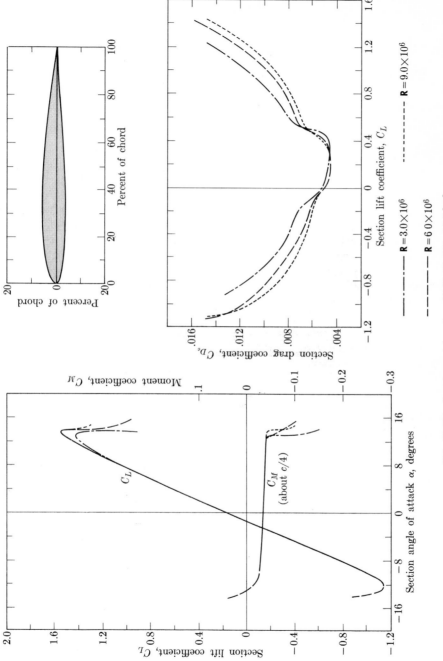

FIG. 15–17. Experimental data for a NACA 63–210 airfoil [18].

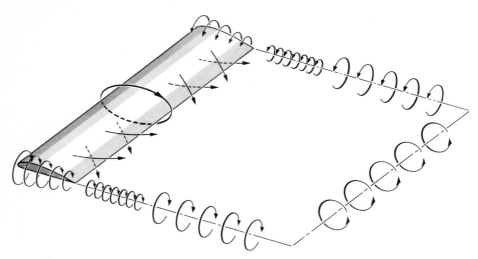

FIG. 15–18.   Vortex system for airfoil of finite span.

span squared to the area projected in the plane of the chord. For a rectangular wing,

$$\text{AR} = \frac{b^2}{S} = \frac{b^2}{cb} = \frac{b}{c}, \tag{15–33}$$

where b = span of wing, S = wing area, c = chord. Thus an airfoil of finite span is characterized by a given aspect ratio; two-dimensional sections have an infinite aspect ratio.

Basically, the lifting force is produced because the average pressure on the top surface of an airfoil is less than the average pressure on the underside. On a wing of finite span, this pressure differential must disappear at wing tips, so that lateral pressure gradients exist with opposite signs on top and bottom. The result is a tendency for a lateral flow on both surfaces, such that the underside fluid spills over to the topside at the tips. This cross flow gives rise to wing-tip trailing vortices as shown in Fig. 15–18. Actually, the cross flow causes a sheet of trailing vortices along the trailing edge of the airfoil, but the effect is most pronounced at the tip. A simple model of the finite-span airfoil is a vortex system in which trailing vortices join a bound vortex at the foil and a starting vortex far downstream to form a loop of constant circulation.

The spilling over at the tips has the effect of reducing lift from the maximum at midspan to zero at the tips. Lift production corresponds to a change in the vertical momentum of the fluid passing the foil. This can be seen qualitatively in Fig. 15–15(b). For a finite-span foil, the fluid is deflected less and the momentum change and lift are less than for a pure two-dimensional case. The lower lift corresponds to a lower effective angle of attack and lower circulation. The result

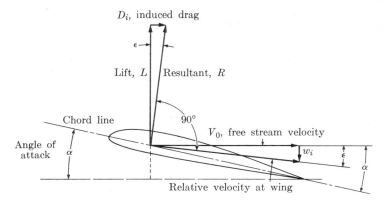

FIG. 15–19.   Resultant force and induced drag on a finite-span airfoil.

can be described in terms of the flow field *induced* by the vortex loop model. In a real fluid, the starting vortex and the downstream portions of the trailing vortices dissipate quickly and can be ignored. Therefore, only the bound and the trailing vortices near the foil need be considered. The bound vortex induces upflow before and downflow behind the foil. The trailing vortices induce a downflow, known as *downwash* w_i, both before and aft. This downwash combines with the bound vortex velocities to cause reduced angle of attack and tangential trailing-edge flow. The local flow relative to the foil is inclined downward at an angle ϵ, as indicated in Fig. 15–19. The magnitude of ϵ is given by

$$\tan \epsilon = w_i/V_0. \tag{15–34}$$

The drag of the finite span is greater than the infinite aspect-ratio foil. Since tip vortices are generated continuously, it is apparent that extra energy is being dissipated. In the ideal-fluid model, this extra energy is stored in the trailing vortices so that energy must be put into the system even though the flow remains irrotational. In the potential-flow model, the resultant force R is inclined aft with respect to the normal to the free-stream velocity V_0 (Fig. 15–19). By definition, the lift L is normal to V_0. The other component of R is a resisting force and is called the *induced drag* D_i.

The term "induced drag" is appropriate because of the similarity of the relationship between vortex filaments and vortex velocities to the relationship between electrical conductors and their magnetic field. From Fig. 15–19 and Eq. (15–34), we have

$$\tan \epsilon = D_i/L = w_i/V_0, \tag{15–35}$$

and

$$D_i = Lw_i/V_0. \tag{15–36}$$

The downwash and the induced drag depends on the spanwise distribution of lift, and hence on the planform of the airfoil. Both downwash and induced drag can be computed for any planform using potential vortex theory. As shown by Prandtl [20], for a planform giving a spanwise elliptical distribution of lift, we have

$$w_i = \frac{2L}{\pi b^2 \rho V_0}.$$

(15–37)

Hence

$$D_i = \frac{L^2}{\pi b^2 \rho V_0^2/2}$$

$$= \frac{C_L^2 S^2 (\rho V_0^2/2)}{\pi b^2},$$

(15–38)

and the induced-drag coefficient is given by

$$(C_{D_c})_i = \frac{C_L^2 S}{\pi b^2}$$

$$= \frac{C_L^2}{\pi(\mathrm{AR})}.$$

(15–39)

This result holds approximately for a rectangular planform.

The above equation shows that for a given wing, the induced drag increases as the aspect ratio becomes smaller. For the infinite aspect-ratio airfoil, the induced drag is zero. Experiments indicate slightly greater values for the induced drag than are predicted by Eq. (15–39) for a wing of essentially rectangular planform. It will be seen later that high-speed considerations lead to the choice of wing planforms other than rectangular.

For a real fluid, the total drag includes surface resistance. The total can be represented closely as the sum of the drag on a two-dimensional airfoil and the induced drag. Then we can write

$$C_D = C_{D\text{profile}} + C_L^2/\pi(\mathrm{AR}),$$

(15–40)

where $C_{D\text{profile}}$ refers to the *profile drag*, that is the total drag on the equivalent two-dimensional airfoil operating at the same lift coefficient C_L as the finite span.

While the foregoing discussion has been largely concerned with the characteristics of objects in subsonic air or gas flow, the principles apply equally well to objects such as hydrofoils and struts immersed in a liquid, the only requirement being that the liquid is of sufficiently large extent that free-surface effects are negligible. There are, however, several important differences between cases of full immersion in liquids and cases where both liquid and gas (or vapor) are present. An important example is cavitation, which is discussed in the next section.

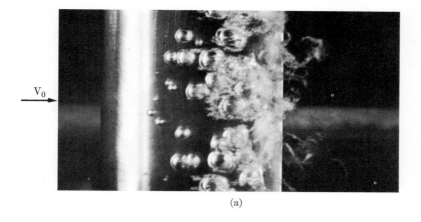

(a)

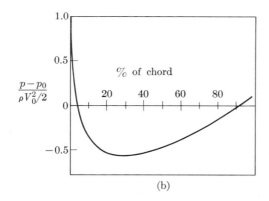

(b)

FIG. 15–20. Traveling cavities on the upper surface of a hydrofoil [21]: (a) photograph at $V_0 = 45$ ft/sec, $\sigma = 0.42$; (b) pressure distribution at inception, $\sigma_i = 0.70$.

15–6　EFFECT OF CAVITATION ON RESISTANCE

Cavitation types. The dynamic reduction of pressure which occurs when liquids flow through constricted or curved passages or around immersed bodies may lead to *cavitation*. Cavitation is a rapid, almost "explosive," change of phase from liquid to vapor which occurs whenever the absolute pressure in a flowing liquid drops by hydrodynamic means to or below a critical value. Under usual circumstances, the critical pressure is the vapor pressure* or slightly below.

Initially, cavitation occurs as minute vapor and gas-filled *traveling* cavities grow rapidly and then collapse as they are swept from the low into a higher pressure region. Figure 15–20(a) [21] is a photograph taken with a 20-μsec exposure which shows cavitation occurring on the low-pressure surface of a hydrofoil. In this example cavities appear, grow to a diameter as large as 0.5 in. and collapse, all in approximately 0.003 sec.

---

* See Section 1–3.1 and Table 1–2 for definition and magnitudes of vapor pressure.

FIG. 15–21.   Cavitation inception in the separation shear layer of a sphere: $V_0 = 40$ ft/sec; $\sigma = 1.75$. (Photograph courtesy California Institute of Technology.)

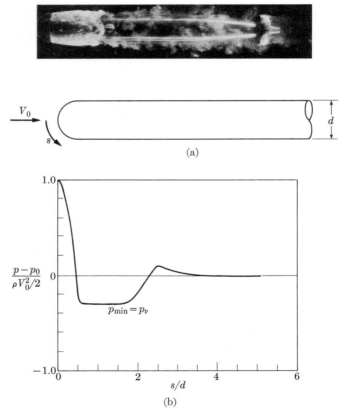

FIG. 15–22.   Body of revolution with an attached cavity: $V_0 = 40$ ft/sec; $\sigma \approx 0.30$: (a) attached cavity; (b) surface-pressure distribution with cavity present. (Photograph courtesy California Institute of Technology.)

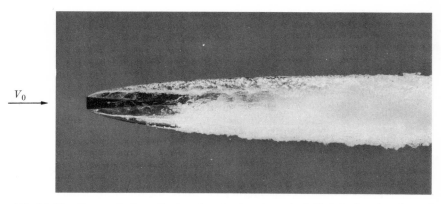

V_0

FIG. 15–23.   Supercavitation.   The two-dimensional cavity wake behind the flat-plate face of a wedge: $V_0 = 30$ ft/sec, $\sigma = 0.5$. (Photograph courtesy California Institute of Technology.)

The first appearance of cavitation is at the location of the lowest pressure in the fluid field. This tends to be at the boundary surface for unseparated flows. Figure 15–20(b) gives the pressure distribution on the hydrofoil at the inception of cavitation. In flows with hydrodynamic separation (boundary-layer separation without change of phase), cavitation will appear first in the fluid space away from the boundary. Figure 15–21 shows inception in the low-pressure cores of turbulent eddies in the shear zone formed by flow separation from the surface of a sphere.

With increased velocity, or a general decrease in the ambient pressure, a large cavitated region will form which appears to be *attached* to the solid boundary. The main liquid springs free of the boundary and flows around the cavity. This is especially pronounced with bluff bodies. Under these circumstances, the pressure distribution is altered and the drag affected. Figure 15–22 shows an attached cavity and the resulting pressure distribution. The limiting minimum pressure is the vapor pressure. Consequently, with attached cavities, the lower limit of the boundary pressure in the cavitated region is independent of the flow dynamics. The maximum pressure on the uncavitated surfaces remains the sum of the velocity pressure and the ambient static pressure. Therefore, the drag of the bodies with cavitation follows a different law than that of bodies without cavitation.

The advanced stages of cavitation tend to form trailing wakes that are void of liquid. This is known as *supercavitation*. An example is shown in Fig. 15–23 by the two-dimensional cavity wake behind a flat plate normal to the flow. The photograph is a view through a window looking at the end of the plate. (The photographs in Figures 15–21, 15–22, and 15–23 were obtained in the High Speed Water Tunnel of the California Institute of Technology, operated under the sponsorship of the Office of Naval Research.)

Cavitation inception. The relation between the minimum absolute pressure p_{min} on an immersed body and a free-stream reference absolute pressure p_0 can be expressed in terms of a dimensionless pressure coefficient

$$C_{p_{min}} = \frac{(p_{min} + \gamma h_{min}) - (p_0 + \gamma h_0)}{\rho V_0^2/2}, \tag{15–41}$$

where

$$h_{min}, h_0 = \text{elevations where the pressures } p_{min} \text{ and } p_0$$
$$\text{are measured,}$$
$$V_0 = \text{reference velocity where the pressure is } p_0.$$

For a well-submerged noncavitating body, $C_{p_{min}}$ depends on body geometry and Reynolds number and is independent of the gravity field.* If the body is near the free surface, $C_{p_{min}}$ may also depend on Froude number. If cavitation occurs when $p_{min} = p_v$, we can define a critical *cavitation number* as

$$\sigma_i = \frac{p_0 - p_v}{\rho V_0^2/2}, \tag{15–42}$$

so that

$$C_{p_{min}} = -\sigma_i + \frac{\gamma(h_{min} - h_0)}{\rho V_0^2/2}. \tag{15–43}$$

For symmetrical bodies, p_0 and h_0 are referred to free-stream values along the line of the body axis and the last term is not zero. However, it is often negligibly small so that the critical cavitation number may be predicted from the pressure coefficient.

Hoerner[16] analyzed measured values of the critical cavitation number as a function of the thickness ratio for several deeply submerged symmetrical strut sections. He found the critical cavitation number to be given closely by $\sigma_i = 2.1(t/2x)$ where $t/2x$ is an effective thickness ratio defined by the terms in Fig. 15–24. For under-water bodies

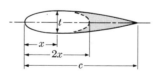

Effective thickness ratio $= \dfrac{t}{2x}$

FIG. 15–24. Critical cavitation number versus thickness ratio for symmetrical struts.

---

* As seen in Eq. (7–20), we can express the pressure as $p = p_d + p_s$, with p_s given by $(p_s + \gamma h) = \text{const}$ throughout the fluid. Then

$$C_{p_{min}} = \frac{p_{d_{min}} - p_{d_0}}{\rho V_0^2/2},$$

which is the form a solution to Eq. (7–22) can take and is therefore independent of gravity.

such as hydrofoils and projectiles, the absolute ambient pressure p_0 depends on the submergence z_0 and the atmosphere pressure p_a. Hence

$$\sigma_i = \frac{p_a + \gamma z_0 - p_v}{\rho V_0^2 / 2}.$$ (15–44)

With σ_i known, it is possible to determine the relation between submergence and velocity. Note that cavitation will occur at different velocities on the top and bottom of symmetrical bodies, since z_0 differs. When $\gamma(z_{0\text{top}} - z_{0\text{bottom}})$ is small compared to $(p_a + \gamma z_0)$, this gravity effect can be neglected.

Drag with supercavitation. A general cavitation number may be defined as

$$\sigma = \frac{p_0 - p_v}{\rho V_0^2 / 2},$$ (15–45)

where σ is evaluated from flow conditions without reference to the body. Then a body having a particular σ_i will show cavitation if $\sigma < \sigma_i$, and be free of cavitation if $\sigma > \sigma_i$.

The occurrence of cavitation on a body affects its drag, and when supercavitation is established, the drag becomes primarily form drag. A simple approximate relation can be derived for the drag coefficient on bodies of revolution as follows.

We let total drag equal form drag and, using the notation of Fig. 8–7, we can write

$$D = D_p = - \int_S p \cos \Phi \, dS = \int_A p \, dA,$$ (15–46)

where $dA = - \cos \Phi \, dS$, which is the surface element projected onto a plane normal to V_0. We then designate the projected separation area as A_s defined by

$$A_s = \int_0^s dA = - \int_s^{\text{TE}} dA,$$

where

$$s = \text{point of liquid separation from body surface,}$$
$$\text{TE} = \text{trailing edge of body.}$$

Now, assuming that the pressure in the attached cavity is constant and equal to the vapor pressure, and introducing the constant free-stream pressure p_0, we obtain

$$D = \int_{A_s} (p - p_0) \, dA - \int_{A_s} (p_v - p_0) \, dA$$

$$= \int_{A_s} (p - p_0) \, dA + (p_0 - p_v) A_s,$$ (15–47)

where $p =$ pressure on wetted surface of the body,

$\quad\quad\quad\quad\quad\quad p_v =$ vapor pressure on the body surface
$\quad\quad\quad\quad\quad\quad\quad\quad$ exposed to the cavity.

The drag coefficient is

$$C_D = \frac{D}{\rho(V_0^2/2)A} = \frac{1}{A}\int_{A_s} C_p \, dA + \sigma\,\frac{A_s}{A}, \tag{15–48}$$

where

$\quad\quad\quad\quad A =$ area of the body projected onto a plane normal to V_0.

The first term in this approximation is the integral of the dynamic effects and is practically a constant equal to the value of C_D when $\sigma = 0$. The second term includes the effect of the limiting p_v. Thus, with supercavitation, C_D is a function of the cavitation number. The coefficient is lowest when $\sigma = 0$. At constant V_0, the effect is to reduce the total resistance as σ is reduced.

For two-dimensional struts, theoretical and experimental results lead to a similar linear expression, namely

$$C_D(\sigma) = C_D(0)(1 + \sigma). \tag{15–49}$$

The pressure distribution on a bluff body having a large separated wake in noncavitating flow will sometimes approximate the distribution for the body with supercavitation. For example, consider the disk normal to the free stream. In this case, the frontal area A and the projected separation area A_s are equal. The noncavitating average pressures on front and back surfaces ($\bar{p}_f$ and $\bar{p}_b$, respectively) are given by

$$\frac{\bar{p}_f - p_0}{\rho V_0^2/2} = 0.76, \tag{15–50}$$

$$\frac{\bar{p}_b - p_0}{\rho V_0^2/2} = -\,0.42. \tag{15–51}$$

The drag coefficient is

$$C_D = \frac{\bar{p}_f - \bar{p}_b}{\rho V_0^2/2}$$

$$= 1.18 \quad \text{(noncavitating)}, \tag{15–52}$$

which agrees with the high Reynolds-number values in Fig. 15–13. Now let us assume that for supercavitating flow the pressure on the back is the vapor pressure and the pressure on the front is unchanged. When we replace $\bar{p}_b$ with $\bar{p}_v$, the cavitating drag coefficient becomes

$$C_D = 0.76 + \sigma. \tag{15–53}$$

The experimental results for disks give a value of $C_D \approx 0.78$ for $\sigma = 0$ (ambient pressure equal to cavity pressure). However, since cavitation first appears in the separation shear layers, $\sigma_i > 0.42$ and reaches values of 2.0 or more as the Reynolds number increases.

15–7 INTERFACIAL EFFECTS ON RESISTANCE

In Sections 15–4 and 15–5 we discussed the fluid resistance of submerged objects in incompressible fluids of large or infinite extent. Under these conditions, the dimensionless force coefficients are functions of the body geometry and Reynolds number. In this section we continue the treatment of incompressible fluids and consider the forces on an object moving at or near an interface between two fluids of different densities. For such cases, energy is expended on the generation of interfacial wave motion. Thus gravity influences the hydrodynamic flow field and the lift and drag on the body. As discussed in Section 15–1, the Froude number must be considered as an additional parameter when gravity affects the flow field.

The total drag on a body is always the sum of a frictional component D_f and a pressure component D_p. The proximity to an interface affects the ratio D_f/D_p as well as the magnitude of the drag. So far as the body is concerned, it experiences a changed distribution of wall shear stress and of pressure. The latter is the primary effect, however, so that wave resistance is essentially an added pressure or form resistance.

Figure 15–25 shows theoretical values of the drag, D_w, due to wave generation for slender ellipsoidal bodies moving beneath an air-water interface. The bodies have different diameter-to-length ratios d/l and different relative submergences z_0/l. The theoretical values were computed from potential theory assuming a nonviscous fluid [22]. The wave drag is essentially equal to the total drag less the

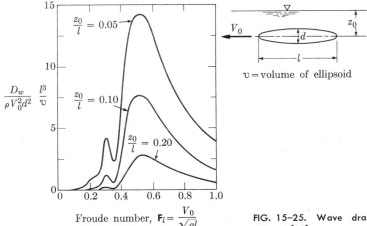

$$\frac{D_w}{\rho V_0^2 d^2} \frac{l^3}{\upsilon}$$

Froude number, $\mathbf{F}_l = \dfrac{V_0}{\sqrt{gl}}$

$\upsilon = $ volume of ellipsoid

FIG. 15–25. Wave drag of a submerged ellipsoid [22].

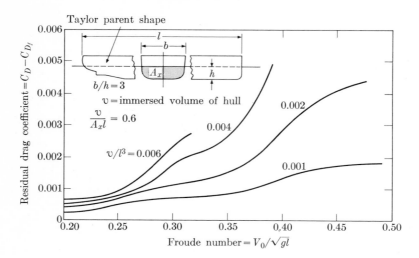

FIG. 15-26. Residual-drag (total drag minus skin friction) coefficient for a ship hull as a function of the Froude number (adapted from Hoerner [16]). (The drag coefficient is based on the wetted-hull surface area, S.) [25]

frictional drag in the absence of waves (i.e., when deeply submerged) [23]. The wave drag is a maximum at the Froude number (based on length l) $\mathbf{F}_l = 0.5$, and the wave drag becomes unimportant if the submergence z_0/l exceeds 0.5. From these results, a reasonable estimate of the depth–wave-drag relation for a submarine may be obtained.

The separation of the total hydrodynamic drag into skin-friction and wave-making components is an important problem in predicting the drag of full-scale ships and submarines. In model tests in a towing tank, equality of Froude and Reynolds numbers cannot be obtained since essentially the same liquid is used in model and prototype (see Section 7–4). The common procedure is to simulate the dynamic wave pattern correctly by testing the model at the prototype Froude number. Then the Reynolds number of the model test may be several hundred times less than that for the prototype so that the skin friction is not properly modeled. However, the frictional drag $(D_f)_M$ of the model hull can be calculated from boundary-layer theory (using the wetted-surface area of the hull as the equivalent "plate").* When subtracted from the measured-model total drag $(D_{\text{tot}})_M$, the *residual* drag is taken to be the wave drag of the model $(D_w)_M$. Then $(D_w)_M$ is scaled according to the Froudian force ratio to obtain the prototype residual drag. Thus, for the model, we have

$$(D_{\text{tot}})_M - (D_f)_M = (D_{\text{resid}})_M \approx (D_w)_M \qquad (15\text{-}54)$$

---

* The Reynolds-number range of model tests is such that a transition from a laminar to a turbulent boundary layer normally would occur somewhere along the hull. This would make the model skin-friction drag calculation highly indeterminate; therefore, a turbulence stimulator (for example, a strip of sandpaper) is used routinely on the bow of a model ship.

FIG. 15–27. Bow waves generated by moving ship. (Official photograph, U. S. Navy.)

and for the prototype,

$$(D_w)_P \approx \frac{(D_w)_M}{F_r},\tag{15–55}$$

where

$$F_r = \gamma L_r^3 = \text{the Froudian force ratio (see Eq. 7–31).}$$

Finally, the prototype frictional drag is computed for the prototype Reynolds number by means of boundary-layer theory and added to the extrapolated residual drag. Thus we get

$$(D_{\text{tot}})_P \approx (D_f)_P + (D_w)_P.\tag{15–56}$$

The quantity $(D_f)_P$ must include the effects of prototype roughness, and additions for the aerodynamic resistance of the superstructure must also be made [24].

A plot of the residual drag coefficients from model tests of a family of ship hulls is shown in Fig. 15–26 [25]. As in the case of the submerged ellipsoids, all curves tend toward a maximum residual drag near a Froude number of 0.5. Figure 15–27 shows the bow waves produced by a ship underway in a relatively calm sea.

15–8 COMPRESSIBILITY EFFECTS ON DRAG

Fluid compressibility influences the resistance of immersed bodies through the propagation of energy away from the body by elastic waves. The Mach number **M** is the significant parameter for elastic or acoustic wave motion, as is shown in Sections 7–5, 13–7, and 14–3. The effects of compressibility on drag in liquids can be dismissed by observing that the acoustic velocity in water is of the order of 4700 ft/sec, which is much in excess of practical speeds. Hence only very low Mach-number flows are possible in water under steady conditions.

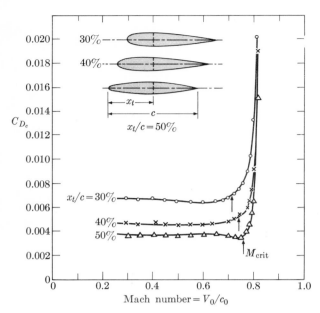

FIG. 15–28. Effect of compressibility on drag of symmetrical airfoils of infinite aspect ratio (adapted from Hoerner [16]).

The effect of compressibility on two-dimensional symmetric airfoils designed for subsonic flow is demonstrated by Fig. 15–28 [16]. This gives drag coefficients at zero lift versus the Mach number of the free-stream flow. The data are for a family of symmetrical airfoils with a constant thickness ratio but with the maximum thickness at different positions along the chord. Each foil has an essentially constant drag coefficient for $M < M_{crit} = 0.75$. Above the critical Mach number, the drag increases sharply (known as drag divergence), indicating the formation of elastic waves at some point aft of the leading edge where the local Mach number is equal to or greater than unity. At 30,000 ft in standard atmosphere (Table 1–7), the free-stream velocity corresponding to a Mach number of 0.75 is 750 ft/sec or 500 mph. This speed has been considerably exceeded by subsonic jet aircraft by the use of swept wings. The basic advantage of the swept wing in avoiding the region of drag divergence (that is, $M > M_{crit}$) is shown in Fig. 15–29.

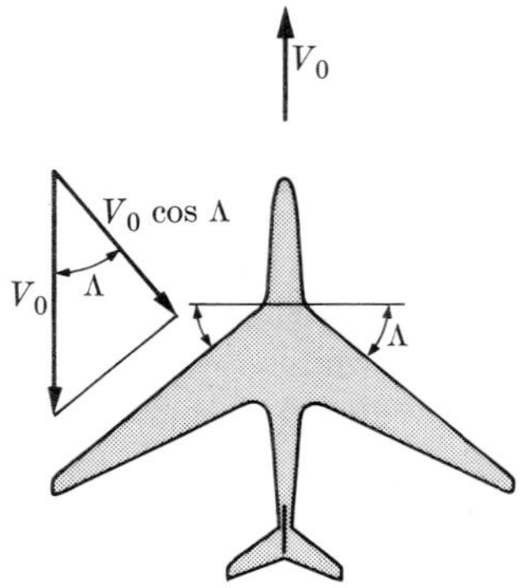

FIG. 15–29. Reduction of effective free-stream velocity for a swept wing.

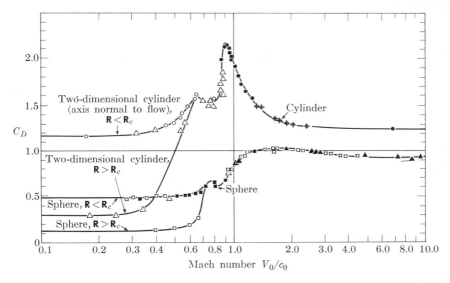

FIG. 15–30. Drag coefficients for cylinders and spheres as functions of Mach numbers and critical Reynolds numbers (adapted from Hoerner [16]).

The vector diagram for the wing, which is swept back at angle Λ, shows the relative free-stream velocity in terms of components normal and parallel to the wing. Since the component parallel to the leading edge of the wing does no useful work, the component $V_0 \cos \Lambda$ normal to the leading edge is the effective free-stream velocity which determines the aerodynamic characteristics of the wing. Thus the Mach number for the swept wing is

$$\mathbf{M}_s = \frac{V_0 \cos \Lambda}{c}, \qquad (15\text{–}57)$$

where c is the local speed of sound.

For the above example of flight at 30,000 ft, a sweptback angle of 37° will permit the forward-speed Mach number, $\mathbf{M} = V_0/c$, to equal 0.92 when the swept-wing Mach number $\mathbf{M}_s = 0.75$. A discussion of finite aspect-ratio effects and other aspects of the high subsonic range may be found in more advanced texts [26].

The resistance of bodies in the transonic, supersonic, and hypersonic ranges represents a broad area of technology which cannot readily be treated within the time limitations of an introductory course. Therefore, only a few experimental results on basic shapes and some references to more advanced sources of information will be given. The variation in the drag coefficient of spheres and cylinders as a function of the free-stream Mach number in the range 0.1 to 10 is shown in Fig. 15–30. This shows the compressibility effect for Reynolds numbers both above and below that necessary for boundary-layer transition. For Mach

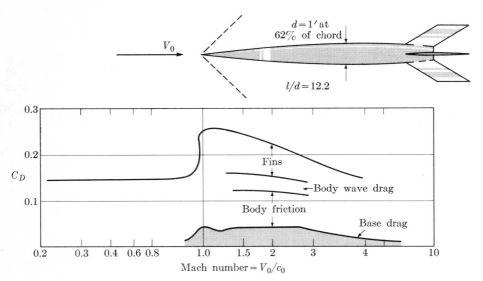

FIG. 15–31.   Drag characteristics of a NACA RM-10 experimental missile (adapted from Hoerner [16]).

numbers greater than 0.7, the viscous effect becomes small and the curves merge. By contrast, the drag characteristic of a slender missile, consisting of a pointed body of revolution with stabilizing fins, is shown in Fig. 15–31 [16]. This body has a very high critical Mach number ($M_c \approx 0.95$), and, at $M = 3$, its drag force is approximately one-fifth that of a sphere of equal maximum diameter. Conventional streamlining, in the subsonic sense of well-rounded leading edges, leads to extremely high drag forces in supersonic flow in comparison with thin pointed sections.

At transonic and supersonic speeds, the swept wing of the high subsonic range gives way to the "delta" wings of essentially triangular planform. These lifting surfaces have small aspect ratios and cannot be analyzed by two-dimensional theories.

The term "hypersonic" indicates flow at very high supersonic Mach numbers, usually within the continuum phase of gas dynamic theory. Heat-transfer effects are very important in this range. At the upper limits of the atmosphere of the earth, the assumption of a fluid continuum is no longer valid. This is the realm of rarefied gas dynamics and free molecule flows, and conventional definitions of viscosity and sonic speed are no longer physically significant [27].

PROBLEMS

15-1. Plot the velocity distribution v_θ along the radial line at $\theta = 90°$ (Fig. 15-1) for an irrotational flow past a circular cylinder of radius a.

15-2. Determine the angle θ (Fig. 15-1) at which a small piezometer opening should be placed on the surface of a cylinder in order to measure the free-stream pressure. Explain how a small cylindrical probe having one or more piezometer openings might be used as an instrument for determining the angular direction of the velocity vector in a flow field.

15-3. Plot several streamlines defined by Eq. (15-8) for values of $r < a$. Does this flow have any physical significance?

15-4. A steel sphere (SG = 7.0) 6 in. in diameter is accelerating horizontally at a constant rate of 2 ft/sec² through a large body of water which was initially at rest. Find the force acting on the sphere due to its acceleration through the water.

15-5. Water is accelerating at a constant rate of 2 ft/sec². A steel sphere (SG = 7.0) 6 in. in diameter is to be held stationary within the flow. Find the force acting on the sphere due to the acceleration of the water. Compare this force with that obtained in Problem 15-4 and comment on the difference, if any.

15-6. Compute the total drag per unit width of a thin flat plate 1 ft long moving with a steady velocity of 0.01 ft/sec through glycerin at 68°F.

15-7. The optical components of a submarine periscope are contained in a cylinder 12 in. in diameter. (a) Specify the chord length of a strut designed to enclose the cylinder and result in minimum drag. (b) When the velocity of the submarine is 50 fps, calculate the saving in power which results from the streamlining of the cylinder.

15-8. (a) Determine the water velocity at which a cable of 0.1 in. diameter will vibrate with a period of 0.02 sec. (b) Determine the air velocity at which the same cable will vibrate with the same period.

15-9. The behavior of the drag coefficient for cylinders at high Reynolds numbers between 2×10^5 and 5×10^5 has been offered as a possible explanation of the anomalous failures of masts and poles during high winds. Check this hypothesis by calculating the drag forces acting upon each of two cylindrical masts 50 ft high and 6 in. and 12 in. in diameter, assuming for both, uniform wind velocities of: (a) 30 mph; (b) 60 mph; and air with $\rho = 0.0023$ slug/ft³; $\nu = 1.6 \times 10^{-4}$ ft²/sec. Comment on the significance of these *four* results.

15-10. (a) When a cylindrical chimney 3 ft in diameter is exposed to a 36-mph wind, what is the bending moment at the bottom of the chimney? Neglect end effects. Express the answer in terms of chimney height. (b) Approximately where on the cylinder profile will flow separation occur for this velocity?

15-11. A torpedo is shaped like a cylinder 21 in. in diameter and 18 ft long. Assume that its normal speed is 70 ft/sec. When a 1 : 10 scale model is tested in a water tunnel at a velocity of 30 ft/sec, the total drag (surface resistance plus form drag) is 3.0 lb. (a) Calculate the approximate form drag for the model by estimating the skin-friction drag. (b) Assuming that the form-drag coefficient is independent of Reynolds number, calculate the form drag for the prototype at a velocity of 60 ft/sec. Assume completely turbulent boundary layers for both model and prototype.

15–12.   Two long cylinders of diameter 0.1 ft are towed through water (68°F) with their axes normal to the direction of motion. At a towing velocity of 8.7 fps the drag of one cylinder was 81 lb/ft of length, and the other was 44 lb/ft. Estimate the roughness of the two cylinders.

15–13.   In wind tunnels for testing aircraft, a turbulent air stream moves past a stationary model. The drag on the model depends upon the boundary-layer development, which in turn is affected by the degree of turbulence in the air stream. Explain briefly how increasing the turbulence (by the use of screens, for example) will affect the drag of (a) a relatively long, well-streamlined, smooth surface body, and (b) a bluff body such as a sphere (also smooth surface). Assume that only the turbulence is changed and that the mean velocity and all other conditions of flow remain unchanged.

15–14.   Two wind tunnels which are quite similar in design produce the same airstream velocity. In one air stream, however, the turbulence is appreciable while in the other it is negligible. The drag of a smooth sphere is measured in each tunnel with the results that in tunnel A, $C_D = 0.5$, while in tunnel B, $C_D = 0.2$. In both tests the Reynolds number for the sphere is 10^5. From these tests determine which stream is the more turbulent. Explain why the tests will indicate the relative degree of turbulence.

15–15.   Explain why the drag coefficient for a square plate (held normal to the direction of the fluid stream) is lower than that of a long plate (length/breadth $= 20$) of the same width also held normal to direction of the fluid stream. What is the drag-force ratio for the two plates for a Reynolds number greater than 10^3?

15–16.   An instrument is to be supported on the bed of a stream by a vertical strut which is elliptical in cross section with a $1:3$ ratio of minor to major axes. The major axis (aligned parallel to the flow) is 6 ft. The drag coefficient for this elliptical section depends on the Reynolds number as follows:

Reynolds number (characteristic length = minor axis)	10^4	10^5	10^6	10^7
C_D	0.5	0.12	0.12	0.14

Determine the drag force on the strut in 20 ft of 60°F water flowing at 6 fps.

15–17.   A rectangular billboard 10 ft high by 50 ft long is erected on an open supporting framework so as to be freely exposed to the wind. Estimate the total-pressure force due to a 45-mph wind, using (a) the pressure distribution curve of Fig. 15–10, (b) the drag coefficient from Table 15–2. Explain why the two answers are not the same.

15–18.   Determine the drag of a 5×5 ft plate with roughened edges when held at right angles to a stream of air (60°F) having a velocity of 30 fps. By what percentage will the drag be reduced when the plate is turned parallel to the stream?

15–19.   A sphere 1 ft in diameter with a density 1.5 times that of water is dropped into a large body of water (60°F). Determine the terminal velocity of the sphere.

15–20.   A fluid-mixing device consists of two thin circular disks 0.1 ft in diameter mounted on a vertical rod, as shown in Fig. 15–32. The mixer is rotated at a constant speed of 50 rpm in water at 60°F. (a) Estimate the torque in foot-pounds necessary to rotate this device. (b) If the mixer is in a tank 2 ft in diameter, would the torque be expected to increase, decrease, or remain constant with time? Explain.

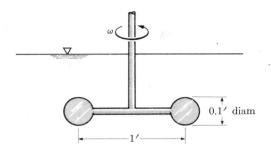

0.1′ diam

1′

FIGURE 15-32

15-21. When a rotating cylinder 6 in. in diameter is moving through water with a velocity of 20 ft/sec, it is subject to a lift force of 400 lb/ft of length. Determine the lift coefficient and the theoretical rotational speed of the cylinder in revolutions per minute.

15-22. An airfoil section having the profile and characteristics shown in Fig. 15-17 is to be used in wings having an equivalent span of 48 ft and an 8-ft chord. The wings are to support an airplane weighing 16,000 lb. (a) Determine the necessary angle of attack of the wing for level flight at 180 mph at an elevation of 15,000 ft. (b) Determine the power necessary to overcome wing drag under these conditions.

15-23. A rectangular wing has a 50-ft span and a 10-ft chord. (a) Using the two-dimensional air-foil characteristics given in Fig. 15-17, determine the profile drag force for $\alpha = 2°$ at a speed of 300 mph and an elevation of 10,000 ft. (b) Determine the magnitude of the induced drag.

15-24. In a laboratory wind tunnel a two-dimensional airfoil produces a lift coefficient of 0.5 at an angle of attack of 4°. How much greater must the angle of attack be for a wing of the same section having a span of 100 ft and a chord of 12.5 ft to produce the same lift coefficient?

15-25. Assume that the velocity distribution around the forward half of a sphere in a real fluid can be approximated by $v_\theta = \frac{3}{2}V_0 \sin \theta$. Assume that boundary-layer separation occurs at $\theta = 90°$. (a) Determine the magnitude of the drag coefficient if the pressure in the resulting wake is p_0, the static pressure in the undisturbed flow. Also $V_0 = 20$ ft/sec in water (68°F) and the sphere diameter is 1 ft. (b) Determine the pressure p_0 at which cavitation can be expected on the sphere for the same V_0.

15-26. A hemisphere 2 ft in diameter is to form the nose of a torpedo. If the torpedo is to travel at a 15-ft submergence, determine the maximum speed without cavitation occurring. Let the velocity and pressure over the hemisphere be predicted by the potential-flow solution for flow past a sphere where $\psi = U \sin^2 \theta [a^3/2r - r^2/2]$.

15-27. If the torpedo in the previous problem is accelerating, what effect will there be on the maximum cavitation-free velocity?

15-28. A 200-lb bomb with a hemispherical nose 1 ft in diameter is dropped into the sea from a high altitude. (a) At impact its velocity is 100 ft/sec and a supercavitation condition results with the cavity filled with air at atmospheric pressure. Assume that the cavity separation occurs at the maximum diameter ($\theta = 90°$) and that pressure distribution over the hemisphere is the same as that over a sphere in potential flow. Compute the drag on the bomb. (b) Compare this drag with the magnitude that would exist at the same velocity if the cavity were void of air and hence at the vapor pressure.

15–29. An oceanographic research instrument has the shape of an ellipsoid with a maximum diameter of 0.5 ft and a length of 4 ft. The instrument is to be towed in the direction of its major axis at speeds ranging from 2 to 8 fps at a depth of 0.4 ft (from the undisturbed water surface to the centerline of the ellipsoid). Wind-tunnel tests on an object of similar shape and Reynolds-number range indicate a drag coefficient (based on the maximum cross-sectional area) of 0.10. Plot the towing force as a function of towing speed in sea water at 68°F.

15–30. The shape of a submarine may be approximated by an ellipsoid having a length of 100 ft and a maximum diameter of 10 ft. Determine the optimum cruising speed when the submarine is operating near the surface if the maximum possible speed is 35 fps.

15–31. A ship having a Taylor standard hull shape (Fig. 15–26) has a length $l = 200$ ft, a beam $b = 24$ ft, a wetted hull surface area of 40,000 ft^2 and a section area $A_x = 180$ ft^2. (a) Determine the residual drag coefficient for a speed of 25 fps. (b) Estimate the total drag of the ship at the same speed.

15–32. A proposed ship having a length of 140 ft and a hull area equal to 4700 ft^2 below the water line is to be tested by means of a geometrically similar model in a towing tank. The length of the model is 5 ft. When the model is towed at a velocity of 4 knots, a total drag of 1.95 lb is measured. The towing tank contains fresh water at 60°F. (a) Calculate the total drag of the prototype ship when moving in sea water at 68°F. (b) To what prototype velocity does this correspond? Assume that the boundary layer on the model is completely turbulent.

15–33. Drag tests on an ellipsoid 0.1 ft in diameter have been made in a high-speed wind tunnel operating with air at atmospheric pressure and standard temperature. The following table summarizes the experimental results for the coefficient of drag at various Reynolds numbers for this ellipsoid:

C_D	0.50	0.50	0.18	0.20	0.80	1.00
R	6×10^4	2×10^5	3×10^5	5×10^5	6.6×10^5	1×10^6

(a) Determine the drag in pounds for a geometrically similar ellipsoid 0.4 ft in diameter traveling in standard air at a velocity of 200 ft/sec. (b) Make a rough plot of the data given above and explain the physical significance of the shape of various portions of the drag-coefficient curve. (c) Estimate the drag in pounds for a similar ellipsoid 0.4 ft in diameter traveling in water (standard temperature) with a velocity of 18 ft/sec. The depth of submergence is very large compared to the diameter of the ellipsoid.

15–34. (a) Determine the drag force on a spherical object 1 ft in diameter moving at a speed of 1000 fps through air at an altitude of 25,000 ft. (b) Estimate the percentage of the total drag force which is due to compressibility effects.

15–35. Determine the terminal velocity of a missile having the shape shown in Fig. 15–31 dropped vertically from an altitude of 50,000 ft under standard atmospheric conditions. The missile weighs 2800 lb and has a maximum diameter of 1 ft. Compare the terminal velocity with the free-fall velocity (at zero elevation), neglecting air resistance. Explain the difference in the two results.

REFERENCES

1. MILNE-THOMPSON, L. M., *Theoretical Hydrodynamics*, 4th ed., The Macmillan Company, New York, 1960, Art. 6.22.

2. STELSON, T. E., and F. T. MAVIS, "Virtual Mass and Acceleration in Fluids," *Trans. ASCE*, **122** (1957), pp. 518–525.

3. KUO, Y. H., "On the Flow of an Incompressible Viscous Fluid Past a Flat Plate at Moderate Reynolds Numbers," *J. Math. Phys.*, **32** (1953), pp. 83–101.

4. TOMOTIKA, S., and T. AOI, "The Steady Flow of a Viscous Fluid Past an Elliptic Cylinder and Flat Plate at Small Reynolds Numbers," *Quart. J. Mech. Appl. Math.*, **6** (1953), pp. 290–312.

5. MILNE-THOMPSON, L. M., *op. cit.*

6. FAGE, A., V. M. FALKNER, and W. S. WALKER, "Experiments on a Series of Symmetrical Joukowski Sections," *ARC Reports and Memoranda*, No. 1241 (1929).

7. HOMANN, F., "Einfluss grosser Zähigkeit bei Strömung um Zylinder und Kugel," *Forsch. Gebiete Ingenieurw.*, **7,** 1–10 (1936).

8. KÁRMÁN, Th.V., "Flüssigkeits- und Luftwiderstand," *Physik. Z.*, **13** 49, (1911). See also F. H. ABERNATHY, and R. E. KRONAUER, "The Formation of Vortex Streets," *J. Fluid Mech.*, **13**, Part 1 (May, 1962), pp. 1–20.

9. ROSHKO, A., "On the Development of Turbulent Wakes from Vortex Streets," *Nat. Advisory Comm. Aeron.*, Rep. 1191 (1954).

10. SCHLICTING, H., Boundary Layer Theory, 4th ed., McGraw-Hill Book Company, New York, 1960. Figure 21–20 gives data from A. FAGE and J. H. WARSAP, *ARC Reports and Memoranda*, No. 1283 (1930).

11. BROWN, F. N. M., University of Notre Dame, provided photographs.

12. DRYDEN, H. L., and A. M. KUETHE, "Effect of Turbulence in Wind-Tunnel Measurements," *Nat. Advisory Comm. Aeron.*, Rep. 342 (1929).

13. WATTENDORF, F., and A. M. KUETHE, *Physics* (1934), pp. 153–164.

14. MILLIKAN, C. B., and A. L. KLEIN, "The Effect of Turbulence," *Aircraft Eng.*, **169** (Aug., 1933).

15. EISNER, "Das Widerstandsproblem," *Intern. Congr. Appl. Mech. 3rd*, Stockholm (1930).

16. HOERNER, S. F., *Fluid Dynamic Drag*, 2nd ed. (published privately).

17. MILNE-THOMPSON, L. M., *Theoretical Aerodynamics*, The Macmillan Company, London, 1958.

18. ABBOTT, I. H., A. E. von DOENHOFF, and L. S. STIVERS, Jr., "Summary of Airfoil Data," *Nat. Advisory Comm. Aeron.*, Tech. Rept. 824, p. 162.

19. DOMMASCH, D. O., S. S. SHERBY, and T. F. CONNOLLY, *Airplane Aerodynamics*, 2nd ed., Pitman Publishing Corp., New York, 1958.

20. PRANDTL, L., *Essentials of Fluid Dynamics*, Hafner Publishing Co., Inc., New York, 1952.

21. DAILY, J. W., "Cavitation Characteristics and Infinite Aspect Ratio Characteristics of a Hydrofoil Section," *Trans. A.S.M.E.*, **71**, 269 (1949).

22. WIGLEY, W. C. S., "Water Forces on Submerged Bodies in Motion," *Trans. Inst. Naval Arch.*, 95 (1953), p. 268.

23. WEINBLUM, G., H. AMTSBERG, and W. BOCK, "Tests on Wave Resistance of Immersed Bodies of Revolution," *Mitteilungen der Preussischen Versuchsanstalt für Wasserbau u. Schiffbau*, Berlin, 1936. English Translation *DTMB Transl.* 234 (1950).

24. ROSSELL, H. E., and L. B. CHAPMAN, eds., *Principles of Naval Architecture*, Vol. II, Soc. Naval Arch. and Mar. Engrs., New York, 1958, p. 86.

25. GERTLER, M., "A Reanalysis of the Original Test Data for the Taylor Standard Series," *D. W. Taylor Model Basin*, Rept. No. 806 (March, 1954).

26. DONOVAN, A. F., and H. R. LAWRENCE, ed., *Aerodynamic Components of Aircraft at High Speeds*, Princeton University Press, Princeton, New Jersey, 1957. (Vol. VII, High Speed Aerodynamics and Jet Propulsion.)

27. MASLACH, G. S., and S. A. SCHAAF, "Cylinder Drag in the Transition from Continuum to Free-Molecule Flow," *Phys. Fluids*, **6**, 3 (March, 1963).

CHAPTER 16

Turbulent Jets and Diffusion Processes

16–1   FREE TURBULENCE

The turbulence phenomena discussed in Chapters 11 through 14 are primarily concerned with fluid motions which are constrained by one or more solid boundaries. Hence the term *wall turbulence* is used to describe turbulence generated in velocity gradients caused by the no-slip condition. The term *free turbulence*, on the other hand, describes turbulent motions which are not affected by the presence of solid boundaries. Some examples of free turbulent flows are shown in Fig. 16–1: (a) the spreading of the edge of a plane jet; (b) a round jet issuing from a slot into a surrounding fluid of the same phase (water into water or air into air); and (c) the flow in the wake of an immersed body. In all cases velocity gradients are generated. If the Reynolds numbers are sufficiently high, the flow is unstable and zones of turbulent mixing are developed.

The analysis of free turbulent motions is in general somewhat easier than that of wall turbulence. In free turbulence, the viscous (molecular) shear stresses usually can be neglected, in comparison with the turbulent eddy stresses, throughout the entire flow field, whereas in conduit flow, due to the damping of turbulence by the wall, the existence of the viscous stresses in the laminar sublayer must be considered. In addition, in jets and wakes in large bodies of fluid, it is usually found that the pressure gradient in the direction of motion is zero.

The free turbulent motions shown in Fig. 16–1 have an important property in common with the boundary-layer motions previously discussed. In all cases, the width b of the mixing zone is small compared to x, and the velocity gradient in the y-direction is large in comparison with the x-direction. These are precisely the assumptions made by Prandtl in simplifying the equations of motion for both laminar and turbulent boundary layers (see Sections 8–2 and 12–3). Consequently for steady two-dimensional free turbulent flows of homogeneous incompressible fluids, the equations of motion and continuity are the same as Prandtl's boundary-layer equations with zero-pressure gradient, namely

$$\bar{u}\,\frac{\partial \bar{u}}{\partial x} + \bar{v}\,\frac{\partial \bar{u}}{\partial y} = \frac{1}{\rho}\,\frac{\partial \tau}{\partial y}, \tag{16–1}$$

$$\frac{\partial \bar{u}}{\partial x} + \frac{\partial \bar{v}}{\partial y} = 0. \tag{16–2}$$

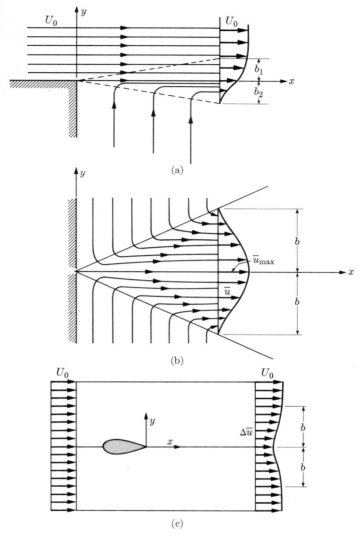

FIG. 16–1. Free turbulent flows: (a) spreading at the edge of a plane jet; (b) velocity distribution in a fully developed immersed jet; (c) velocity distribution in the wake of an immersed object.

The shear stress in Eq. (16–1) is the turbulent stress given by Eq. (11–10), neglecting the molecular-viscosity term. Hence

$$\tau = \eta\, \frac{\partial \bar{u}}{\partial y}.$$ (16–3)

The boundary conditions necessary for the solution of the equations for free turbulent flows differ from those appropriate to the boundary-layer problem.

The analysis of various flows shown in Fig. 16–1 are considered in the following sections for both plane and axisymmetric jets and wakes.

16–1.1   Dynamics of a plane jet

The two-dimensional spreading shown in Fig. 16–1(a) represents the initial conditions in the zone-of-flow establishment of a plane jet issuing from a long slot. If $2b_0$ is the height of the slot, then, as shown in Fig. 16–2, there will be a potential core of length L_0 in which the centerline velocity U_0 remains constant. The outer dashed lines represent the nominal boundaries of the jet, that is, the point where the horizontal velocity $\bar{u}$ is some arbitrarily small fraction of the velocity on the centerline of the jet. For values of x greater than L_0, the jet is said to be fully developed and the velocity profiles are as shown in Fig. 16–1(b), where $\bar{u}_{\max}$ is less than U_0.

The surrounding fluid is entrained and mixed with the fluid issuing from the slot. At large values of y, the ambient fluid approaches the jet in the vertical direction. Due to the entrainment, the volume rate of flow past any section in the jet increases in the x-direction. Certain basic properties of the momentum diffusion of the jet in the fully developed region can be deduced from the simplified equation of motion in the following manner:

The equation of motion (16–1) may be integrated with respect to y,

$$\rho \int_{-\infty}^{\infty} \bar{u}\, \frac{\partial \bar{u}}{\partial x}\, dy + \rho \int_{-\infty}^{\infty} \bar{v}\, \frac{\partial \bar{u}}{\partial y}\, dy = \int_{-\infty}^{\infty} \frac{\partial \tau}{\partial y}\, dy. \qquad (16\text{–}4)$$

The order of differentiating and integrating the first term may be interchanged by noting that $\bar{u}(\partial \bar{u}/\partial x) = (\partial/\partial x)(\bar{u}^2/2)$. In addition, the second term may be

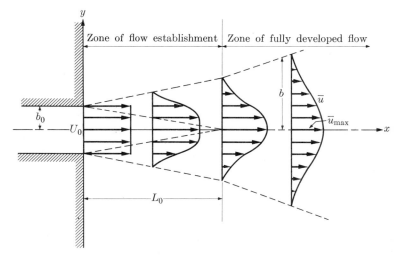

FIG. 16–2.   Development of similar velocity profiles in a plane jet.

integrated by parts,

$$\frac{\rho}{2}\frac{\partial}{\partial x}\int_{-\infty}^{\infty}(\bar{u})^2\,dy + \rho\bar{u}\bar{v}\Big|_{-\infty}^{\infty} - \rho\int_{-\infty}^{\infty}\bar{u}\frac{\partial\bar{v}}{\partial y}\,dy = \tau\Big|_{-\infty}^{\infty}. \tag{16-5}$$

If we use the continuity equation (16–2), the third term on the left-hand side becomes identical with the first term. As $y \to \pm\infty$, $\bar{u} = 0$ and $\partial\bar{u}/\partial y = 0$, and hence $\tau = 0$ from Eq. (16–3). Equation (16–5) may then be written as

$$\rho\frac{\partial}{\partial x}\int_{-\infty}^{\infty}(\bar{u})^2\,dy = 0 \quad\text{and}\quad \int_{-\infty}^{\infty}\rho(\bar{u})^2\,dy = \text{const.} \tag{16-6}$$

The physical significance of Eq. (16–6) becomes apparent when it is remembered that the product $\rho\bar{u}$ is momentum per unit volume, and that $\bar{u}\,dy$ is volume per unit time. Hence the integral of the product $(\rho\bar{u})(\bar{u}\,dy)$ is the total momentum per unit time passing any section of the jet. Equation (16–6) is equivalent to the statement that the flux of momentum of the jet is constant and independent of x. This is a consequence of the assumption of constant pressure since in that case, the net force on any control volume encompassing the jet is zero, and there can be no change in the longitudinal momentum flux.

The constant in Eq. (16–6) may be evaluated from the momentum influx at $x = 0$, and hence, since the momentum per unit volume is ρU_0 and the volume per unit time is $U_0 2b_0$, we have

$$\int_{-\infty}^{\infty}\rho(\bar{u})^2\,dy = 2\rho U_0^2 b_0. \tag{16-7}$$

In the fully developed region of the jet, both the width b and centerline velocity $\bar{u}_{max}$ may be expressed in terms of x to unknown exponents. Hence

$$b \sim x^m, \qquad \bar{u}_{max} \sim x^{-n}.$$

The order of magnitude of the various terms in the equation of motion (16–1) may be obtained in the following manner:

$$(1)\quad \bar{u}\frac{\partial\bar{u}}{\partial x} \sim \frac{(\bar{u})_{max}^2}{x} \sim \frac{x^{-2n}}{x} \sim x^{-2n-1}.$$

$$(2)\quad \bar{v} = -\int\frac{\partial\bar{u}}{\partial x}\,dy \quad\text{(from the continuity Eq. 16–2)}.$$

Hence

$$\bar{v} \sim \frac{\bar{u}_{max}b}{x} \quad\text{and}\quad (\bar{v})\left(\frac{\partial\bar{u}}{\partial y}\right) \sim \left(\frac{\bar{u}_{max}b}{x}\right)\left(\frac{\bar{u}_{max}}{b}\right) \sim x^{-2n-1}.$$

$$(3)\quad \frac{\tau}{\rho} \sim (\bar{u})_{max}^2 \quad\text{and}\quad \frac{\partial}{\partial y}\left(\frac{\tau}{\rho}\right) \sim \frac{(\bar{u})_{max}^2}{b} \sim x^{-2n-m}.$$

Since the left-hand side of Eq. (16–1) is of the order x^{-2n-1} and the right-hand side is of the order x^{-2n-m}, it follows that $2n + 1 = 2n + m$, and hence $m = 1$. When we follow the same reasoning, Eq. (16–6), which expresses the constant momentum flux, is of the order $U_{max}^2 b \sim x^{-2n+m}$. Since $m = 1$, Eq. (16–6) can be independent of x only if the exponent is zero, and hence $n = \frac{1}{2}$. A plane jet, as shown in Fig. 16–2, therefore expands as a linear function of distance from the origin, and the centerline velocity decreases as $1/\sqrt{x}$. The jet Reynolds number may be defined in terms of the local width and centerline velocity as $\mathbf{R}_j = \bar{u}_{max} b/\nu$. Since this is of the order of x^{-n+m} or $\sqrt{x}$, the Reynolds number increases with distance. In practical applications, the Reynolds number can increase only until the dimensions of the jet approach that of the surrounding fluid into which it is discharged.

While the above discussion has given some insight into the gross characteristics of jet diffusion, it has not answered the basic questions of velocity distribution, rate of entrainment of the surrounding fluid, or actual jet dimensions. Several semiempirical approaches based on the assumption of geometric similarity of velocity profiles in the fully developed region of the jet have been developed. The similarity condition implies that

$$\bar{u}/\bar{u}_{max} = f(y/x) = f(\xi), \tag{16–8}$$

where

$$\xi = y/x.$$

The following development for the plane jet illustrates one method of analysis which assumes that the similar velocity profiles are Gaussian curves of the form

$$\frac{\bar{u}}{\bar{u}_{max}} = f(\xi) = \exp\left(-\frac{y^2}{2C_1^2 x^2}\right), \tag{16–9}$$

where C_1 is a constant to be determined experimentally. From Eqs. (16–7) and (16–8), the condition of constant momentum flux becomes

$$2\rho U_0^2 b_0 = \rho(\bar{u})_{max}^2 \int_{-\infty}^{\infty} f^2(\xi)\, dy, \tag{16–10}$$

or

$$2U_0^2 b_0 = (\bar{u})_{max}^2 x I_2, \tag{16–11}$$

where

$$I_2 = \int_{-\infty}^{\infty} f^2(\xi)\, d\xi. \tag{16–11a}$$

Therefore the ratio of the centerline velocity to the initial jet velocity may be expressed in the following manner:

$$\bar{u}_{max}/U_0 = \sqrt{2b_0/x I_2}. \tag{16–12}$$

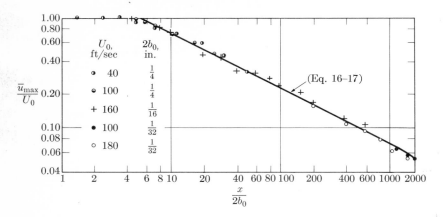

FIG. 16–3. Distribution of centerline velocity in flow from slot [1].

The length of the potential core, L_0, shown in Fig. 16–2, is found by noting that $\bar{u}_{max}/U_0 = 1$ when $x = L_0$, and therefore

$$L_0 = 2b_0/I_2. \tag{16–13}$$

The total discharge per unit width of the jet for $x > L_0$ is given by the integration of the local velocity across a section of the jet,

$$Q = \int_{-\infty}^{\infty} \bar{u}\, dy = \bar{u}_{max} \int_{-\infty}^{\infty} f(\xi)\, dy. \tag{16–14}$$

Since the initial discharge per unit width is

$$Q_0 = 2b_0 U_0, \tag{16–15}$$

the ratio of the total rate of flow to the initial discharge may also be expressed in terms of ξ,

$$\frac{Q}{Q_0} = \frac{\bar{u}_{max}}{U_0} \frac{x}{2b_0} \int_{-\infty}^{\infty} f(\xi)\, d\xi.$$

Making use of the velocity ratio given by Eq. (16–12) and letting

$$I_1 = \int_{-\infty}^{\infty} f(\xi)\, d\xi,$$

we have

$$Q/Q_0 = \sqrt{xI_1^2/2b_0 I_2}. \tag{16–16}$$

Experimental results by Albertson *et al.* [1] show that the exponential velocity distribution of Eq. (16–9) fits the measured velocity distributions in a turbulent jet wn $C_1 = 0.109$, whence $I_1 = 0.272$, $I_2 = 0.192$, and

$$\bar{u}_{\max}/U_0 = 2.28\sqrt{2b_0/x} \qquad (x > L_0), \tag{16–17}$$

$$L_0 = 10.4b_0, \tag{16–18}$$

$$Q/Q_0 = 0.62\sqrt{x/2b_0} \qquad (x > L_0). \tag{16–19}$$

The perimental confirmation of Eqs. (16–17) and (16–18) is shown in Fig. 16–3. The plane jet may be assumed to be turbulent if the Reynolds number baseon the initial velocity and slot height is greater than 30.

16–1. Axially symmetric jets

The case of a round jet issuing from a circular hole and mixing with the surrouding fluid is treated by considering that the motion is symmetrical about he longitudinal axis of the jet. Employing the same boundary-layer approximations. as in the plane jet, we find that the equations of motion in cylindrical coodinates (6–29) can be reduced to a single equation for the longitudinal o z-omponent,

$$\rho v_r \frac{\partial v_z}{\partial r} + \rho v_z \frac{\partial v_z}{\partial z} = \mu\left[\frac{1}{r}\frac{\partial}{\partial r}\left(r\frac{\partial v_z}{\partial r}\right)\right]. \tag{16–20}$$

Since $\tau = u(\partial v_z/\partial r)$, Eq. (16–20) can be written as

$$v_r \frac{\partial v_z}{\partial r} + v_z \frac{\partial v_z}{\partial z} = \frac{1}{\rho r}\frac{\partial(r\tau)}{\partial r}. \tag{16–21}$$

The continuity equation (6–30) becomes

$$\frac{1}{r}\frac{\partial}{\partial r}(rv_r) + \frac{\partial}{\partial z}(v_z) = 0. \tag{16–22}$$

If the shear stress τ is interpreted as that due to the turbulent motion of the fluid, it is proper to accept Eq. (16–21) for the turbulent round jet. Thus Eqs. (16–21) and (16–22) are the axially symmetric counterparts of Eqs. (16–1) and (16–2) for the plane jet. Assuming, as in the plane-jet analysis, that the local jet diameter $d \sim z^m$ and $(\bar{v}_z)_{\max} \sim z^{-n}$, we may again evaluate the order of magnitude of the terms in the equation of motion. The result is the same as in the plane jet, and hence $m = 1$.

The equivalent expression for constant-momentum flux in the round jet is

$$2\pi\rho\int_0^\infty (\bar{v})_z^2 r\, dr = \text{const} = \rho V_0^2(\pi\, d_0^2/4), \tag{16–23}$$

where V_0 and d_0 are the initial velocity and diameter of the jet. The order of magnitude of the momentum-flux expression is $(\bar{v}_z)_{\max}^2\, d^2 \sim z^{-2n+2m}$, and,

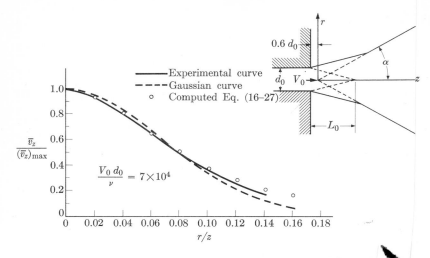

FIG. 16–4. Radial velocity distribution in a round jet [3].

since the flux must be independent of z, we have $-2n + 2n = 0$. Since $m = 1$, it follows that $n = 1$. The jet diameter increases linearly with z, and the centerline velocity decreases inversely with z. Hence the round jet has the interesting property that the Reynolds number is a constant throughout the region of flow. In terms of the concept of an eddy viscosity applicable to the turbulent motion, we have, since η is proportional to a mixing length times a velocity,

$$\eta \sim d(\bar{v}_z)_{\max} \sim z^{m-n} \sim z^0 \sim \text{const.}$$

Hence the eddy viscosity is constant throughout the mixing region of the jet. This observation implies that the original equation of motion (16–20) is valid for a round turbulent jet if the constant molecular viscosity μ is replaced by the constant eddy viscosity η. Equation (16–20) becomes

$$\bar{v}_r \frac{\partial \bar{v}_z}{\partial r} + \bar{v}_z \frac{\partial \bar{v}_z}{\partial z} = \frac{\epsilon}{r} \frac{\partial}{\partial r} \left(r \frac{\partial \bar{v}_z}{\partial r} \right), \tag{16–24}$$

where ϵ is the kinematic eddy viscosity,

$$\epsilon = \eta/\rho.$$

The equation of motion (16–24) and the continuity equation (16–22) may be solved by assuming geometrically similar velocity profiles and the following boundary conditions:

$$\text{at } r = \infty, \quad \bar{v}_z = 0;$$
$$\text{at } r = 0, \quad \bar{v}_r = 0, \quad \text{and} \quad \partial \bar{v}_z/\partial r = 0.$$

The solution [2] for the velocity distribution is

$$\frac{\bar{v}_z}{(\bar{v}_z)_{\max}} = \frac{1}{[1 + (\bar{v}_z)_{\max} r^2/8\epsilon z]^2}. \tag{16–25}$$

This equation compares extremely well with experimental observations in a round turbulent jet, as shown in Fig. 16–4 [3], when ϵ is given by

$$\epsilon = 0.00196 z (\bar{v}_z)_{\max}. \tag{16–26}$$

Hence the velocity distribution in the zone of established flow becomes

$$\frac{\bar{v}_z}{(\bar{v}_z)_{\max}} = \frac{1}{[1 + r^2/0.016z^2]^2}. \tag{16–27}$$

The longitudinal variation of the centerline velocity is

$$(\bar{v}_z)_{\max}/V_0 = 6.4(d_0/z). \tag{16–28}$$

In the above equations the longitudinal distance z is measured from the geo- metrical origin of similarity which, as shown in Fig. 16–4, is a distance $0.6d_0$ from the actual origin of the jet. From Eq. (16–28), the jet is fully developed when $z = 6.4d_0$, and hence the distance $L_0 = 6.4d_0 + 0.6d_0 = 7d_0$. Figure 16–4 also shows that the velocity distribution is reasonably well approximated by a Gaussian curve as was assumed in the analysis of the plane jet.

If Eqs. (16–26) and (16–28) are combined, the constant value of the eddy viscosity is given by

$$\epsilon = 0.013 V_0 \, d_0. \tag{16–29}$$

Therefore

$$\epsilon/\nu = 0.013(V_0 \, d_0/\nu) = 0.013 \mathbf{R}_0. \tag{16–30}$$

For example, if an air jet is discharging into the air at standard conditions with

$$V_0 = 100 \text{ ft/sec} \qquad \text{and} \qquad d_0 = 0.1 \text{ ft},$$

$$\mathbf{R}_0 = 64{,}000 \qquad \text{and} \qquad \epsilon/\nu = 830.$$

Thus the effective eddy viscosity is of the order of one thousand times larger than the molecular viscosity. Apparently, the motion of a round jet becomes laminar ($\epsilon = \nu$) when the Reynolds number is approximately 80.

Since the lateral spread of jets is linear, the angle of spread may be given in terms of an arbitrarily defined width. For the line along which $\bar{v}_z/(\bar{v}_z)_{\max} = 0.5$, the half angle is 6.5° for the plane jet and 5° for the round jet. However, it should not be concluded that the so-called jet boundaries are precisely defined. Due to the turbulent nature of the flow, the actual jet limits are only statistically

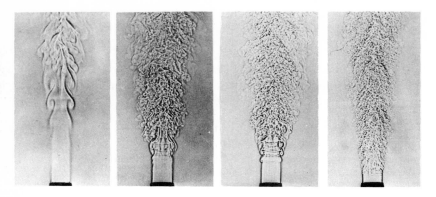

FIG. 16-5.   Development of a turbulent jet [4].

meaningful as shown by the photographs in Fig. 16–5 [4].   In quantitative terms, an intermittency factor Ω may be defined as the ratio of the time during which the flow at a point is turbulent to the total elapsed time of measurement. In a fully turbulent region, $\Omega = 1$, and in the nonturbulent region, $\Omega = 0$. Figure 16–6 [5] shows the intermittency factor across a round turbulent jet. When the velocity is one-tenth the centerline value, the flow is turbulent about one-half of the time.

The amount of fluid entrained by the round jet can be determined by integrating the velocity profiles in the zone of established flow.   The volume rate of flow is given by the simple equation

$$Q = 8\pi\epsilon z, \tag{16-31}$$

with ϵ given by Eq. (16–29), and since $Q_0 = (\pi\, d_0^2/4)V_0$, the ratio

$$Q/Q_0 = 0.42(z/d_0). \tag{16-32}$$

A similar calculation based on the Gaussian curve gives

$$Q/Q_0 = 0.28(z/d_0). \tag{16-33}$$

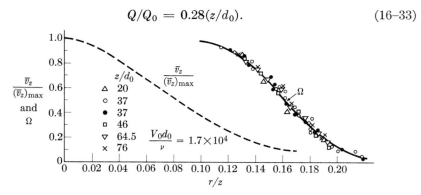

FIG. 16-6.   Radial distribution of intermittency factor in a round jet [5].

However, as shown in Fig. 16–4, the latter yields velocities which are too small near the edge of the round jet.

16–1.3 Free turbulence in wakes

An object placed in a field of flow generates a wake or disturbance in the otherwise nonturbulent flow which persists for large distances downstream, as shown in Fig. 16–1(c). If the Reynolds number is sufficiently high, the downstream wake will become a zone of free turbulence. This condition exists at Reynolds numbers beyond the range of periodic vortex shedding discussed in Chapter 15. In the wake produced by a cylinder, similarity of the downstream velocity profiles is reached only after 100 or more diameters. This is large in comparison with the much shorter distances (5 to 8 diameters) for the plane and round turbulent jets.

For the plane and round wakes, the boundary-layer type of approximation may again be used to simplify the equation of motion, and similarity assumptions are employed to obtain analytical solutions. The velocity distribution in the wake may be expressed in terms of the velocity defect Δu (see Fig. 16–1c), and this is seen to resemble the velocity distribution in jets. When the laws for the rates of spreading and velocity decay for wakes are derived in a manner corresponding to the methods developed for jets, the results [6] may be summarized as follows:

	Plane jet	Round jet	Plane wake	Round wake
m	1	1	$\frac{1}{2}$	$\frac{1}{3}$
n	$\frac{1}{2}$	1	$\frac{1}{2}$	$\frac{2}{3}$

where, as before, the rate of spreading is proportional to the exponent m of the longitudinal distance, and the decay of the maximum velocity (or velocity defect) is proportional to the exponent $-n$ of the longitudinal distance from the origin. In both the plane turbulent wake and the round turbulent jet, the exponents for spreading and velocity decay are the same, and hence the Reynolds number (which is proportional to x^{m-n}) is constant. By the same reasoning, the Reynolds number for the round wake decreases, since the decay of velocity occurs at a faster rate than the increase of wake width.

An important difference between jet and wake motions is in the application of the momentum-flux equation. In the jet, the momentum flux at any section is a constant equal to the initial momentum flux of the entering jet. In the wake, there is a change in the longitudinal flux of momentum, written for the control volume of Fig. 16–1(c), which must be equal to the drag force exerted by the moving fluid on the object. This is due to the fact that an external force must be applied to the object to hold it stationary in the flow. This method of wake surveys has been used to determine the drag of immersed bodies.

16–2 DIFFUSION PROCESSES IN NONHOMOGENEOUS FLUIDS

In the previous sections, various mixing processes in homogeneous fluids have been treated by means of momentum-transfer concepts. Their analysis is based upon the equations of motion and continuity. Homogeneity implies that both the jet fluid and the ambient fluid which is entrained are identical. In situations in which the two fluids are different but approximately of the same density, the volumetric flux relations obtained in Section 16–1 may be used to determine the *average* dilution of one fluid discharging into another. However, in the larger group of technological problems which are concerned with fluid mixtures, it is in general necessary to determine the relative amount of a given substance at a specified point in time and space. Thus a new variable, that of concentration, is introduced into the analysis. This requires that the conservation-of-mass equation (known as the continuity equation in a homogeneous fluid) be reformulated for a nonhomogeneous fluid.

As stated in Section 3–2, the conservation-of-mass principle must be satisfied for each component or species in a heterogeneous fluid. For simplicity, the following discussions are concerned with a binary or two-component system. Thus, if we are interested in the mixing between two different gases, one gas may be designated as species A and the other as species B. More complex cases such as mixing between sea water and fresh water may also be treated with reasonable approximation as a binary system even though sea water is a multicomponent liquid. For example, if it is assumed that the concentration of sodium chloride is representative of the relative concentration of sea water, the mixing process may be considered a binary one with sodium chloride as component A and fresh water as component B.

16–2.1 Molecular diffusion in a binary system

Molecular diffusion is the process by which matter is transported by molecular mobility. The gradual blurring of an originally sharp interface between dissimilar fluids is a common example of ordinary or molecular diffusion. It is recognized that temperature gradients, pressure gradients, and external force fields also contribute to mass flux on a molecular scale. These effects are usually small, although it is easy to find examples in which this is not the case. Examples include the separation of compounds by high-speed centrifuges and the settling of solid-particle suspensions in liquids, where the gravitational field produces a fall velocity of the solids relative to the liquid phase. If the fluid is in a state of convective motion, we must also be careful to distinguish between laminar and turbulent motions. For, if the flow is turbulent, the macroscopic exchange of fluid particles will usually completely overshadow the molecular-exchange process. Ordinary molecular diffusion is often called gradient diffusion, because it may be described by an observational law in which the rate of mass transfer of a substance per unit area is proportional to the gradient of concentration of the substance. This is known as Fick's first law and is analogous to Newton's law of viscosity or Fourier's law of heat conduction as discussed in Section 3–5. Before continuing with a quantitative discussion of diffusion processes, it is necessary to define certain fluid properties and kinematic quantities for a binary system.

Density:

$$\rho_A = \frac{\text{mass of component } A}{\text{volume of mixture of } A \text{ and } B}, \qquad \rho_B = \frac{\text{mass of component } B}{\text{volume of mixture of } A \text{ and } B},$$

$$(16\text{–}34)$$

$$\rho = \frac{\text{mass of } A + \text{mass of } B}{\text{volume of mixture of } A \text{ and } B} = \rho_A + \rho_B. \qquad (16\text{–}35)$$

Concentration:

$$c_A = \frac{\text{mass of component } A}{\text{mass of mixture of } A \text{ and } B} = \frac{\rho_A}{\rho},$$

$$(16\text{–}36)$$

$$c_B = \frac{\text{mass of component } B}{\text{mass of mixture of } A \text{ and } B} = \frac{\rho_B}{\rho}.$$

It then follows that

$$c_A + c_B = 1. \qquad (16\text{–}37)$$

In a mixture, the various components are moving with different velocities. However, we do not refer to the velocity of any one molecule but rather to the average velocity of all the molecules of a given component within a small volume. For component A, we call this velocity $\mathbf{q}_A$ when it is measured with respect to a fixed coordinate system. The mass flux of component A per unit area (mass per second per unit area) is therefore a vector equal to the product $\rho_A \mathbf{q}_A$. The unit area is normal to the direction of the velocity vector. Denoting the flux vector (relative to fixed coordinates) by $\mathbf{N}_A$, we have

$$\mathbf{N}_A = \rho_A \mathbf{q}_A. \qquad (16\text{–}38)$$

The mass flux $\mathbf{N}_A$ may also be interpreted as the momentum of component A per unit volume of the mixture. In a similar manner,

$$\mathbf{N}_B = \rho_B \mathbf{q}_B. \qquad (16\text{–}39)$$

We are now in a position to define the local hydrodynamic velocity $\mathbf{q}$ as that which would be measured by a Pitot tube. This must equal the total momentum per unit mass of the mixture, and hence

$$\frac{\text{total momentum}}{\text{volume of mixture}} = \mathbf{N}_A + \mathbf{N}_B \quad \text{and} \quad \frac{\text{total mass of mixture}}{\text{volume of mixture}} = \rho_A + \rho_B = \rho.$$

Therefore

$$q = \frac{N_A + N_B}{\rho_A + \rho_B} = \frac{\rho_A q_A + \rho_B q_B}{\rho}, \tag{16–40}$$

and also, using Eq. (16–36), we have

$$q = c_A q_A + c_B q_B. \tag{16–41}$$

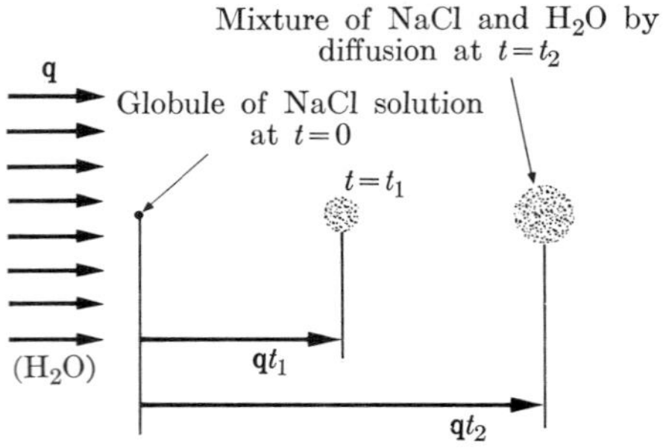

In dealing with diffusion processes, we are interested in the mass transfer relative to the hydrodynamic or convective velocity of the fluid. For example, consider a small globule of sodium-chloride solution injected into water flowing uniformly in a channel, as shown in Fig. 16–7. The center of mass of the globule moves downstream with the hydrodynamic velocity q. Because of molecular diffusion, the salt tends to spread, and hence the velocity q_A of the salt differs from the hydrodynamic velocity q. Denoting the mass flux per unit area of component A (relative to the hydrodynamic velocity) as J_A, we have

$$J_A = \rho_A(q_A - q). \tag{16–42}$$

It is this flux which is proportional to the local concentration gradient. Hence Fick's first law may be stated in vector form as

$$J_A = -\rho D_{AB} \nabla c_A, \tag{16–43}$$

where D_{AB} is the molecular-diffusion coefficient or diffusivity for the binary system. Equation (3–11) is therefore the y-component of Eq. (16–43). Like molecular viscosity, D_{AB} is a fluid property depending on the components A and B, their relative concentration, and the temperature and pressure of the

FIG. 16–7. Molecular diffusion relative to the convective motion of the fluid.

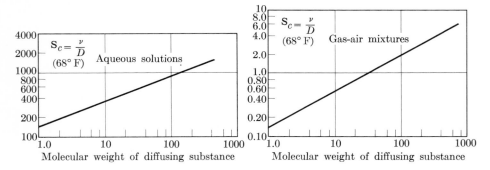

FIG. 16–8. Schmidt number as a function of molecular weight for gas-air mixtures and dilute aqueous solutions [10].

TABLE 16–1

MOLECULAR DIFFUSIVITY OF BINARY MIXTURES

Components A–B	Temperature, °C	D_{AB}, cm^2/sec
Gases [7] (at 1-atm pressure)		
CO_2–N_2O	0	0.096
CO_2–N_2	0	0.144
CO_2–N_2	25	0.165
H_2–CH_4	25	0.726
Liquids		
NaCl–H_2O [8]	0	0.784×10^{-5}
	25	$1.61 \ \times 10^{-5}$
	50	$2.63 \ \times 10^{-5}$
Glycerol–H_2O [9]	10	$0.63 \ \times 10^{-5}$

[*Note:* Diffusivities in liquids depend on the concentration of the diffusing substance. The above values are for dilute aqueous solutions.]

system. Table 16–1 gives some values of the diffusivity of certain binary mixtures (component A diffusing into component B).

The molecular diffusivities of gases and liquids differ by a factor of approximately 10^6, which indicates the relative molecular mobilities of these two phases of matter. Experimental data on molecular diffusivity are frequently given in terms of the Schmidt number, $S_c = \nu/D$. This dimensionless number is the ratio of the kinematic viscosity to the molecular diffusivity. An empirical correlation of the Schmidt number for gas-air mixtures and dilute aqueous solutions as a function of the molecular weight of the diffusing substance is given in Fig. 16–8 [10]. Values obtained in this manner are accurate within ±20% and are useful for estimating purposes. The Schmidt number for liquids is characteristically of the order of 1000; for gases it is near unity.

16–2.2 Convective-diffusion equation

The expression for the conservation of mass of one component in a binary system is obtained by summing the mass fluxes for a differential control volume. A convenient expression for the mass flux relative to a fixed-coordinate system is found by combining Eqs. (16–38) and (16–42). Thus

$$\mathbf{N}_A = \mathbf{J}_A + \rho_A \mathbf{q}.$$

When we use Fick's first law, Eq. (16–43), and introduce the concentration by means of Eq. (16–36), the mass flux becomes

$$\mathbf{N}_A = -\rho D_{AB} \nabla c_A + \rho c_A \mathbf{q}. \tag{16–44}$$

The latter has an obvious advantage over the original expression for the mass-flux vector (Eq. 16–38), since $\mathbf{N}_A$ is now in terms of measurable quantities: local density ρ, concentration c_A, and the hydrodynamic velocity $\mathbf{q}$.

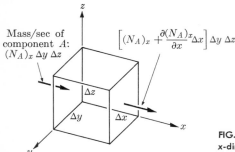

Mass/sec of component A:
$(N_A)_x \, \Delta y \, \Delta z$

$\left[(N_A)_x + \dfrac{\partial (N_A)_x}{\partial x} \Delta x \right] \Delta y \, \Delta z$

FIG. 16–9. Mass flux of component A in x-direction.

We may now form the conservation-of-mass expression for component A in Cartesian coordinates by referring to Fig. 16–9. This may be stated as follows:

$$
\begin{bmatrix}
\text{net flux of mass } A \\
\text{through fluid ele-} \\
\text{ment (inflow minus} \\
\text{outflow)}
\end{bmatrix}
+
\begin{bmatrix}
\text{time rate of produc-} \\
\text{tion at mass } A \text{ by} \\
\text{chemical and/or bi-} \\
\text{ological reaction in-} \\
\text{side fluid element}
\end{bmatrix}
=
\begin{bmatrix}
\text{time rate of accumu-} \\
\text{lation of mass } A \text{ in-} \\
\text{side fluid element}
\end{bmatrix}.
$$

In equation form, we have

$$
-\frac{\partial (N_A)_x}{\partial x} \Delta x \, \Delta y \, \Delta z - \frac{\partial (N_A)_y}{\partial y} \Delta x \, \Delta y \, \Delta z - \frac{\partial (N_A)_z}{\partial z} \Delta x \, \Delta y \, \Delta z + r_A \, \Delta x \, \Delta y \, \Delta z
$$

$$
= \frac{\partial \rho_A}{\partial t} \Delta x \, \Delta y \, \Delta z. \tag{16–45}
$$

Upon dividing by the element of volume, $\Delta x \, \Delta y \, \Delta z$, and rearranging, we obtain

$$
\frac{\partial \rho_A}{\partial t} + \frac{\partial (N_A)_x}{\partial x} + \frac{\partial (N_A)_y}{\partial y} + \frac{\partial (N_A)_z}{\partial z} = r_A, \tag{16–46}
$$

or in vector form,

$$
\frac{\partial \rho_A}{\partial t} + \nabla \cdot \mathbf{N}_A = r_A. \tag{16–47}
$$

In the above equations, r_A represents the mass of A produced per unit volume per unit time. Conversely, if the component disappears in a reaction process, r_A would be negative.

In a similar manner, the conservation of mass of substance B may be written as

$$\partial \rho_B / \partial t + \nabla \cdot \mathbf{N}_B = r_B. \tag{16–48}$$

In a binary system, the production of A can only come about at the expense of B since the total mass of the system must also be conserved. Hence

$$r_A = -r_B. \tag{16–49}$$

Therefore, adding Eqs. (16–47) and (16–48), we obtain

$$\frac{\partial \rho_A}{\partial t} + \frac{\partial \rho_B}{\partial t} + \nabla \cdot \mathbf{N}_A + \nabla \cdot \mathbf{N}_B = 0,$$

or, using Eqs. (16–34) and (16–40), we have

$$\frac{\partial \rho}{\partial t} + \nabla \cdot (\rho \mathbf{q}) = 0, \tag{16–50}$$

which, not surprisingly, is the general continuity equation for the fluid mixture (Eq. 6–16).

Returning to the single-component conservation equation (16–47), we replace the flux vector $\mathbf{N}_A$ by its equivalent from Eq. (16–44), and since $\rho_A = \rho c_A$, we get

$$\partial (\rho c_A) / \partial t + \nabla \cdot (-\rho D_{AB} \nabla c_A + \rho c_A \mathbf{q}) = r_A. \tag{16–51}$$

Expanding and rearranging, we obtain

$$\partial (\rho c_A)/\partial t + \rho c_A \nabla \cdot \mathbf{q} + \mathbf{q} \cdot \nabla \rho c_A = \rho D_{AB} \nabla^2 c_A + (\nabla c_A) \cdot (\nabla \rho D_{AB}) + r_A. \tag{16–52}$$

If the fluid is considered to be incompressible, $\nabla \cdot \mathbf{q} = 0$. If, in addition, the solution is dilute, both the total density of the mixture (ρ) and the diffusion coefficient are essentially constant. Hence after division by ρ, Eq. (16–52) becomes

$$\partial c_A / \partial t + \mathbf{q} \cdot \nabla c_A = D_{AB} \nabla^2 c_A + r_A / \rho. \tag{16–53}$$

In Cartesian form, we have

$$\frac{\partial c_A}{\partial t} + u \frac{\partial c_A}{\partial x} + v \frac{\partial c_A}{\partial y} + w \frac{\partial c_A}{\partial z} = D_{AB} \left[\frac{\partial^2 c_A}{\partial x^2} + \frac{\partial^2 c_A}{\partial y^2} + \frac{\partial^2 c_A}{\partial z^2} \right] + \frac{r_A}{\rho}, \tag{16–54}$$

and in cylindrical coordinates,

$$\frac{\partial c_A}{\partial t} + v_r \frac{\partial c_A}{\partial r} + \frac{v_\theta}{r} \frac{\partial c_A}{\partial \theta} + v_z \frac{\partial c_A}{\partial z}$$

$$= D_{AB} \left[\frac{1}{r} \frac{\partial}{\partial r} \left(r \frac{\partial c_A}{\partial r} \right) + \frac{1}{r^2} \frac{\partial^2 c_A}{\partial \theta^2} + \frac{\partial^2 c_A}{\partial z^2} \right] + \frac{r_A}{\rho}. \tag{16–54a}$$

This is known as the *convective-diffusion equation;* it is the expression for the conservation of mass of substance A undergoing ordinary diffusion in an incompressible laminar flow.

If the convective velocity and production rate are zero, Eq. (16–53) reduces to

$$\partial c_A / \partial t = D_{AB} \nabla^2 c_A. \tag{16–55}$$

This form of the equation is often called Fick's second law and is a well-known second-order differential equation. If the concentration is replaced by the temperature T, and D_{AB} is replaced by the thermal diffusivity α (see Eq. 3–9), we have the "heat equation"

$$\partial T / \partial t = \alpha \nabla^2 T, \tag{16–56}$$

which is one of the basic equations of heat-transfer theory. A large number of solutions of these equations for various initial and boundary conditions have been obtained. The Prandtl number is analogous to the Schmidt number in that it expresses the ratio between the molecular viscosity and the thermal diffusivity,

$$\mathbf{P}_r = \nu / \alpha. \tag{16–57}$$

The Prandtl number [11] for common gases is approximately 0.7 at atmospheric pressure. For liquids the variation is large; the Prandtl number for water at standard temperature is approximately 8.0.

16–2.3 Diffusion and convection in turbulent flow

The turbulent-flow counterpart of the convective-diffusion equation is obtained by making use of an analogy between molecular and turbulent diffusion. The development is similar to that used to obtain the Reynolds equations for turbulent motion (Eq. 11–22) from the Navier-Stokes equation. As in Eqs. (11–17), we represent the instantaneous hydrodynamic velocity components in terms of the sum of a time-averaged and a fluctuating velocity. Thus

$$u = \bar{u} + u', \qquad v = \bar{v} + v', \qquad w = \bar{w} + w'.$$

In addition, the concentration of the diffusing substance may be represented in a similar manner as

$$c_A = \bar{c}_A + c'_A. \tag{16–58}$$

Assuming D_{AB} is a constant, one substitutes the mean and fluctuating velocities and concentrations into the convective-diffusion equation (16–54). Then each term is time averaged according to the rules of Section 11–2. Assuming that the total-mass density ρ is constant and making use of the continuity condition,

$$\frac{\partial \bar{u}}{\partial x} + \frac{\partial \bar{v}}{\partial y} + \frac{\partial \bar{w}}{\partial z} = 0,$$

we obtain

$$\frac{\partial \bar{c}_A}{\partial t} + \bar{u}\frac{\partial \bar{c}_A}{\partial x} + \bar{v}\frac{\partial \bar{c}_A}{\partial y} + \bar{w}\frac{\partial \bar{c}_A}{\partial z} = -\frac{\partial}{\partial x}\overline{(u'c'_A)} - \frac{\partial}{\partial y}\overline{(v'c'_A)} - \frac{\partial}{\partial z}\overline{(w'c'_A)}$$

$$+ D_{AB}\left[\frac{\partial^2 \bar{c}_A}{\partial x^2} + \frac{\partial^2 \bar{c}_A}{\partial y^2} + \frac{\partial^2 \bar{c}_A}{\partial z^2}\right] + \frac{r_A}{\rho}. \qquad (16\text{–}59)$$

The time-averaged cross products $\overline{u'c'_A}$, etc., represent the volume fluxes per unit area of substance A due to the fluid turbulence. The mass flux of the substance per unit area is therefore given by the product of these terms and the mean density,

$$\text{turbulent mass flux per unit area} = \rho\overline{(u'c'_A)}.$$

By analogy with Fick's first law (Eq. 16–43), we assume that the turbulent flux is proportional to the gradient of the time-averaged concentration,

$$\rho\overline{(u'c'_A)} = -\rho E_x\frac{\partial \bar{c}_A}{\partial x}, \qquad \rho\overline{(v'c'_A)} = -\rho E_y\frac{\partial \bar{c}_A}{\partial y}, \qquad \rho\overline{(w'c'_A)} = -\rho E_z\frac{\partial \bar{c}_A}{\partial z}.$$

$$(16\text{–}60)$$

The coefficients E_x, E_y, and E_z are the turbulent-diffusion coefficients. Only for isotropic turbulence are they the same in the three coordinate directions. Substituting Eqs. (16–60) into (16–59), we have

$$\frac{\partial \bar{c}_A}{\partial t} + \bar{u}\frac{\partial \bar{c}_A}{\partial x} + \bar{v}\frac{\partial \bar{c}_A}{\partial y} + \bar{w}\frac{\partial \bar{c}_A}{\partial z} = \frac{\partial}{\partial x}\left(E_x\frac{\partial \bar{c}_A}{\partial x}\right) + \frac{\partial}{\partial y}\left(E_y\frac{\partial \bar{c}_A}{\partial y}\right) + \frac{\partial}{\partial z}\left(E_z\frac{\partial \bar{c}_A}{\partial z}\right)$$

$$+ D_{AB}\left[\frac{\partial^2 \bar{c}_A}{\partial x^2} + \frac{\partial^2 \bar{c}_A}{\partial y^2} + \frac{\partial^2 \bar{c}_A}{\partial z^2}\right] + \frac{r_A}{\rho}. \qquad (16\text{–}61)$$

In turbulent motion, the turbulent-diffusion coefficients are many orders of magnitude larger than the molecular-diffusion coefficients. Hence, unless we are concerned with diffusion near a solid boundary (where the turbulence is damped), it is usually permissible to neglect the molecular diffusion entirely. The turbulent counterparts of the Prandtl and Schmidt numbers are defined as the ratio of the kinematic eddy viscosity to the turbulent thermal and mass diffusivities, respectively. The numerical values are based on measurements of the spread of velocity, temperature, and mass concentration in turbulent mixing processes. The turbulent Prandtl and Schmidt numbers are approximately equal for both liquids and gases. The numerical value is about 0.7, which indicates that in turbulent flow, mass and heat diffuse at equal rates and that this rate is greater than the rate of turbulent diffusion of momentum [12].

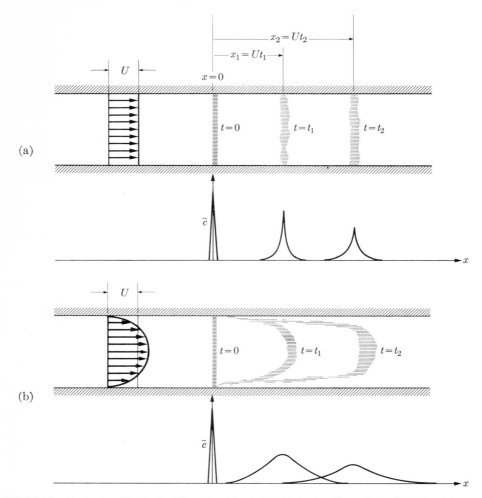

FIG. 16–10. Mechanism of longitudinal dispersion: (a) turbulent diffusion in uniform velocity flow; (b) turbulent dispersion due to nonuniform velocity distribution.

16–2.4 Dispersion in one-dimensional motion

The general three-dimensional convective-diffusion equation is extremely difficult to solve, even when the molecular diffusion is ignored, due to the variable velocity and diffusion coefficients. Many diffusion and mixing problems may therefore be treated by assuming a one-dimensional motion in a region of constant cross-sectional area. In this case, Eq. (16–61) reduces to

$$\frac{\partial \bar{c}_A}{\partial t} + U \frac{\partial \bar{c}_A}{\partial x} = \frac{\partial}{\partial x}\left(E_T \frac{\partial \bar{c}_A}{\partial x}\right) + \frac{r_A}{\rho}. \tag{16–62}$$

The mean convective motion of the fluid is now represented by the average velocity U (the volume rate of flow divided by the cross-sectional area), and the convective mass transfer due to the mean motion in the x-direction is therefore given by the term $U\partial \bar{c}_A/\partial x$. It is also implied that the concentration $\bar{c}_A$ is the average value at a cross section. In a shear flow such as would occur in a conduit or open channel, the velocity distribution is not uniform. The difference in the longitudinal convective mass transfer which is associated with the actual velocity distribution and that which is accounted for by the mean velocity must therefore be incorporated into the diffusion term. This effect is known as *longitudinal dispersion*, and the symbol E_T is used to distinguish between turbulent diffusivity E_x and longitudinal dispersion. The phenomenon of dispersion is shown schematically in Fig. 16–10. At time $t = 0$, a finite quantity of a tracer substance is introduced instantaneously and uniformly across the conduit. If the tracer moves down the conduit with a uniform velocity U, it will spread longitudinally due to turbulent mixing, as shown in the plot of concentration versus x in Fig. 16–10(a). In the actual shear flow shown in Fig. 16–10(b), adjacent layers of fluid are moving with different longitudinal velocities while mixing laterally due to turbulence. This results in a much greater longitudinal spread, as shown in the accompanying concentration plot.

Taylor [13] has demonstrated that the longitudinal dispersion coefficient in a uniform turbulent flow in a straight conduit is given by

$$E_T = 10.1 r_0 \sqrt{\tau_0/\rho}, \tag{16–63}$$

where r_0 is the pipe radius, and τ_0 is the shear stress at the wall. This value of E_T is about 200 times larger than the mean value of the turbulent diffusivity $\bar{E}_x$.

We will consider several examples of the use of the one-dimensional convective-diffusion (or dispersion) equation for a conservative substance in which the rate of production r_A is equal to zero. If, in addition, the flow is both steady and uniform, the longitudinal-dispersion coefficient is a constant, and Eq. (16–62) becomes

$$\frac{\partial \bar{c}_A}{\partial t} + U \frac{\partial \bar{c}_A}{\partial x} = E_T \frac{\partial^2 \bar{c}_A}{\partial x^2}. \tag{16–64}$$

The solution of this equation depends upon the initial and boundary conditions which are specified in the following examples.

Example 16–1: Constant concentration at origin

A constant concentration $(\bar{c}_A)_0$ is maintained at $x = 0$. Initially, the concentration of substance A is zero in the region of positive values of x. This condition may be realized in a conduit leading from a large tank or reservoir with x equal to zero at the junction of the conduit and tank. The diffusing substance A is contained in the tank at constant concentration, and fluid B fills the conduit. At time $t = 0$, the fluid in the conduit begins to move with a mean velocity U. The initial and boundary

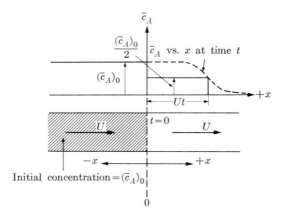

FIG. 16–11.   Unsteady diffusion with constant concentration at x = 0.

conditions, shown schematically in Fig. 16–11, are

$$\bar{c}_A(0, t) = (\bar{c}_A)_0, \quad t \geq 0; \qquad \bar{c}_A(x, 0) = 0, \quad x > 0; \qquad \bar{c}_A(+\infty, t) = 0, \quad t \geq 0.$$

The solution is given by

$$\frac{\bar{c}_A}{(\bar{c}_A)_0} = \frac{1}{2} e^{Ux/E_T} \operatorname{erfc}\left[\frac{x + Ut}{2\sqrt{E_T t}}\right] + \frac{1}{2} \operatorname{erfc}\left[\frac{x - Ut}{2\sqrt{E_T t}}\right], \qquad (16\text{–}65)$$

where the complementary error function erfc $(\alpha) = 1 - \operatorname{erf}(\alpha)$, and the error function

$$\operatorname{erf}(\alpha) = 2/\sqrt{\pi} \int_0^\alpha e^{-\zeta^2} d\zeta$$

is a tabulated integral found in standard mathematical tables.   The mean point on the concentration curve, $\bar{c}_A = (\bar{c}_{A0}/2)$, moves with the speed U, as shown in the upper part of Fig. 16–11.

For the special case in which $U = 0$, Eq. (16–65) reduces to

$$\frac{\bar{c}_A}{(\bar{c}_A)_0} = \operatorname{erfc}\left(\frac{x}{2\sqrt{E_T t}}\right) = 1 - \operatorname{erf}\left(\frac{x}{2\sqrt{E_T t}}\right). \qquad (16\text{–}66)$$

If values of $\bar{c}_A/(\bar{c}_A)_0$ are plotted as an ordinate on an arithmetic-probability scale versus $x/\sqrt{t}$ as an abscissa on a linear scale, Eq. (16–66) becomes a straight line if E_T is a constant.   Figure 16–12 shows some experimental data plotted in this manner.   In these experiments [14] turbulence was generated in a long tank of water by a stack of screens moving vertically in simple harmonic motion, and therefore, since there is no velocity distribution, $E_T = E_x$.   At the midpoint of the tank ($x = 0$), a barrier separated water containing a small amount of a tracer (NaCl) in the negative x-region from pure water in the positive x-region.   At time $t = 0$, the barrier was removed and concentration measurements were made at various times and distances in the

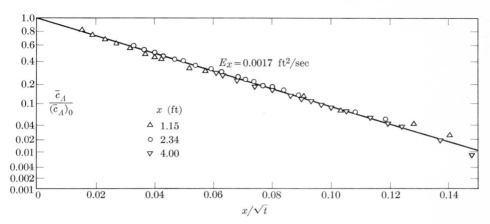

FIG. 16–12.   Concentration versus $x/\sqrt{t}$ for unsteady diffusion process in Fig. 16–11.

positive x-region. The data are seen to clearly define a constant value of the turbulent diffusion coefficient E_x. The numerical value of E_x is found by noting that there must be a point on the abscissa (as yet unknown) at which $x/\sqrt{t} = \sqrt{\pi E_x}$. From Eq. (16–66), we have

$$\frac{\bar{c}_A}{(\bar{c}_A)_0} = 1 - \mathrm{erf}\left(\frac{\sqrt{\pi E_x}}{2\sqrt{E_x}}\right) = 1 - \mathrm{erf}\left(\frac{\sqrt{\pi}}{2}\right).$$

Since $1 - \mathrm{erf}\,(\sqrt{\pi}/2) = 1 - \mathrm{erf}\,(0.89) = 1 - 0.79 = 0.21$, we get

$$\bar{c}_A/(\bar{c}_A)_0 = 0.21 \qquad \text{when} \qquad x/\sqrt{t} = \sqrt{\pi E_x}.$$

Therefore, by reading the value of the abscissa corresponding to $\bar{c}_A/(\bar{c}_A)_0 = 0.21$, one can calculate the numerical value of E_x.

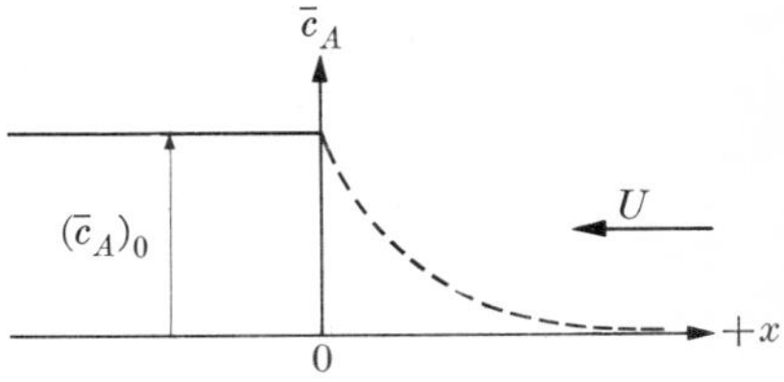

FIG. 16–13.   Steady-state concentration distribution for a constant concentration at $x = 0$.

A steady-state solution of Eq. (16–65) is obtained by letting $t \to \infty$. For a velocity U in the positive x-direction (Fig. 16–11), $\bar{c}_A/(\bar{c}_A)_0 = 1$ for all x since erfc $(+\infty) = 0$ and erfc $(-\infty) = 2$. If the flow is in the negative x-direction, U is negative, and as $t \to \infty$, the steady-state solution of Eq. (16–65) is an exponential curve as shown in Fig. 16–13,

$$\bar{c}_A/(\bar{c}_A)_0 = e^{-Ux/E_T}. \tag{16–67}$$

Physically this steady-state distribution represents a mass-transfer balance between the dispersive flux of mass in the direction of decreasing concentration $(+x)$ and the convective flux in the direction of flow $(-x)$.

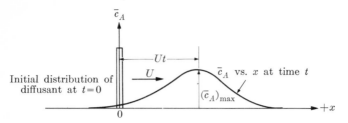

FIG. 16-14. Unsteady concentration distribution for instantaneous injection of a tracer at x = 0.

Example 16-2: Instantaneous injection at origin

Consider a steady uniform turbulent flow in a long conduit. A finite quantity of tracer material is introduced instantaneously and uniformly over a cross section of the flow at $x = 0$ and $t = 0$. The material is dispersed and convected downstream, as shown in Fig. 16-14. The initial condition may be expressed in terms of the Dirac delta function $\delta(x)$. This is a mathematical function for an initial impulse of concentration introduced at $x = 0$ and $t = 0$ which is generated by a finite quantity of matter. At the initial condition, the concentration approaches infinity while the thickness of the impulse approaches zero in such a way that the area under the impulse function (which represents the total mass of the tracer) is a constant. This property of the delta function is given by

$$\int_{-\infty}^{+\infty} \delta(x)\, dx = 1.$$

Thus the initial condition is represented by

$$\bar{c}_A(x, 0) = (M/\rho A)\, \delta(x),$$

where

$$M = \text{mass of tracer material introduced,}$$
$$A = \text{cross-sectional area of flow.}$$

Since the total mass of the tracer material is conserved at all times, we have

$$\int_{-\infty}^{+\infty} \bar{c}_A(x, t)\, dx = \int_{-\infty}^{+\infty} \bar{c}_A(x, 0)\, dx = (M/\rho A) \int_{-\infty}^{+\infty} \delta(x)\, dx = M/\rho A.$$

The boundary condition is obtained by specifying that the concentration at $x = \pm\infty$ is zero at all times,

$$\bar{c}_A(\pm\infty, t) = 0, \qquad t \geq 0.$$

The solution of Eq. (16-64) subject to the above conditions is

$$\bar{c}_A = \frac{M}{\rho A \sqrt{4\pi E_T t}}\, e^{-(x - U t)^2 / 4 E_T t}. \tag{16-68}$$

The tracer material moves downstream and spreads as a Gaussian curve which is symmetrical in the x-direction at any given time. The peak concentration moves with the speed U, as shown in Fig. 16–14. The magnitude of the peak concentration decreases in the positive x-direction.

The convective-diffusion equation has applications in many fields. Most of the basic processes of chemical technology are based upon mass-transfer concepts. The simultaneous transport of various fluids in pipelines and the displacement and dispersion of miscible fluids in porous media and ground-water aquifers also involve solutions of either the multidimensional or one-dimensional conservation-of-mass equation. Oceanographers, meteorologists, and engineers concerned with waste disposal into the air, rivers, or oceans employ similar concepts for the analysis of mixing and diffusion processes. A more thorough understanding of complex sedimentation and salinity-intrusion problems in estuaries has also been gained through mathematical models based on the convective-diffusion equation.

PROBLEMS

16–1. Consider the following examples of jet discharge into a surrounding fluid: (i) Water jet discharging into air. (ii) Water jet discharging into water. (iii) Air jet discharging into air. (iv) Air jet discharging into water. (a) Which of these may be analyzed by the methods of Section 16–1? (b) Which pairs may exhibit dynamically similar behavior?

16–2. A plane turbulent jet of air discharges from a slot into surrounding air having the same temperature. In the fully developed zone the outer jet boundary may be defined as a line connecting a series of points along which the local velocity component $\bar{u}$ is 10% of the local centerline velocity. Determine the angle between the line defining the outer jet boundary and the x-axis of the jet and the point of intersection of this line with the x-axis. Use the notation of Fig. 16–2.

16–3. A two-dimensional plane jet has an initial width $b_0 = 0.1$ ft and an efflux velocity of 50 fps. Plot the velocity distribution along the centerline of the jet for a distance $z = 4$ ft and the distribution of longitudinal velocity in the lateral direction at $z = 2$ ft.

16–4. A circular jet having a diameter of 0.5 ft and an initial velocity of 5 ft/sec discharges waste water vertically upward from a pipeline on the bottom of a lake. The density of the effluent and that of the lake water are approximately equal. The initial concentration of a certain chemical in the waste water is one part per thousand parts of water. (a) Determine the necessary depth, at the point of discharge of the effluent, so that the surface concentration of the chemical in the lake is 1% of its initial value. (b) Keeping the initial jet velocity at 5 ft/sec, make a plot of the depth necessary for the specified dilution of 1% versus the nozzle diameter. (c) Repeat part (b), keeping the nozzle diameter at 0.5 ft and varying the initial velocity of the jet.

16–5. Essentially uniform turbulence is to be produced at the free surface of a tank of 60°F water in order to study mass transfer of a gas across the free surface. It is proposed to create the surface turbulence by means of vertical jets of water discharging from an array of nozzles in the bottom of the tank. An eddy viscosity of approximately 0.001 ft²/sec is required at the water surface and the depth of water in the tank is to be at least 1 ft. Assume that the outer boundary of the jet is represented by the radius at which the local velocity is one-tenth of the local centerline velocity. The center-to-center spacing of the nozzles is to be 1 ft. Determine the nozzle diameter and initial jet velocity required to meet these specifications.

16–6. Air is discharged into surrounding air from an annular nozzle formed by two circular plates a distance $2b_0$ apart, as shown in Fig. 16–15. Derive the equation for constant momentum flux for the annular jet.

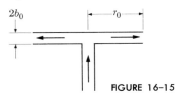

FIGURE 16–15

16–7. A circular jet of air discharges into the atmosphere from a nozzle of diameter d_0 at a velocity V_0. After the jet becomes fully established, if the flow is turbulent, the distribution of axial velocity in the radial direction is very nearly given by the Gaussian curve

$$v_z/(v_z)_{\max} = e^{-(r/0.1z)^2}.$$

Determine the ratio of jet discharge at any distance z to the initial discharge, Q/Q_0.

16–8. A nonturbulent fluid stream has a uniform velocity V_0 in the z-direction. At a point in the stream, $x = y = z = 0$, a small amount of tracer material is injected continuously at a rate W_A (mass of tracer per second). The axially symmetric, steady-state concentration distribution is given by the following solution of Eq. (16–54a):

$$c_A = \frac{W_A}{4\pi R D_{AB}} \exp\left\{-\left[\frac{V_0}{2D_{AB}}(R - z)\right]\right\},$$

where $R = \sqrt{x^2 + y^2 + z^2}$.

If for a certain V_0 and W_A, measurements of c_A as a function of $r = \sqrt{x^2 + y^2}$ at a given value of z are made, show how the data may be plotted to determine the magnitude of D_{AB}.

16–9. Express the conservation-of-mass equation (16–54) in dimensionless form by using reference values of velocity V_0, length L, and concentration $(c_A)_0$. Let $r_A = 0$ and show that the resulting equation contains a single dimensionless group equal to the product of the Reynolds and Schmidt numbers.

16–10. Discuss the conditions under which the convective-diffusion equation reduces to the Laplace equation $\nabla^2 c_A = 0$. What analytical techniques are available for solving this type of problem?

16–11. A large tank of liquid contains solid particles of uniform size in suspension. Turbulence is generated by a stack of screens oscillating harmonically in the vertical direction so that the turbulent diffusivity E_{z0} is constant. The equation for the vertical concentration distribution of the suspended material is

$$w_f c = -E_{z0}(dc/dz),$$

where w_f is the fall velocity of a particle in still liquid, z is the vertical coordinate measured from the bottom of the tank, and c is the particle concentration. If the particle concentration is c_0 at $z = a$, find the concentration distribution in the vertical direction for $z > a$.

16–12.   The equation for the vertical distribution of suspended sediment in a turbulent free-surface stream is as follows:

$$w_f c = -E_z(dc/dz),$$

where z is the vertical coordinate measured from the stream bed, c is the concentration of suspended sediment in the water, E_z is the turbulent diffusion coefficient in the z-direction which is a function of z, and w_f is the fall velocity of the sediment in water at rest. If it is assumed that the turbulent diffusion coefficient is equal to the turbulent eddy viscosity, the following relation for E_z is obtained:

$$E_z = 0.4\sqrt{\tau_0/\rho}\, z(1 - z/z_0),$$

where τ_0 is the shear stress at the bed of the stream, and z_0 is the depth of the stream. (a) Determine the vertical concentration of suspended sediment in terms of a known concentration c_0 at $z = 0.1\, z_0$. (b) Explain why the concentration distribution should not be expressed in terms of a known concentration at $z = 0$.

16–13.   Determine the longitudinal dispersion coefficient E_T for a galvanized iron pipe 2 in. in diameter carrying 60°F water at a mean velocity of 5 ft/sec. Compare the value of E_T with the value of the kinematic eddy viscosity ϵ averaged over the cross section of the pipe.

16–14.   Develop the equation for longitudinal dispersion in a uniform free-surface flow of arbitrary cross section. Start with the dispersion equation for turbulent flow in a circular pipe (Eq. 16–63) and express E_T in terms of the hydraulic radius and the channel slope in the direction of flow.

16–15.   The following data were obtained in a laboratory study of one-dimensional turbulent diffusion without convection. A dilute solution of NaCl in water diffused into an equal-density solution of sugar in water. Time $t = 0$ is the start of the test in which a constant concentration c_0 of NaCl is maintained at the point $x = 0$. Also c is the concentration of NaCl at any x and t. Determine the turbulent diffusion coefficient for this test.

$t = 500$ sec		$t = 1000$ sec		$t = 2000$ sec	
x	c/c_0	x	c/c_0	x	c/c_0
1.0	0.47	1.0	0.61	1.0	0.74
2.0	0.18	2.0	0.36	2.0	0.51
3.0	0.05	3.0	0.17	3.0	0.30
		4.0	0.04	4.0	0.14

REFERENCES

1. ALBERTSON, M. L., Y. B. DAI, R. A. JENSEN, and H. ROUSE, "Diffusion of Submerged Jets," *Trans. ASCE*, **115**, 639 (1950).

2. HINZE, J. O., *Turbulence*, McGraw-Hill Book Co., New York, 1959, pp. 411, 424–426.

3. HINZE, J. O., *op. cit.*, p. 424.

4. SHAPIRO, A. H., "Turbulent Transfer Processes in Parallel Jets," *Symposium on Hydrodynamics in Modern Technology*, M.I.T. Hydrodynamics Laboratory, 1951, pp. 141–143.

5. CORRSIN, S., and A. L. KISTLER, "The Free Stream Boundaries of Turbulent Flows," *Nat. Advisory Comm. Aeron.*, Tech. Note 3133 (1954).

6. SCHUBAUER, G. B., and C. M. TCHEN, *Turbulent Flow*, No. 9., Princeton Aeronautical Paperbacks, Princeton University Press, 1961, p. 90.

7. HIRSCHFELDER, J. O., C. F. CURTISS, and R. B. BIRD, *Molecular Theory of Gases and Liquids*, John Wiley and Sons, Inc., New York, 1954.

8. DORSEY, N. E., *Properties of Ordinary Water Substance*, Reinhold Publishing Corp., New York, 1940.

9. REID, R. C., and T. K. SHERWOOD, *The Properties of Gases and Liquids*, McGraw-Hill Book Co., New York, 1958. Also C. N. SATTERFIELD and T. K. SHERWOOD, *The Role of Diffusion in Catalysis*, Addison-Wesley Publishing Co., Inc., Reading, Mass., 1963.

10. SPALDING, D. B., *Convective Mass Transfer*, McGraw-Hill Book Co., New York, 1963, pp. 139, 143.

11. ROHSENOW, W. M., and H. Y. CHOI, *Heat, Mass and Momentum Transfer*, Prentice-Hall, Inc., Englewood Cliffs, N.J., 1961.

12. ABRAMOVICH, G. N., *The Theory of Turbulent Jets*, The M.I.T. Press, Cambridge, Mass., 1963, p. 31.

13. TAYLOR, G. I., "The Dispersion of Matter in Turbulent Flow Through a Pipe," *Proc. Roy. Soc. (London) (A)*, **223** (1954).

14. HARLEMAN, D. R. F., J. M. JORDAAN, and J. D. LIN, "The Diffusion of Two Fluids of Different Density in a Homogeneous Turbulent Field," *M.I.T. Hydrodynamics Lab.*, Tech. Rept. 31 (Feb., 1959).

Answers to Odd-Numbered Problems

CHAPTER 1

1-1. (b) 2.5×10^{-5} lb/sec-ft^2; 1-3. (a) 1200 rad/sec;
(c) 7.5×10^{-5} lb/sec-ft^2 (b) 0.06 lb/sec-ft^2

1-5. $k = 1.39$; $R = 52.9$ ft/lb-lbm °R

1-7. 10.8 Btu/lbm; 112.5 Btu/lbm 1-9. 1115 ft/sec; 1520 ft/sec

1-11. 3.70×10^{-7} lb/sec-ft^2 at 1000'; 3.31×10^{-7} lb/sec-ft^2 at 20,000'

1-13. 25 psig 1-15. 41.3 psia 1-17. 2° 18'

1-19. 5.87 ft 1-21. 15,380 lb 1-23. 70.2 lb

1-25. 232,000 lb; 0.066 ft

CHAPTER 2

2-3. (b) $\mathbf{v}_t - (\mathbf{\Omega} \times \mathbf{r})$; $-2\mathbf{\Omega} \times \mathbf{v}_t + \mathbf{\Omega} \times (\mathbf{\Omega} \times \mathbf{r}) - \dfrac{d\mathbf{\Omega}}{dt} \times \mathbf{r}$

2-7. 29.2×10^{-4} ft/sec^2

CHAPTER 4

4-1. 483 ft^3/sec 4-3. 22.4 ft^3/sec 4-5. -140 Btu/sec

4-7. 2075 ft/sec 4-9. 20.8 psia 4-11. $y_1 c_c b \left(\dfrac{2g}{y_1 + c_c b} \right)^{1/2}$

4-13. 0.696 ft 4-17. 10.7 hp 4-19. 17.1 lb/sec

4-21. 935 lb 4-23. 2.1 hp 4-25. (b) 432,000 hp

4-27. 384 hp 4-29. $p_0 - p_1 = \rho_a V_0 [V_1 - \tfrac{5}{9} V_0]$

4-31. (a) 990 ft-lb (b) 31.7° 4-33. (a) 37° 25'

CHAPTER 6

6-3. $-3Ax^2y$

6-5. (a) $u = 10z - 20(z - 10z^2)$; (b) to the right
(c) $\tau_{zx} = \mu[U/a - (a/2\mu)(1 - 2z/a)(dp/dx)]$

6-7. $V = -(\gamma/3\mu)a^2 \, \partial h/\partial x$

6-9. 1.83 ft/sec; 0.707 in; 0.29 in.

6-11. $v_z = -\dfrac{1}{4\mu} \left[r_2^2 - r^2 - \dfrac{r_2^2 - r_1^2}{\ln (r_2/r_1)} \ln \dfrac{r}{r_2} \right] \partial(p + \gamma h)/\partial z$

6–13. 7.7 ft/lb-sec; 6510 ft/lb-sec; 54.2 ft/lb-sec
6–15. (a) $\mathbf{q} = \mathbf{i}ax + \mathbf{j}2a - \mathbf{k}az$
6–17. 54.5 psig   6–19. 5.67 rad/sec
6–21. $\dfrac{\rho}{2}\left[\dfrac{Q}{2b\ln(R+b)/(R-b)}\right]^2\left[\dfrac{1}{(R-b)^2} - \dfrac{1}{(R+b)^2}\right]$

CHAPTER 7

7–1. (a) 196 cfs; (b) 0.33 psi   7–3. (c) 5.4 cfs
7–5. 11.75 in   7–7. 750 knots;   7–9. (b) 147 rpm
7–11. Wind tunnel   7–13. (1) 0.245 rpm
7–15. $\nu_M = 11.2 \times 10^{-6}$ ft^2/sec   7–17. (b) $1/\sqrt{3}$
7–19. 4890 ft/sec; 489 ft/sec   7–21. 2.4 atmospheres

CHAPTER 9

9–1. (a) 0.007 in; (b) 0.0015 in
9–3. $V_0 = \dfrac{4\mu}{3\rho a}\left[1 + \sqrt{1 + \dfrac{\rho a^3}{3\mu^2}(\gamma_s - \gamma)}\right]$
9–5. (a) 18.5 psia   9–7. ratio = 0.07   9–13. 122 ft
9–15. (a) 3.75×10^{-7} cfs/ft
9–19. (b) $-\dfrac{a^2\gamma}{3\mu}\dfrac{\partial H}{\partial x}$; $-\dfrac{a^2\gamma}{3\mu}\dfrac{\partial H}{\partial z}$

CHAPTER 10

10–1. $u = 1.26$; 1.91 ft/sec
       $v = 1.1 \times 10^{-3}$ ft/sec; 2.76×10^{-3} ft/sec
10–3. 0.024 $x\frac{1}{2}$ lb   10–5. $\delta^*/\delta = 0.333$; $\theta/\delta = 0.133$
10–13. 14.2 ft of oil
10–15. (a) $\tau_0/\rho = \dfrac{ABU^{3/2}}{x^{1/2}}$; (b) $c_f = \dfrac{2AB}{\sqrt{Ux}}$
10–17. (a) $\theta = \frac{39}{280}\delta$; $c_f = 0.645/\mathbf{R}_x^{1/2}$; (b) $\delta = 4.65x/\mathbf{R}_x^{1/2}$

CHAPTER 11

11–1. $0.54 < x < 1.59$ ft   11–3. 3.1 ft/sec
11–5. $\bar{p}_0 + \rho_2(\bar{u}^2 + \overline{u'^2})$

CHAPTER 12

12–1. (a) 2.5×10^{-5} ft; (b) 6.4 ft/sec; (c) 39.5 ft/sec; (d) 0.2 ft
12–5. (a) 0.15 lb; (b) 0.0087 lb/ft^2; (c) 0.15 ft, 0.27 ft
12–7. 422 hp   12–9. (a) 10 lb; (b) 0.10 ft; (c) 0.08 ft
12–11. 300,000 lb   12–13. (a) 1.6×10^{-4} ft
12–15. 2.6×10^{-6} ft (model); 2.5×10^{-5} ft (prototype)

CHAPTER 13

13–1. (a) 0.014 ft; (b) 0.37 lb/ft^2; (c) 0.0017 ft
13–3. 24,800 hp vs. 22,200 hp   13–5. 0.07 ft^3/sec
13–9. -27 psi   13–11. 0.92 ft^3/sec
13–13. 4×10^{-6} ft for comm. roughness
13–15. (a) 44.4 psig; (b) 0.0021 ft   13–17. (a) 94.3 lb; (b) 1.8 lb/ft^2
13–19. $k_s/D = 0.0048$ 13–21. 60.78 °F; 189 lb/in^2
13–23. 0.35 ft (use 5 in. pipe)
13–25. (a) 417; (b) 1000/sec; (c) 3.1 ft/sec; (d) -0.04 lb/ft^2
13–27. 0.31; 3.9×10^6 13–29. 0.018 Btu/sec
13–31. 0.082 lbm/sec   13–33. 60 ft
13–35. 231 ft^3/sec   13–37. (a) 395 ft^3/sec; (b) 13 ft wide; (c) 1.31
13–39. 0.0042   13–41. (a) 0.0092 (b) 0.0041

CHAPTER 14

14–5. $\alpha = 15°$ 14–7. 2.37 ft^3/sec; 2.1 ft^3/sec; 1.6 ft^3/sec
14–11. 5.98 ft^3/sec   14–15. 94 ft   14–17. 1.6 ft^3/sec
14–19. 1.2   14–21. 65°F
14–23. 0.045 ft^2; 0.055 ft^2 14–25. 738 ft/sec; 421.4°R; 1.07
14–27. 222 slugs/(sec-ft^2)   14–29. 10.4; 2.7; 0.26
14–31. 2700 ft/sec; 6.5 psia   14–35. $Q = C_d \sqrt{2g}\, h_w{}^{5/2}$
14–37. 7.2 ft   14–39. $5.67 y_c{}^{3/2}$
14–41. (a) 191 ft^3/sec; (b) 1.25 ft fall
14–45. (a) 109 ft   14–47. 43 ft
14–49. 11.7 ft   14–51. 40°

CHAPTER 15

15–5. 0.38 lb   15–7 (a) 4 ft; (b) 60 hp/ft of periscope
15–9. (a) 72 lb, 222 lb; (b) 111 lb, 133 lb
15–11. (a) 1.28 lb; (b) 513 lb   15–15. 0.037
15–17. (a) 5670 lb; (b) 3100 lb   15–19. 10.7 ft/sec
15–21. 2.06; 250 rpm   15–23. (a) 470 lb; (b) 870 lb
15–25. (a) 9/8; (b) 8.73 psia   15–31. (a) 0.0016; (b) 44,000 lb
15–33. (a) 1.19 lb; (6) 7.9 lb $(C_D = 0.20)$
15–35. 4320 ft/sec; 1790 ft/sec

CHAPTER 16

16–5. 0.083 ft; 0.93 ft/sec   16–7. 0.128 z/d_0
16–11. $c = c_0 e^{-(w_f/E_{z0})(z-a)}$ 16–13. $E_T = 0.25$ ft^2/sec;$^{E_T/\bar\epsilon = 150}$
16–15. 0.0065 ft^2/sec

Index

A

Absolute pressure, 5, 19, 286
Acceleration, centrifugal, 45
  convective, 44
  Coriolis, 45
  gravitational, 3
  local, 44
Accelerated motion of bodies, 374–376
Accelerating frame of reference, 45,
  77–78
Added mass, 374, 376
Additional apparent stress, 219, 221, 222
Adiabatic bulk modulus, 12
Adiabatic flow (*see* isentropic flow)
Adiabatic flow with friction in pipes,
  290, 292
Adiabatic process, 15
Airfoils, lift and drag, 170, 171, 388–394
Analogy, shock wave, 358
  transport (heat, mass, and
    momentum), 54–57
Anemometer, hot-wire, 217
Angular deformation, rate of, 1, 6, 7, 99
Angular deformation in solids, 100
Angular momentum, 84–87
Anisotropic porous media, 183, 187
Apparent gravity (gravitational force),
  46, 111
Apparent stress, 219, 221
Archimedes' principle, 24–25
Aspect ratio, 385, 392–394
Atmosphere, standard, 5, 26, 36

B

Backwater curves, 353–357
Bends, losses in, 309, 320–321
Bernoulli equation along a streamline,
  125
Bernoulli equation, comparison of forms,
  125
  irrotational motion, 121
  one-dimensional, 71

Bingham plastic, 2
Blasius' boundary layer solution,
  196–200
Blasius' turbulent flow friction formula,
  271
Blower, 86
Body forces, 16, 77, 111
Borda mouthpiece, 347
Bound vortex, 387, 392
Boundary layer, Blasius' solution (flat
    plate), 196–200
  definition of, 162
  integral momentum equation
    (Kármán's), 168
  laminar, 193–211
  laminar sublayer, 231, 267, 272
  Prandtl's equations, 166
  pressure gradient effects, 201,
  252–255
  resistance, laminar, 198, 199
    turbulent, 237–245, 248, 249, 254
  Reynold's equations, 223
  rotating disk, 205
  roughness effects, 246–251
  skewed, 205
  three-dimensional, 204–208
  turbulent, 227–255
  velocity distributions, 196, 197, 207,
    232–236, 245
Boundary layer thickness definitions,
  167, 168
Boundary layer thickness relations,
  195, 196, 198, 205, 228, 233, 245
Branching pipes, 323
Buffer zone, 232, 233, 238
Bulk modulus of elasticity, 11
Buoyant force, 25, 370, 375

C

Capillarity, 27–29
Cauchy-Riemann equations, 120

444